AF553825

BIOMOLECULES

BIOMOLECULES

By

Dr. Shubhrata R. Mishra

Department of Botany
Vikram University
Ujjain (M.P)

DISCOVERY PUBLISHING HOUSE
NEW DELHI-110002

Published by:
Namit Wasan
DISCOVERY PUBLISHING HOUSE PVT. LTD.
4383/4B, Ansari Road, Darya Ganj
New Delhi-110 002 (India)
Phone : +91-11-23279245; 23253475; 43596065
E-mail : discoverybooksindia@gmail.com
discoverypublishinghouse@gmail.com
namitwasan9@gmail.com
web : www.discoverypublishinggroup.com

***Edition:* 2020**

ISBN: 978-81-7141-729-2

Biomolecules

Printed at:
Infinity Imaging Systems
Delhi

PREFACE

The present title "Biomolecules" has been carefully compiled and edited to meet the long felt needs of increasingly number of students and researchers who have to deal with the different aspects of biochemistry and physiology. It encompasses all the chemical reactions carried out by cells. It provides a balanced and integrated treatment of the entire field of molecular biology. The material presented in the following pages will afford the student and researcher in the subject a comprehensive view and appropriate background material for focused studies. The method of presentation is very clear and lucid, which can be easily followed by an educated layman.

It is our earnest hope that this book will be of great value to all our students.

Author

CONTENTS

CHAPTER 1
BIOMOLECULES

Introduction

What is life?

Zoology is a science that deals with animals and animal life, and the student of zoology should have an understanding of the meaning of the term *life*. Unfortunately, there is no good definition of life. Although one has no difficulty distinguishing between a familiar plant or animal and a rock, other distinctions may not be so obvious. For instance, after a frog has had its brain destroyed, it will die eventually. At what precise point does the frog cease to live? Directly related to this seemingly pedestrian question is the medical legal question of human death: when is a human legally dead, and therefore, at what point may human organs be taken for transplantation? The processes that are agreed generally to be characteristic of all living organisms are metabolism, growth, reproduction, irritability, and evolution. The consideration of these characteristics illuminates the concept of life more than any single definition can.

Characteristics of Life

Life is easier to recognize than define. Well all recognize that a dog is alive, a stone is not. What properties of the dog distinguish it from the stone?

In 1976 on July 20 at 11:53 Greenwich Mean Time (GMT) and on September 3 at 10:39 GMT, identical lenders from two unnamed spacecraft, *Viking 1 and Viking 2*, settled gently on the surface of the planet Mars. Contained within each lander were a number of experiments designed to help us learn more about the planet. Five of these experiments sought evidence that would bear on the question, Is there anglify on Mars? The experiment taking most direct approach to the problem was the television camera. If the eye of the camera looked out and saw the eye of a Martian looking in, the question would be answered. However, that did not happen. Nor did the camera reveal any signs of the activities of living things. There were plenty of stones but nothing resembling a dog. So what other kinds of evidence of life could the landers look for?

The Chemistry of Life

One of the most characteristic features of living things on our planet is that they are constructed from molecules containing carbon atoms. This feature of living organisms is so characteristic, in fact, that the chemistry of carbon compounds is called **organic chemistry**. Some rocks also contain carbon atoms, but unless life processes have put them there, they do not contain organic molecules. Since life as we know it does not exist without organic molecules, one of the experiments incorporated in the *Viking* landers was designed to test for the presence of organic molecules in a sample of Martian soil. (None was found.)

The molecules of which living matter is composed are extraordinarily varied and complex. In addition to carbon, some two dozen of the 90-odd other elements found on earth participate in their construction.

Fig. : 1.1 Elemental composition of the earth's crust (the lithosphere) and human body.

Construction of Lithosphere		Composition of Human Body	
Oxygen	47	Hydrogen	63
Silicon	28	Oxygen	25.5
Aluminium	7.9	Carbon	9.5
Iron	4.5	Nitrogen	1.4
Calcium	3.5	Calcium	0.31
Sodium	2.5	Phosphorus	0.22
Potassium	2.5	Chlorine	0.03
Magnesium	2.2	Potassium	0.06
Titanium	0.46	Sulfur	0.05
Hydrogen	0.22	Sodium	0.03
Carbon	0.19	Magnesium	0.01
All others	<0.1	All others	<0.01

Two points should be made about these lists. First, living matter use only a fraction of the elements available to it, and the relative proportions of the elements it acquires from its nonliving surroundings are quite different from their proportions in the environment. Put another way, one of the properties of life is to take up certain elements that are scarce in the nonliving world and concentrate them within living cells. Hydrogen, carbon, and nitrogen together represent less than 1% of the atoms in the earth's crust but some 74% of the atoms in living matter. The ability of life to concentrate rare elements within itself is really quite remarkable. Some sea animals accumulate elements like vanadium and iodine within their cells to concentrations a thousand or more times as great as in the surrounding seawater.

A second point about the elemental composition of living matter is our uncertainty about the exact number of elements required. Some elements (e.g., selenium and aluminum) are found in minute concentrations within living things. Does such an element, despite its scarcity, play a role essential to life, or is it simply an accidental acquisition from the organism's surroundings (e.g., in its food)?

To prove the necessity the following experiment can be performed. Young rats are reared in plastic trace-element isolators. All dust is filtered out of the air the animals receive, and their food is prepared with the utmost care so that contaminating elements are reduced to a minimum. Such studies have demonstrated that rats cannot grow normally unless they receive minute quantities of vanadium, selenium, and fluorine in their diet. With respect to fluorine, 0.5 parts of fluorine per million parts of food are able to restore the animals to normal health and growth. There is an irony and an important lesson is this finding. Certain compounds of fluorine are used as rat poisons and, at the concentrations the animal ingests, they do their job well. So here is an element that in large doses is lethal but in small doses is absolutely essential to life.

The organization of the atoms and molecules in a living organism is far more dynamic than that in a nonliving object like a rock. The rock we see today contains the same atoms it contained last year. By contrast, the dog we see today is largely constructed of atoms acquired since we saw it last year. The atoms it has lost have returned to the environment. This rapid turnover of material in living things is part of their **metabolism**.

Metabolism

A dog continually exchanges material with its surroundings. It eats, drinks, urinates, and defecates. It breathes. With the appropriate equipment, we can show that the air the dog exhales differs in composition from the air it in hales: Oxygen has been removed and carbon dioxide added.

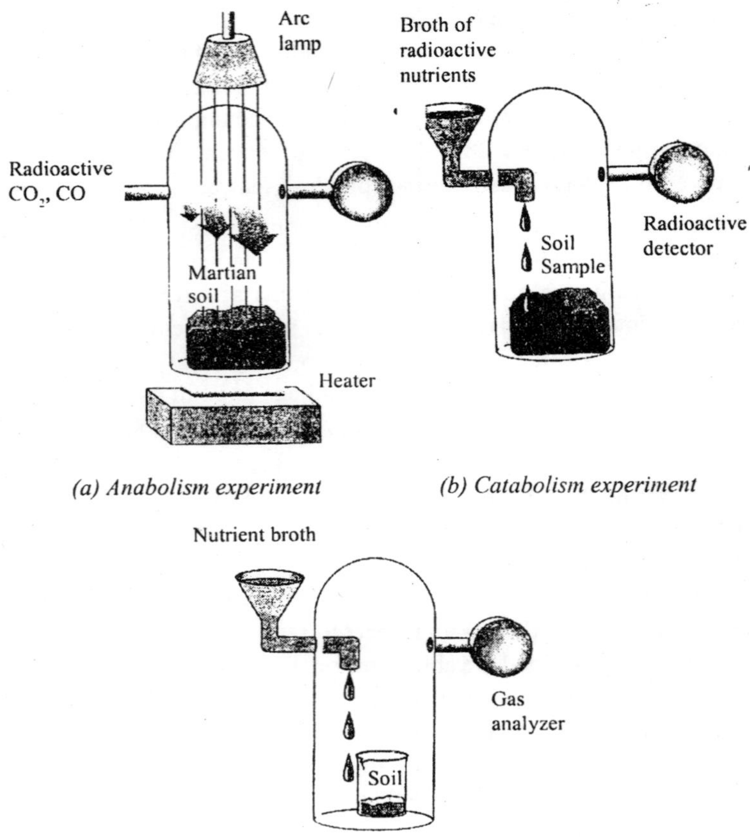

(a) Anabolism experiment *(b) Catabolism experiment*

(c) Gas exchange experiment

Fig. 1.2 : The three life detection experiments performed by the Viking landers

All organisms take in material from, and give material back to, the environment as they engage in metabolism. Three of the experiments included in the *Viking* landers were designed to look for evidence of metabolism occurring within samples of Martian soil.

One of the experiments looked for evidence that some thing in the soil could synthesize complex organic molecules from such simple carbon compounds as carbon dioxide (CO_2) and carbon monoxide (CO). In this experiment a mixture of radioactive CO_2 and CO was introduced

In this experiment a mixture of radioactive CO_2 and CO was introduced into a vessel containing a small sample of Martian soil and Martian atmosphere. The synthesis of organic molecules from simple precursors like CO_2 requires energy, and the chief form of that energy used here on earth is sunlight (used in photosynthesis). Therefore, the incubation mixture was illuminated by a bright arc lamp. After 5 days the soil sample was heated to drive off any radioactive organic molecules that might have been synthesized from the radioactive CO_2 (and/or CO) by organisms in the soil (figure 1.2a). This experiment was thus looking for evidence of metabolism involving the synthesis of complex substances from simple ones. Such metabolism is called **anabolism**.

A second experiment was designed to look for signs of **catabolism**—that is, destructive metabolism. In this experiment a soil sample was incubated with a dilute soup of radioactive organic molecules. Over the next 10 days, the atmosphere above the sample was regularly monitored for the appearance of radioactive gases such as carbon dioxide. Their appearance would show that the radioactive organic molecules with which the soil samples were soaked had decomposed, and this *might* have been caused by the metabolism of living things.

The third experiment designed to look for evidence of metabolism in Martian soil was the gas exchange experiment. In this experiment a known mixture of gases was placed above the soil sample and then periodically analyzed to see if any gases (e.g., CO_2) had disappeared from or been added to the mixture. If soil alone should fail to bring about a change in gas composition (it did not—large amounts of oxygen were given off at first), then a nutrient soup would be added to the soil. If organisms were there to catabolize the organic molecules in the soup, some of the products of catabolism should be gases and their appearance detected by the equipment.

The underlying assumption in these three experiments, which were run in an apparatus weighing 35 lb (16 kg), is that a fundamental property of life is metabolism—and if there is life on Mars, even of the simplest and most primitive kind, it should reveal its presence by its capacity to carry on metabolism.

Growth

The essence of reproduction in a living organism is the *self-controlled duplication* of its characteristic structures. It occurs when the organism takes in more material from its environment than it gives back *and* organizes these materials into its own structures. We call this kind of reproduction activity **growth**.

To do this, the organism must expend some of the energy released during its metabolism, and it must contain the **information**—encoded in **genes**—that will guide the building of its structures. If you have studied chemistry, you might argue that the crystals in a rock, once at least, grew in the same way. A crystal placed in a solution for its units slowly organizes these units according to its pattern and thus grows. A basic difference, however, exists between growth of a crystal and growth of a living organism. The crystal grows by accumulating from its surroundings units that are already identical to those in the crystal. The living organism, on the other hand, grows by transforming materials that are not unique to it into those that are. A dog can thrive on the same diet that we eat. The dog, however, converts the food into more dog, whereas we convert it into more human. In each case the characteristic pattern into which the nutrients are assembled is governed by the information encoded in the genes.

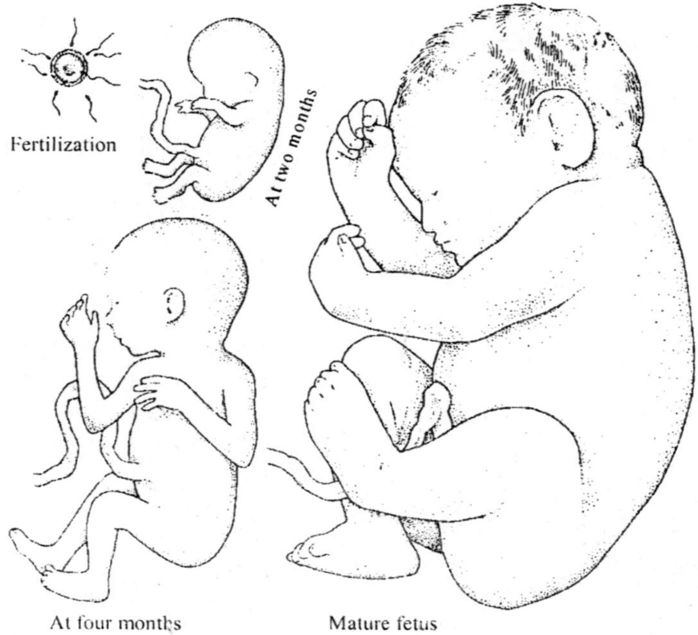

Fig 1.3 : Showing growth in human fetus

Genes are made of DNA. The DNA in our cells carries the information that makes us what we are. Every structure in your body and every function carried out in your body results from the expression of information encoded in your DNA. Your sex, blood type, the colour of your eyes, all the thousands of physical traits you are born with are encoded in DNA. The amount of DNA needed to encode a single attribute or function, say the information needed to make the hormone insulin, is called a **gene**. The oxygen-carrying haemoglobin in our red blood cells, the antibodies that protect us against disease, the proteins from which our muscles are made are all the products of genes. Although our bodies contain many substances that are not proteins (e.g.,glycogen), the synthesis of these depends on enzymes, and enzymes are proteins. Genes make proteins in two steps. First, the gene is copied into a molecule of RNA, a process called, appropriately, **transcription**. This is folowed by the **translation** of the RNA transcript into protein.

Reproduction

The third characteristic of organisms is the potential for *reproduction*. Each living thing reproduces offspring of its own kind. The new individual may be produced from a part of the old one (asexual reproduction), or it may be produced by a union of two sex cells formed by two individuals normally of the same species (sexual reproduction).

What has been said about reproduction as a criterion of life does not always apply. In the highly organized societies of the social insects some individuals—the workers—are naturally sterile. For example, a honeybee colony consists of a queen, thousands of sterile female workers, and a small number of male drones. A queen mates only once in her lifetime of 10 to 15 years, and she lays both fertilized and unfertilized eggs. The unfertilized eggs develop into female. If the female *larvae* are fed a rich diet they develop into fertile queens, but if they are fed a restricted diet they develop into sterile workers. The developmental environment of these workers has negated their potential of reproduction. The same is true for humans who are sterile because of some environmental accident or some inherited malady, for example, castration caused by an accident, or sterility resulting from a hormonal imbalance.

Responsiveness

All living things are capable of responding to certain changes (stimuli) in their surroundings. Changes in light, heat, gravity, sound, mechanical contact, and chemicals in the surroundings are common stimuli to which organisms respond. In order to respond to these stimuli, the organism must have a means of detecting them. The eyes, ears and nose of a dog are effective stimulus detectors as anyone who has watched these animals can testify.

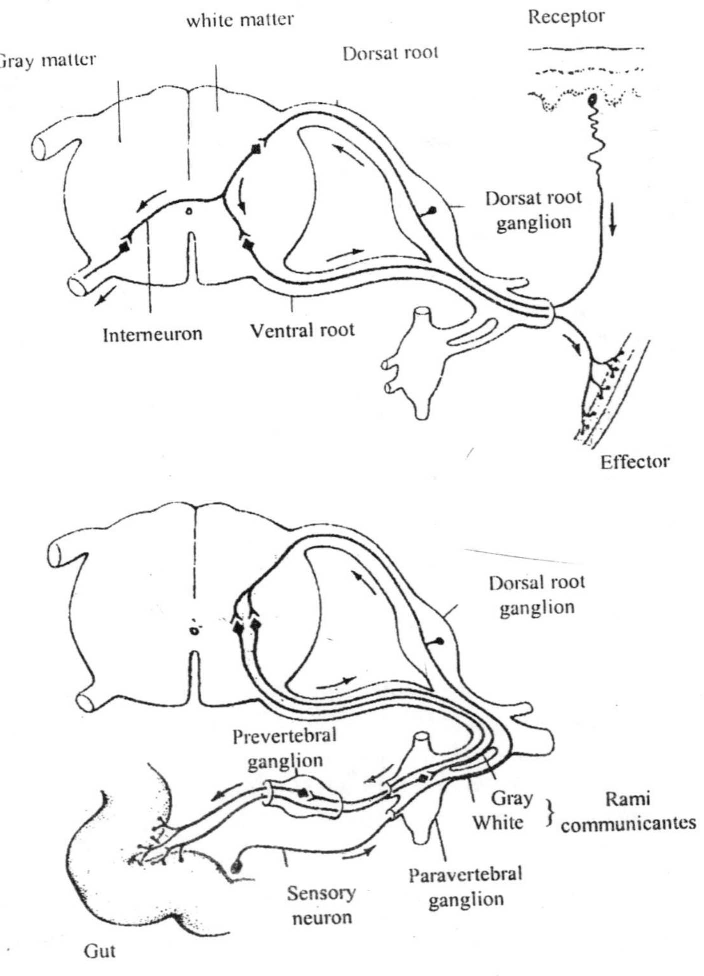

Fig. 1.4 : Autonomic reflex arc.

In order to effective, responses to changes in the environment must be coordinated. Even the simplest organisms consist of many parts, and each of these must do the right thing at the right time for an appropriate action to be carried out. When a dog is called to supper, some of its muscles contract, other relax, its digestive glands begin to secrete, and so on. Each part must work in harmony with the others. A system of **nerves** and a system of chemical regulators called **hormones** coordinate the actions of the dog and many other animals. Plants have no nervous systems and rely mainly on hormones for their coordination.

The action carried out by an organism in response to a stimulus is accomplished by its **effectors**. Muscles and glands are the most important effectors in animals. Like all other parts of the coordinating system, they consume energy.

Organisms respond to changes in their environment by altering their relationship to it. In running to its supper dish, the dog is changing its position in response to your signal. These responses, which often occur in definite patterns, make up the organism's **behaviour**. It is an active not a passive thing. A dislodged stone rolling downhill is not behaving; it is simply being carried along passively by the force of gravity. A hungry dog, on the other hand, is *creating* the change in its relationship to its surroundings.

Table 1.1 : The Vertical Method of Subdividing Biology

Discipline	Word Origin	Meaning of prefix	The Study of	
Anthropology	Gk. *anthrōpos*	Man	Natural history of man	Zoology (Gk. *zōon*, animal)
Mammalogy	L. *mamma*	Breast	Mammals	
Ornithology	Gk. *ornis*	Bird	Birds	
Herpetology	Gk. *herpeton*	Reptile	Reptiles	
Ichthyology	Gk. *ichthys*	Fish	Fishes	
Entomology	Gk. *entomon*	Insects	Insects	
Helminthology	Gk. *helmins*	Worms	Parasitic worms	
Protozoology	Gk. *protos* *zōon*	First Animal	Protozoa	
Bacteriology	Gk. *baktērion*	Small rod	Bacteria	Microbiology
Virology	L. *virum*	Poison	Viruses	
Phycology	Gk. *phykos*	Seaweed	Algae	
(Algology)	L. *alga*	Seaweed	Algae)	Botany (Gk. *Botane*, pasture) or Phytology
Mycology	Gk. *mykēs*	Fungus	Fungi	
Bryology	Gk. *bryon*	Moss	Mosses and liverworts	
Pteridology	Gk. *pteris*	Fern	Ferns	

Evolution

The final characteristic of living organisms is *evolution*, that is, living organisms change in the characteristics that specify their forms and functions. Organisms undergo changes in their genetic material (mutate); these changes are passed on to the offspring and to succeeding generations. Such changes are the raw materials for evolution. Those mutations which provide an adaptive advantages to an organism will result in the organism contributing a greater number of offspring to the next generation than other members of the population. Thus the property of evolution is a function of populations of organisms and not of individual organisms.

The evolutionary changes they occur within a line of descent are often adaptive—that is, the changes enable the descendants to live in their environment more efficiently than their ancestors could have. One of the best-studied cases of adaptive evolution is the industrial melanism.This phenomenon occurred over the course of a century (approximately 1850-1950), when many of the moths found in the industrial regions of England became very dark in colour. This change coincided with the widespread destruction of lichens because of air

Table 1.2 The Horizontal Method of Subdividing Biology

Discipline	Word Origin		Meaning of prefix	The Study of
Morphology	Gk.	*morphē*	Form	Form and structure
Anatomy	Gk.	*ana*	Up	Organisms as determined by dissection
		tomē	Cutting	
Histology	Gk.	*histos*	Tissue	Tissues by microscopy
Cytology	Gk.	*kytos*	Hollow vessesl	Cells
Physiology	Gk.	*physis*	Nature	Functions ans activities of cells and organisms
Taxonomy	Gk.	*taxis*	Arrangement	Classification
Ecology	Gk.	*oikos*	Household	Relationship between organisms and their environment
Genetics	Gk.	*genesis*	Descent	Heredity and variation
Embryology	Gk.	*embryon*	Embryo	Formation and development of organisms
Paleontology	Gk.	*Palaios* on	Ancient Being	Fossils and fossil impression
Evolution	L.	*evolvere*	To unroll	Descent of species

Table 1.2 Contd.

Discipline	Word	Origin	Meaning of prefix	The Study of
Zoogeography	Gk.	zōon	Animal	Distribution of animals
		gē	Earth	
		graphein	To write	
Phytogeography	Gk.	phyton	Plant	Distribution of plants
		gē	Earth	
		graphein	To write	
Biochemistry	Gk.	bios	Life	Chemistry of organisms
		chemeia	Transmutation	
Biophysics	Gk.	bios	Life	Biological phenomena in terms of physical principles
		physis	Nature	
Immunology	L.	immunis	Free	Resistance of organisms to infection
Radiobiology	L.	radius	Ray	Effects of radioactivity on biological material
	Gk.	bios	Life	
Endocrinology	Gk.	endon	Within	Study of hormones and their effects
		krinein	To separate	
Psychobiology	Gk.	psyche	Soul	Related areas of psychology and biology
Limnology	Gk.	limne	Lake or marsh	Freshwater lakes, ponds, and streams

pollution. Lichen-encrusted tree trunks provide excellent camouflage for light-colored moths, whereas bare tree trunks provide camouflage for dark moths. The widespread adoption of air pollution controls in recent decades has allowed the lichens to return and, at the same time, has led to the gradual reappearance of the light-coloured moths. Here is a case of evolutionary change occurring in a remarkably brief interval of time. Like all evolutionary changes, it involved changes in the genes carried by the individuals in the evolving population.

Evolution involves something more than gradual change within a single line of descent. It also involves a proliferation of kinds of organisms, as single groups of organisms give rise to two or more distinct kinds of descendants. The number of kinds of organisms that now live on the earth is far greater than the number that were present during the first billion years that life existed on earth. For example, five subfamilies of present-day ants, encompassing many hundreds of species, are thought to be descended from early ants like *Sphecomyrma*. The study of interrelationships between organisms and their environment is called **ecology**.

In summary, living things differ from nonliving things in several ways: (1) their structural organization is dazzlingly intricate and complex; (2) they metabolize; (3) they reproduce themselves; (4) they respond to change in their environment; and (5) all of these features are shaped over time by the process of evolution.

In addition to the five functional characteristics, living organisms are characterized by their structure (body forms), their organization, and their composition. The structure of each organism is no haphazard array of materials. The organism is put together in quite a precise manner, and deviations from the precise order often result in severe incapacitation or death. This text discusses the organization of various organisms at several levels; indeed, there is a system of progressive organization called a *hierarchy* of *organization*. This hierarchy runs froms *atoms* and *molecules* to *cells*, the units of structure aand function of organisms. Cells are organized into *tissues*, which are groups of like cells that are structurally modified to perform a particular function. Tissues are combined into *organs*, and a number of organs functioning together constitute an *organ system*.

An organism, then, is composed of a certain number of organ systems that are integrated in the structural and functional organization of the individual. The hierarchy of organization of living things can be extened to the *supraorganismic* level. Organisms affect and are affected by each other; the relationships among organisms depend on both physical and biotic factors. Organisms of the same kind constitute *populations*; several specific populations living in the same environment are called *communities*. The community with its nonliving environment constitutes an *ecosystem*. As mentioned earlier the large, easily recognizable regions of the world (such as grasslands, coniferous forests, and deserts) are called biomes.

What Is Biology?
What Is Zoology?

Biology is the science of living things. The word biology is derived from the greek *bios*, meaning "life," and *logos*, "the study of." Historically, knowledge about living things was developed somewhat independently by students of plants and by students of animals. As a result, many biologists think of two main subdivisions of biology – *botany*, the study of plants, and *zoology*, the study of animals. Other biologists feel that there are really three types of organisms plants, animals, and micro-organisms -and consider *microbiology* to be a major subdivision of biology. Table 1.1 gives some areas of study according to the taxonomic-based division of biology. This list is not comprehensive, but it gives some idea of the "natural areas" that have arisen according to the classification scheme. Included in Table 1.1 are the origins of these terms. This way of subdividing biology is sometimes called the *vertical method.*

The other major method of subdividing biology into disciplines is based on an operational or functional approach which cuts across taxonomic lines. Some of these are listed in Table 1.2 This type of scheme is sometimes called the *horizontal method* of organization. Since the scope of biology has practically no boundaries, many of the important advances, particularly in the past few years, have been made by workers who defy categorization into a particular branch of biology. How does one classify an individual who studies conditions for the origin of life on the primitive earth – biochemist, evolutionist, microbiologist, philosopher, or jack-of-all-trades? Because workers have become specialized in various aspects of biology, many investigations are carried on by *team research* in which zoologists, botanists, microbiologists, chemists, physicists, physicians, and mathematicians collaborate.

With the advent of many new research techniques and a resultant explosion of knowledge, there has been a tendency to divide biology into a new set of horizontal categories consisting of *molecular biology, cellular biology, developmental biology, organismal biology,* and *population* and *community biology.* This method of division will be called *modified horizontal* because the categories are more inclusive than those in the horizontal method of division. Several college biology curricula have been organized along the lines of the modified horizontal method. Molecular biology encompasses biochemistry and biophysics, and presumably can be said to include all aspects of biology which take a molecular approach to the problems and their solution. Cellular biology simply may include all approaches to structure and function of cells. Included are the chemical and physical organization, energetics, transport, mobility and stabilizing mechanisms of cells, growth and division, differentiation, and development. Developmental biology is much broader than the traditional embryology; it includes development from the molecular level to gross structural levels. The phenomena of regeneration, wound repair, and aging are included in developmental biology.

It is obvious that there is considerable overlap among molecular biology, cellular biology, and developmental biology. All of them involve investigations at the molecular level, and all involve the flow of information in biological systems; the term molecular genetics is often used to refer to these aspects of biology.

Organism biology focuses on whole organisms. It is concerned with such matters as functional and developmental anatomy, phylogeny, comparative physiology, psychobiology, and ecology. Population and community biology involve a part of traditional ecology. In this field the primary concern is with the structure and dynamics of populations, communities, ecosystems, and with the ecological aspects of natural selection.

The three methods of dividing biology – vertical, horizontal, and modified horizontal – do not delineate many aspects of applied biology. Biology interacts with the space science, earth sciences, mathematics, physical sciences, social sciences, and humanities. Among these applied fields are animal husbandry, plant breeding, medicine, agriculture, pharmacology, conservation, horticulture, bioclimatology, and many others.

Concepts of Zoology

This introductory chapter has touched on several areas of zoology as a discipline; the vastness and complexity of the discipline are apparent. It is reasonable for one to wonder at the outset if the concepts of zoology can be stated in a brief outline. These concepts should be fundamental ideas that serve as the skeleton of zoology. The following twelve concepts would be suggested by most professional zoologists.

1. Living systems obey the laws of chemistry and physics.
2. All organisms are composed of cells and cell products (the cell concept).
3. Living cells are able to convert energy from one form to another.
4. Metabolism is catalyzed by enzymes which are under genetic control.
5. Many vitamins are related structurally to coenzymes.
6. Hormones regulate many cellular and organismal activities.
7. As a result of evolution, living systems are organized; structurally

and functionally, from atom to biome to carry out a variety of specific functions.

8. Control or regulatory mechanisms exist at all levels of organization.
9. Life comes only for living things by reproduction (biogenesis rather than abiogenesis).
10. DNA is the genetic material (RNA serves this role in certain viruses), and genetic potential is transmitted from generation to generation through genes (gene theory).
11. The primary factor in organic evolution is the number of offspring which survive and become parents of the next generation.
12. Organisms interact with their biological and physical environments.

CHAPTER 2
ATOMS AND MOLECULES

Because every organism, as well as its environment, is made up of atoms and molecules, biologists use their knowledge of chemistry every day. In fact, much of modern biology is concerned with **molecular biology**—the chemistry and physics of the molecules that constitute living things. A molecular biologist might study how the structure of a DNA molecule allows it to store and transmit genetic information, or might investigate the precise interactions among a cell's atoms and molecules that maintain the energy flow essential to life. However, an understanding of chemistry is essential to *all* biologists. An evolutionary biologist might study evolutionary relationships by comparing proteins produced by different types of organisms. An ecologist might study the biological effects of changes in the salinity of the water in an estuary. An ecologist might study the biological effects of changes in the salinity of the water in an estuary. A botanist might be a "chemical prospector," seeking new sources of medicines from plants.

In recent years biologists have learned that, although organisms are strikingly diverse, their chemical composition and fundamental metabolic processes are remarkably alike. These chemical similarities provide strong evidence for the evolution of all organisms from a common ancestor, and explain why much of what biologists learn from studying bacteria or rats in laboratories can be applied to other organisms, including humans. Furthermore, the physical and chemical principles governing organisms are not unique. They are the same as those governing nonliving systems.

In this chapter we lay a foundation for understanding how the structure of atoms determines the way they form chemical bonds to produce compounds of increasing complexity. Most of our discussion will center around small, simple substances known as **inorganic compounds**. Among the biologically important groups of inorganic compounds are water, simple acids and bases, and simple salts. We pay particular attention to water, the most abundant substance on Earth's surface, and examine how its unique properties affect organisms as well as their nonliving environment.

Atoms

All matter, living and nonliving, is composed of **atoms**. More than a mi million atoms could fit in a single layer over the period at the end of this sentence. Each atom consists of a dense, positively charged nucleus, around which one or more negatively charged electrons move. The nucleus contains one or more protons and may contain one or more neutrons. Electrons, protons, and neutrons are not indivisible particles. Each has a substructure, but that level of organization (the world of quarks) has no known consequence for biology.

The weight, or mass, of a proton serves as a standard unit: the atomic mass unit (amu) or dalton (after the English scientist John Dalton, who studied atoms two centuries ago). A single proton or neutron weighs one dalton, which is 1.7×10^{-24}gram (0.0000000000000000000000017 g), whereas the mass of an electron is 9×10^{-28} (0.0005 dalton). Because they weigh so much less than protons and neutrons, electrons contribute only negligibly to the mass of an atom.

The positive electric charge on a proton is defined as a unit of charge. An electron has a charge equal and opposite to that of a proton. Thus the charge of proton is + 1 unit, that of an electron is – 1. The neutron, as its name suggests, is electrically neutral, so its charge is 0. The number of protons in an atom equals the number of electrons, so the atom itself is electrically neutral.

Table 2.1
Abundance of Some Chemical Elements

Atomic Number	Sym-bol	Element	Atomic Weight	Abundance, as % of Universe	Earth's Crust	Human Body
1	H	Hydrogen	1.01	87	0.14	9.5
2	He	Helium	4.00	12		
3	Li	Lithium	6.94		0.01	
4	Be	Beryllium	9.01			
5	B	Boron	10.81			
6	C	Carbon	12.01	0.03	0.03	18.5
7	N	Nitrogen	14.01	0.01		3.3

Atomic Number	Sym-bol	Element	Atomic Weight	Abundance, as % of Universe	Earth's Crust	Human Body
8	O	Oxygen	16.00	0.06	46.6	65.0
9	F	Fluorine	19.00		0.03	
10	Ne	Neon	20.18	0.02		
11	Na	Sodium	22.99		2.83	0.2
12	Mg	Magnesium	24.31		2.09	0.1
13	Al	Aluminium	26.98		8.13	
14	Si	Silicon	28.09		27.7	
15	P	Phosphorus	30.97		0.12	1.0
16	S	Sulfur	32.06		0.05	0.3
17	Cl	Chlorine	35.45		0.03	0.2
18	A	Argon	39.95			
19	K	Potassium	39.10		2.59	0.4
20	Ca	Calcium	40.08		3.63	1.5
21	Sc	Scandium	44.96			
22	Ti	Titanium	47.90		0.44	
23	V	Vanadium	50.94		0.02	
24	Cr	Chromium	52.00		0.02	
25	Mn	Manganese	54.94		0.1	
26	Fe	Iron	55.85		5.0	
27	Co	Cobalt	58.93			
28	Ni	Nickel	58.70		0.01	
29	Cu	Copper	63.55		0.01	
30	Zn	Zinc	65.38		0.01	
31	Ga	Gallium	69.72			
32	Ge	Germanium	72.59			
33	As	Arsenic	74.92			
34	Se	Selenium	78.96			
35	Br	Bromine	79.90			
36	Kr	Krypton	83.80			
37	Rb	Rubidium	85.47		0.03	
38	Sr	Strontium	87.62		0.03	
39.	Y	Yttrium	88.91			
40	Zr	Zirconium	91.22		0.02	
41	Nb	Niobium	92.91			

Atomic Number	Sym-bol	Element	Atomic Weight	Abundance, as % of Universe	Earth's Crust	Human Body
42	Mo	Molybdenum	95.94			
43	Tc	Technetium	98.91			
44	Ru	Ruthenium	101.07			
45	Rh	Rhodium	102.91			
46	Pd	Palladium	106.40			
47	Ag	Silver	107.87			
48	Cd	Cadmium	112.41			
49	In	Indium	114.82			
50	Sn	Tin	118.69			
51	Sb	Antimony	121.75			
52	Te	Tellurium	127.60			
53	I	Iodine	126.90			
54	Xe	Xenon	131.30			

The list contains only the first 54 of 92 natural elements.

Atoms and Elements, Molecules and Compounds

All matter, whether it exists in an alga or a distant star, is composed of elements. **Elements** are substances that cannot be divided into simpler substances by chemical means. **Atoms** are the smallest indivisible particles of these elements. Atoms *can* be divided into their component parts, but when this is done, the special qualities of that element are lost. Each element, or type of atom, has its own special chemical behaviour by which it is identified. Oxygen, chlorine, and carbon are examples of elements.

Ninety-two elements occur naturally, and about seventeen others have been synthesized in the laboratory. Each has been named and given a letter symbol. For example, the element called hydrogen is designated by the letter H. Sulphur is S. Oxygen is O. Of course, nothing is all that simple, so we find that sodium is designated by Na, from its Latin name *natrium*. The symbol standing alone usually designates on atom. For example, O refers to one atom of oxygen. The symbol 2O refers to two atoms of oxygen. But the symbol O_2 means that two oxygen atoms are joined together into one *molecule* of oxygen, the form in which oxygen usually exists in nature. **Molecules** are formed from combinations of atoms, and **compounds** are substances formed of one kind of molecule. Water, then, is a compound formed of hydrogen and oxygen atoms joined to form molecules that are designated by the symbol H_2O.

Chnops

CHNOPS is an elegant word that might help you to identify the six important elements that make up about 99 percent of living matter. However, if you drop the word casually at your next gathering, it is not likely that everyone will know you are referring to carbon, hydrogen, nitrogen, oxygen, phosphorus, and sulfur. If they do, you are in the company of organic chemists and should leave immediately.

The structure of an Atom

Let us now consider what an atom is supposed to look like. We say "supposed to" because atoms cannot be seen, so all the evidence is circumstantial. Atoms, according to theory, are composed of a dense center around which small particles spin in orbit. The center, called the **nucleus**, consists of **protons** and, usually, **neutrons**. Each proton carries one positive charge, designated by a plus sign (+).

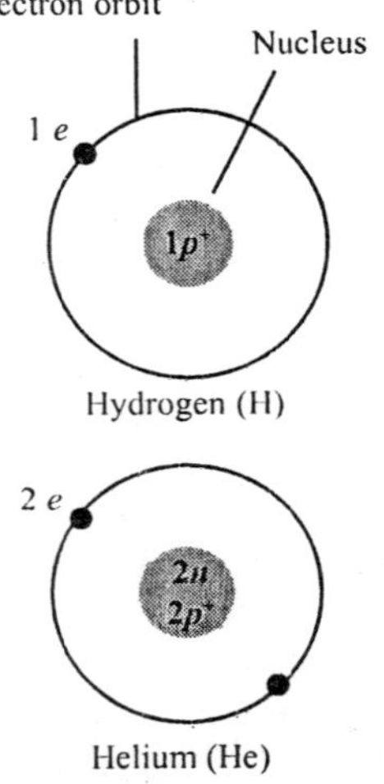

Fig. 2.1 The Hydrogen atom (top) and the helium atom (bottom)

Neutrons, as the name implies, carry no charge. Protons and neutrons are similar in size and are said to weigh one atomic unit each. The particles in orbit around the nucleus are called **electrons**. They are much smaller than protons (having about 1/1835 the mass). Each electron carries one negative charge, designated by a minus sign (–). An atom of hydrogen, the simplest element, consists of one proton, and one electron, with no neutrons.

Helium, the next simplest element, is composed of two protons, two neutrons, and two electrons. Since an atom's **atomic number** refers to the number of protons in the nucleus, the atomic number of hydrogen is one; of helium, two. It is important to know that elements differing by even a single proton may have vastly differing chemical properties.

Table 2.2: Components of the Atom

NAME	WEIGHT	CHARGE
Proton	one atomic unit	positive
Electron	negligible	negative
Neutron	one atomic unit	

You may have noticed in the diagrams of hydrogen and helium that the number of positively charged protons are balanced by the negatively charged electrons, resulting in the atoms having a net neutral charge. If they have equal numbers of protons and electrons, they are said to be electrically balanced. As a final note about our examples here, it should be mentioned that the figures are not drawn to scale. In nature, the electrons are nowhere near the nucleus. In fact, if the period at the end of this sentence were a nucleus. In fact, if the period at the end of this sentence were a nucleus, its nearest electron would probably be across the street somewhere.

Isotopes

Atoms with the same atomic number (that is, the same number of protons) have the same-chemical properties and are said to belong to the same **element**. Formally speaking, an element is any substance that cannot be broken down to any other substance by ordinary chemical means. However, not all atoms of an element have the same number of neutrons. Atoms of an element that possess different numbers of neutrons are said to be **isotopes** of that element. Most elements in nature exist as mixtures of different isotopes. There are, for example, three isotopes of the element carbon (C), all of which possess six protons. Over 99% of the carbon found in nature exists as an isotope with six neutrons. Because its total mass is 12 daltons (6 from protons plus 6 from neutrons), this isotope is referred to as carbon–12, and symbolized 12C. Most of the rest of the naturally occurring carbon is carbon–13, an isotope with seven neutrons. The rarest carbon isotope is carbon–14, with eight neutrons. Unlike the other two isotopes, carbon–14 is unstable: its nucleus tends to break-up, which with lower atomic numbers. This nuclear break-up, which emits a significant amount of energy, is called radioactive decay, and isotopes that decay in this fashion are said to be **radioactive**.

Some radioactive isotopes are more unstable than others, and, therefore, they decay more readily. For any given isotope, however, the rate of decay is constant. This rate is unusually expressed as the **half-life**, the time it takes for one half of the atoms in a sample to decay. Carbon–14, for example, has a half-life of about 5600 years. A sample of carbon containing 1 gram of carbon-14 today would contain 0.5 gram of carbon–14 after 5600 years, 0.25 gram 11,200 years from now, 0.125 gram 16,800 years from now, and so on. By determining the ratios of the different isotopes of carbon and other elements in samples of biological origin and in rocks, scientists are able to make absolute determinations of the times when these materials formed.

While there are many useful applications of radioactivity, there are also harmful side effects that must be considered in any planned use of radioactive substances. Radioactive substances emit energetic subatomic particles that have the potential to severely damage living cells, producing mutations in their genes, and, at high doses, cell death. Consequently, exposure to radiation is now very carefully controlled and regulated.

Isotopes differ in number of neutrons

Most elements consist of a mixture of atoms with different numbers of neutrons and thus different masses. Such atoms are called **isotopes**. Isotopes of the same element have the same number of protons and electrons; only the number of neutrons varies. The three isotopes of hydrogen, ${}^{1}_{1}H$ (ordinary hydrogen), ${}^{2}_{1}H$ (deuterium), and ${}^{3}_{1}H$ (tritium), contain zero, one, and two neutrons, respectively.

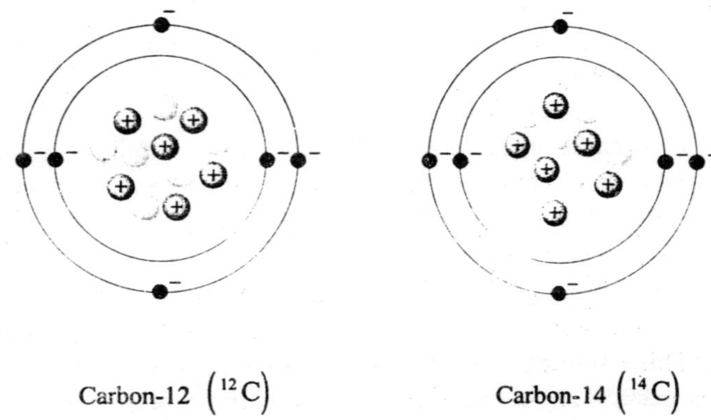

Carbon-12 $\left({}^{12}C\right)$ Carbon-14 $\left({}^{14}C\right)$

Fig.2.2 : Isotopes differ in atomic mass.

The standard for comparing elements is based on assignment of an atomic mass of exactly 12 to $^{12}_{6}C$ the most common isotope of carbon. The atomic mass for an element reflects the masses of the mixtures of isotopes that occur in nature. For example, although more than 99% of the hydrogen atoms in a naturally occurring sample have an atomic mass of 1 amu (to be precise, 1.00000078 amu on the carbon–12 scale), the atomic mass of hydrogen is 1.0079 amu. Likewise, its atomic weight is 1.0079. This reflects the fact that a small amount of deuterium, $^{2}_{1}H$ (mass number 2), and an even smaller amount of tritium, $^{3}_{1}H$ (mass number 3), occur along with the common form of hydrogen, $^{1}_{1}H$.

All the isotopes of a given element have essentially the same chemical characteristics. However, some isotopes with an excess of neutrons are unstable and tend to break down, or decay, to a more stable isotope (usually becoming a different element). Such isotopes are termed **radio-isotopes** because they emit radiation when they decay. Sophisticated instruments allow scientists to detect and measure this radiation.

Raidoisotopes such as ^{3}H (tritium), ^{14}C, and ^{32}P have been extremely valuable research tools, useful in studies ranging from determining the age of fossils to DNA synthesis to sugar transport in plants. This is because, despite the difference in the number of neutrons, the organism treats all isotopes of a given element in a similar way.

In medicine, radioisotopes are used for both diagnosis and treatment. The reactions of a sugar, hormone, or drug can be followed in the body by labeling the substance with a radioisotope such as carbon-14 or tritium. For example, the active component in marijuana (tetrahydrocannabinol) can be labelled and administered intravenously. Then the amount of radioactivity in the blood and urine can be measured at successive intervals. Results of such measurements have determined that this compound remains in the blood for several weeks, and products of its metabolism can be detected in the urine. Radioisotopes are also used to test thyroid gland function, to measure the rate of red blood cell production, and to study many other aspects of body function and chemistry.

Because radiation can interfere with cell division, radioisotopes have been used in the treatment of cancer (a disease often characterized by rapidly dividing cells).

Electron can be involved in chemical changes

Although the mass of electron makes only negligible contribution to the mass of an atom, electrons carry an electrical charge that profoundly affects the chemical properties of the atom. Each electron bears a charge of – 1, exactly equal but opposite to the charge on a proton. The electrons are attracted by the positive charges of the protons. The atom as a whole has no net charge. The positive charges of the protons equal the negative charges of the electrons.

Although the number of protons in an atom does not change during a chemical reaction, the number and relative positions of the electrons usually do change.

Electrons occupy orbitals corresponding to energy levels

Electrons move rapidly through characteristic regions of space termed **orbitals**. Each orbital contains a maximum of two electrons. It is impossible to know an electron's position at any given time. For this reason each orbital represents the region in which the electrons that occupy it are most probably found. This probability can be depicted as a shaded area whose density is proportional to the probability that an electron is present there at any given instant. Such a representation

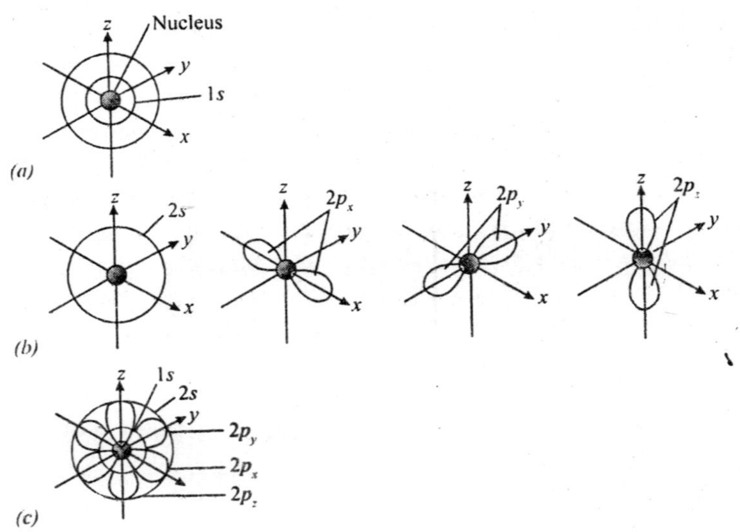

Fig. 2.3 : Atomic orbitals can be represented as probability densities ("electron clouds")

of an orbital is referred to as an "electron cloud".

The energy of an electron depends on the orbital it occupies. The lowest energy orbital, the 1s orbital, is nearest the nucleus and is spherical in shape. The next 2*s*, is also spherical. The electron clouds of the 2*p* orbitals are dumbbell-shaped. These occur in groups of three, arranged at right angles to each other. Other, higher energy electron orbitals are either spherical or dumbbell-shaped or are represented by more complex three-dimensional coordinates.

Electrons in orbitals with similar energies, said to be at the same **principal energy level**, make up an **electron shell.** In general, electrons in a shell distant from the nucleus have greater energy than those in a shell close to the nucleus. For example, the lowest energy electrons, the 1s electrons, occupy the first (innermost) electron shell. Although each orbital may contain no more than two electrons, there may be several orbitals within a given electron shell. The 2*s* electrons and the 2*p* electrons have similar energies and occupy the second electron shell. The number of electrons in the highest energy level (outermost electron shell) determines the chemical properties of an atom.

The way that electrons are arranged around an atom is referred to as the **electron configuration** of that atom. The electron configuration can be conveniently represented by a series of concentric circles around the nucleus, each circle representing an electron shell. It is important to remember, however, that electrons do *not* circle the nucleus in fixed concentric pathways.

Electrons always fill the lower energy orbitals before occupying the higher energy ones. The maximum number of electrons in the innermost shell (1s, which is a single spherical orbital) is two. The second shell has four orbitals (2s, which is spherical , plus the three dumbbell-shaped 2p orbitals) and thus can contain a maximum of eight electrons. The third shell has a maximum of 18 electrons arranged in nine orbitals, and the fourth has 32 electrons in 16 orbitals. Although the third and outer shells can each contain more than eight electrons, they are stable when only eight are present. We may consider the first shell to be complete when it contains two electrons, and the other shells to be complete when they each contain eight electrons.

Remember that each atom is largely empty space. The distance from an electron to the protons and neutrons in the central nucleus may be thousands of times greater than the diameter of the nucleus itself. The tendency of the negatively charged electrons to fly off into space is countered by their attraction to the positive charges of the protons in the nucleus.

Energy is required to move a negatively charged electron farther away from the positively charged nucleus. An electron can move to an orbital farther from the nucleus by receiving more energy, or it can give up energy and sink to a lower energy level in an orbital nearer the nucleus.

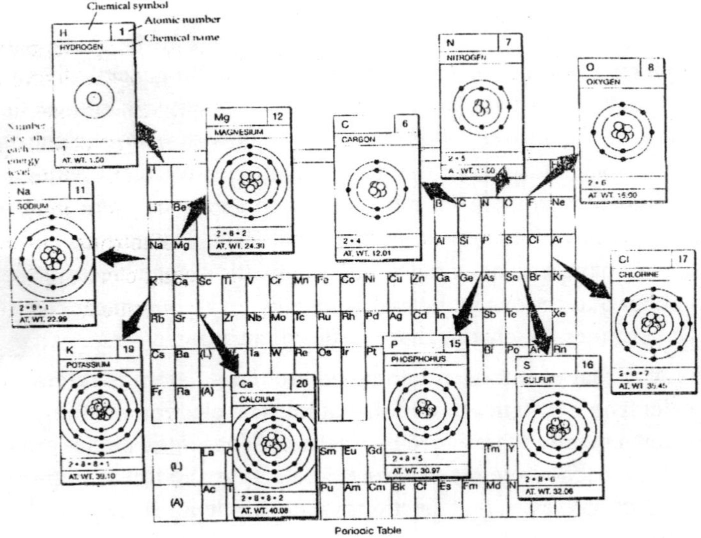

Fig. 2.4 : This representation of the periodic table shows Bohr models of some biologically important atoms.

When energy is added, an electron can jump from one energy level to the next, *but it cannot stop at some intermediate energy level*. To move an electron from one level to the next, the atom must absorb a tiny discrete packet of energy known as a **quantum** (pl. *quanta*), which contains just the right amount of energy for the transition — no more and no less. The term *quantum leap* is used in everyday language to indicate a sudden discontinuous move from one level to another. Changes in electron energy levels are important in energy conversions in organisms. For example, in photosynthesis quanta of light energy absorbed by chlorophyll molecules cause electrons to move to a higher energy level.

The Behavior of electrons

The part of the atom of greatest biological interest is the electron. What happens in cells happens because of the way electrons behave. The characteristic number of electrons in each atom of an element determines how the atom reacts with other atoms. All **chemical reactions**, in cells or anywhere else, are exchange of electrons, or changes in the sharing of electrons between atoms.

Where a given electron in an atom is at any given time is impossible to say. We can only describe a volume of space within the atom where the electron is likely to be. The region of space within which the electron is found at least 90 percent of the time is the electron's **orbital**. An electron spins like a top-or like Earth on its axis-and, like a top, may spin in one of two directions: clockwise or counter-clockwise. In an atom, a given orbital can be occupied by at most two electrons, which must spin in opposite directions. Thus any atom larger than helium (atomic number = 2) must have electrons in two or more orbitals.

The orbitals constitute a series of shells around the nucleus. The innermost shell, called the K shell, consists of only on orbital, an s orbital. The s orbital fills first, and its electrons have the lowest energy. Hydrogen ($_1H$) has one K-shell electron; helium ($_2He$) has two; all other atoms have two K-shell electrons and electrons in other shells as well. The L shell is made up of four orbitals (an *s* orbital and three *p* orbitals) and hence can hold up to eight electrons. The M, N, O, P, and Q shells have different numbers of orbitals.

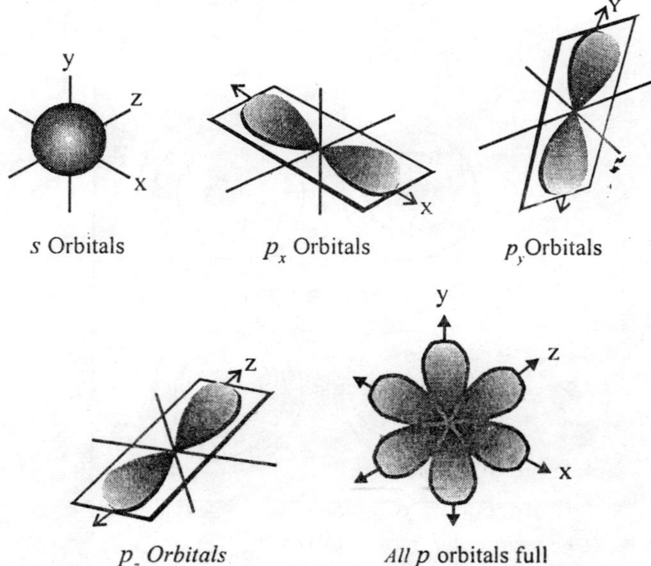

s Orbitals p_x Orbitals p_y Orbitals

p_z *Orbitals* *All p* orbitals full

In any atom, it is the outermost shell of electrons that determines

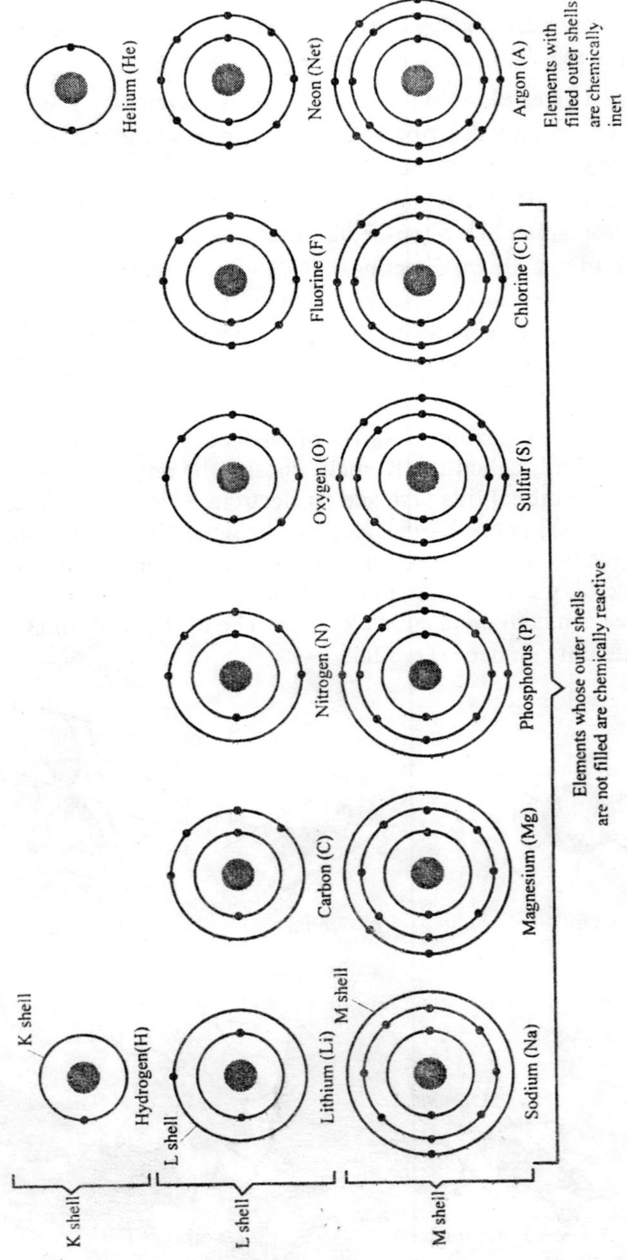

Fig. 2.5 : Electron Shells Determines the Reactivity of Atoms.

what the atom can do chemically. When the outer shell is full, the atom will not react with other atoms. Examples of some chemically inert elements (elements with full outer shells) are helium, neon, and argon. Other elements are reactive in various degrees, depending on the number of electrons in the outermost shell. They are, in a sense, seeking ways to fill their outer shells with electrons by combining with other atoms.

Electrons and the Properties of the Atom

If you remember the numbers two, eight, and eight, you will have at your fingertips the answer to very few questions. But one such questions would be, what is the maximum number of electrons that can exist in the first three shells of an atom? Only a specific number of electrons can occupy any one electron shell; the first (the innermost) shell can accommodate only two, and the second and third hold only eight each. (The fourth can hold thirty-two, and some atoms have even more shells.) However, no matter how many shells an atom has its chemical properties are determined primarily by the outer shell. It is the one most likely to gain or lose electrons, and the atom tends to behave in such a manner as to keep this shell filled. When this outer shell is filled, the atom is said to be **inert**. It is almost impossible to make inert atoms react with any other element. The outer shells of helium and neon, for example, are full. In the case of helium, with only two electrons, there is only one shell, the inner one, and it has all the electrons it can hold. Neon, as you see, has only two shells, both completely occupied. These, then, are "inert" elements that generally do not react with other elements.

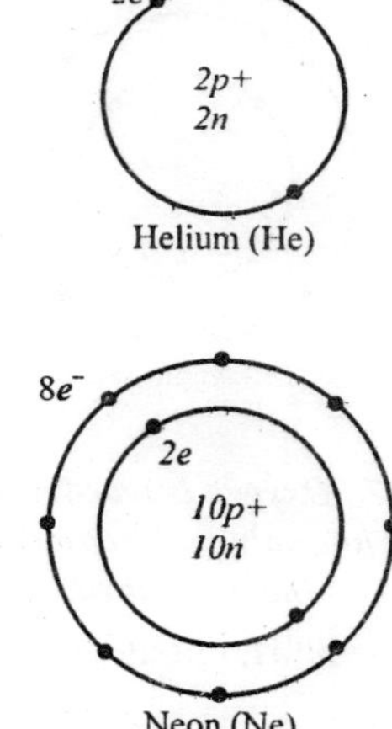

Fig. 2.6 : Helium (top) and neon (bottom) are called inert atoms.

So we see that atoms tend to "satisfy" themselves in two ways. First, *they tend to balance the number of their protons and electrons*, and second, *they tend to keep their outer shells filled*. When these conditions are met, the atoms are inert, unreactive, and not of great interest to us here. With atoms, as perhaps with people, the most active and interesting ones are often the most "dissatisfied".

Why is the number of electrons in the outer shell so important? Perhaps this can best be understood by beginning with a couple of examples. Let us consider the case of oxygen. Oxygen has an atomic number of 8; that is, it has eight protons, and, therefore, needs eight electrons. However since the inner shell will require two electrons, the outer shell is left with only six—two short of its shell requirements. So, oxygen tends to accept two electrons from other atoms.

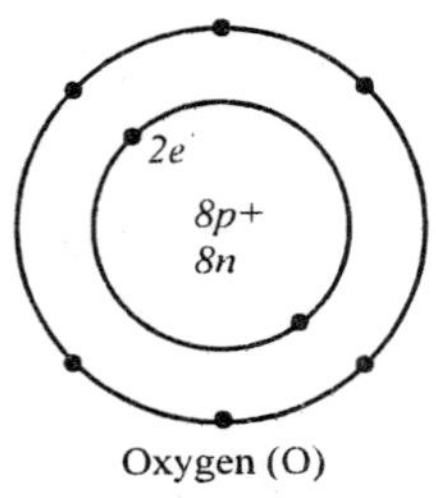

Fig. 2.7 : Oxygen is reactive because it has only six electrons in the outer shell.

On the other hand, note that sodium has an atomic number of 11. This means it has two electrons in its first shell, eight in the second shell, but only one in the third shell. Since it is not energetically feasible for a sodium atom to gain seven electrons, it tends to fulfill its outer shell requirements by losing its single outer electron. This loss means that it now only has two shells, but at least both are satisfied.

Of course, the loss of the electron (with its negative charge) ionizes the sodium. That is, it leaves it with a positive charge. Sodium ionized in this way is written Na^+. Magnesium, which has twelve electrons, is sometimes written as Mg^{++} (do you see why?). The fact that chlorine has an atomic number of 17 may give you a clue to the reason salt (NaCl) is formed so easily. Take a minute here to consider the question.

The Atoms of Life

Of the 92 naturally occurring elements on earth, only 11 are found in organisms in more than trace amounts (0.01% or higher). These 11 elements have atomic numbers less than 21 and, thus, have low atomic masses. The most common elements inside organisms are not the elements that are most abundant in the earth's crust. For example, silicon, aluminium, and iron constitute 39.2% of the earth's crust, but they exist in trace amounts in the human body. On the other hand, carbon atoms make up 18.5% of the human body but only 0.03% of the earth's crust.

Four elements–oxygen, hydrogen, carbon and nitrogen–comprise 96.3% of the weight of the human body and a similarly large proportion of the weight of all other organisms. There are several reasons why these four elements are so abundant in living things.

1. They all form molecules with one another by sharing electrons and so making covalent bonds.
2. These bonds are weak enough to be broken at temperatures compatible with life, enabling the atoms to be rearranged in a variety of molecules.
3. Approximately 90% of the atoms (and about 75% of the weight) of orgnisms are oxygen and hydrogen, reflecting the predominant role of water (H_2O) in living systems.
4. Many of the molecules they form are gases, including water; most of the other gaseous molecules are soluble in water. As water vapour in the primitive earth's atmosphere cooled and fell as rain, it brought the other dissolved molecules into the oceans. Life is thought to have evolved from complex molecules formed by the interaction of these smaller molecules in the oceans and the atmosphere.

Chemical Bonds

The elements differ in the affinity that their atoms have for additional electrons. The greater this affinity for electrons, the greater the electronegativity of the atom. If we distort the periodic table so that the greater an element's electronegativity, the higher it is placed in the table (figure 2.10), two generalizations can be made. First, the lower the atomic number of element in a given column, the greater the electronegativity of its atoms. This makes sense when you remember that each additional shell or energy level is farther away from the attraction of the positively charged protons present in the nucleus. Second, those elements (nonmetals) that need only one or two electrons to fill their outermost shells are more electronegative than those elements (metals) that would need to acquire six or seven electrons to fill a shell. Metals are satisfied, instead, to get rid of the one or two electrons they have in their outermost shell. The relative electronegativity of two interacting atoms determines what kind of chemical bond forms between them.

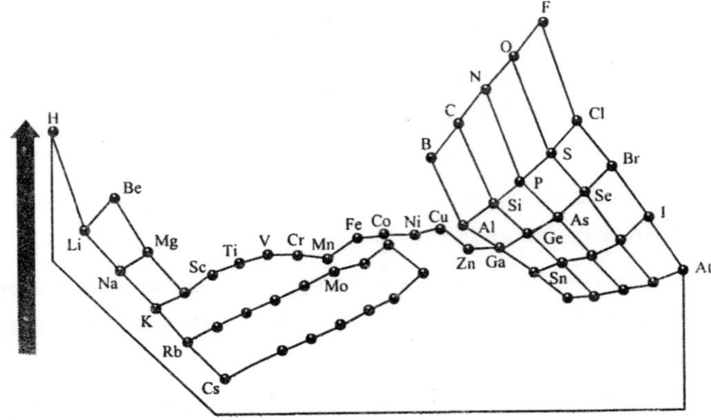

Fig. 2.8 : ***Relative electronegativities. A portion of the periodic table has been drawn to show the relative electronegativities of some of the elements. Although fluorine (F) is the most highly electronegative of the elements, it is the electronegativity of runner-up oxygen that is exploited by living things.***

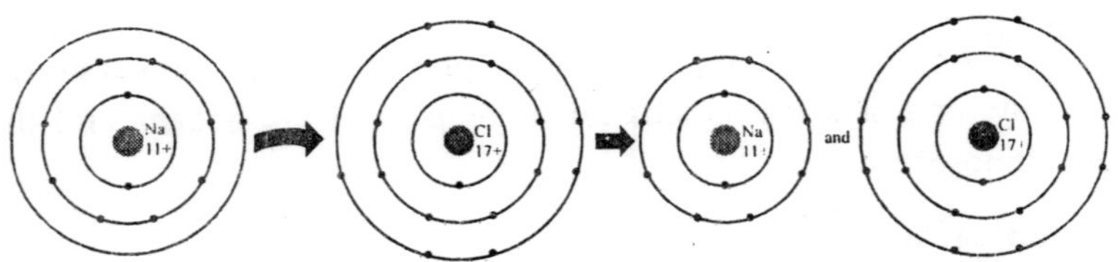

Fig. 2.9 : ***Formation of sodium chloride. In the transfer of an electron from a sodium atom to a chlorine atom, each atom acquires an outer shell of eight electrons and thus attains stability. Each also acquires an electrical charge. Charged atoms are called ions.***

Ionic Bonds

Fluorine is the most electronegative of all the elements. Sodium is only weakly electronegative. When these two elements are brought together, the electron affinity of the fluorine atom is so great that it completely pulls away the single outer electron of the sodium atom. Sodium, like all the metals, does not have a great affinity for its outer electron so it gives its electron up readily. Having lost an electron, however, it now has a single positive charge (11 protons, but only 10 electrons). Similarly, the fluorine atom now has a single negative charge because of its additional electron (9 protons, 10 electrons). These electrically charged atoms are called **ions**. The mutual attraction of opposite electrical charges, holds the ions together by **ionic bonds**. Similarly, the interaction of sodium and chlorine leads to the formation of sodium and chloride ions. In each of these cases, the ratio of the ions is 1:1. However, the ions are not held together in pairs but are stacked in three-dimensional arrays. Each sodium ion is held by six chloride ions (above, below, front, back, left, and right), while each chloride ion is, in turn, held by six sodium ions. The result of this stacking of ions is a cubical crystal of common table salt.

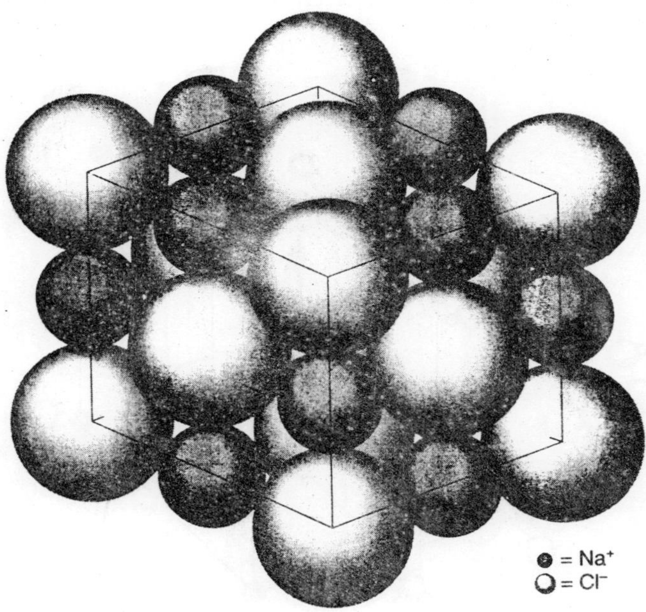

Fig. 2.10 : The structure of a crystal of sodium chloride, NaCl. The orderly stacking of Na^+ and Cl^- ions produces a crystal in the shape of a cube. All crystals are formed by the orderly stacking of their subunits. This order can be analyzed by x-ray crystallography.

Ionic compounds are widespread in nature. Although we naturally associate them with the nonliving world of rocks and minerals, many ionic compounds are essential to life. Salts of sodium potassium, calcium, chlorine, and other elements are found dissolved in, for example, the water of blood and cell fluid. In this condition the positive and negative ions become separated or **dissociated** from each other. The essential chemical properties of the substance are not changed in the process, however. Sodium ions (Na^+) and chloride ions (Cl^-)

dissolved in water still retain the properties of table salt. They do not regain the poisonous properties of the neutral atoms from which these ions were made.

Covalent Bonds

Carbon and hydrogen atoms have similar electronegativities. When brought together, the only way in which each can achieve a stable electronic configuration is to share pairs of electrons between them. Each of carbon's four *L*-shell electrons associates with the single

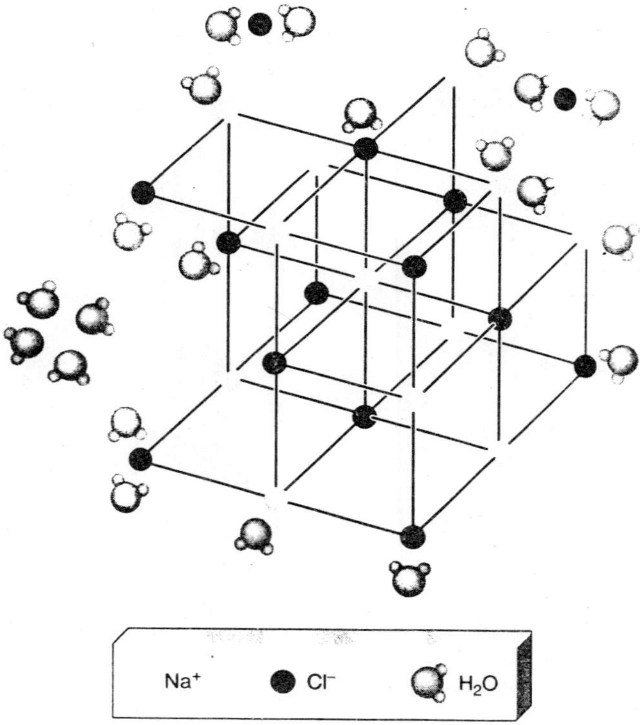

Fig. 2.11 : How water dissolves table salt (NaCl). Although the water molecule as a whole is electrically neutral, the uneven distribution of its electrons causes it to be somewhat polar, that is, negatively charged around the oxygen atom and positively charged around the two hydrogens. Because of this polarity, water is an excellent solvent for ionic compounds. The Cl^- ions are attracted to the positively charged region of water molecules, the Na^+ ions to the negatively charged region. As a result the ions become dissociated and the crystal dissolves.

electron contributed by a hydrogen atom. In this way a molecule of methane is formed. Each shared pair of electrons constitutes a **covalent bond**. It can be shown with the ×'s representing the electrons contributed by the hydrogen atoms and the dots representing the outermost (L) shell electrons of the carbon atom. However, all the electrons are precisely alike, so indicating all of them by dots is less misleading. And if such electron-dot formulas become too cumbersome, we can replace each electron pair by a dash. This gives us the **structural formula** of the molecule. If that is too much bother or if (as may be the case with more complex molecules) we are not sure just where the covalent bonds are located, we can use a molecular

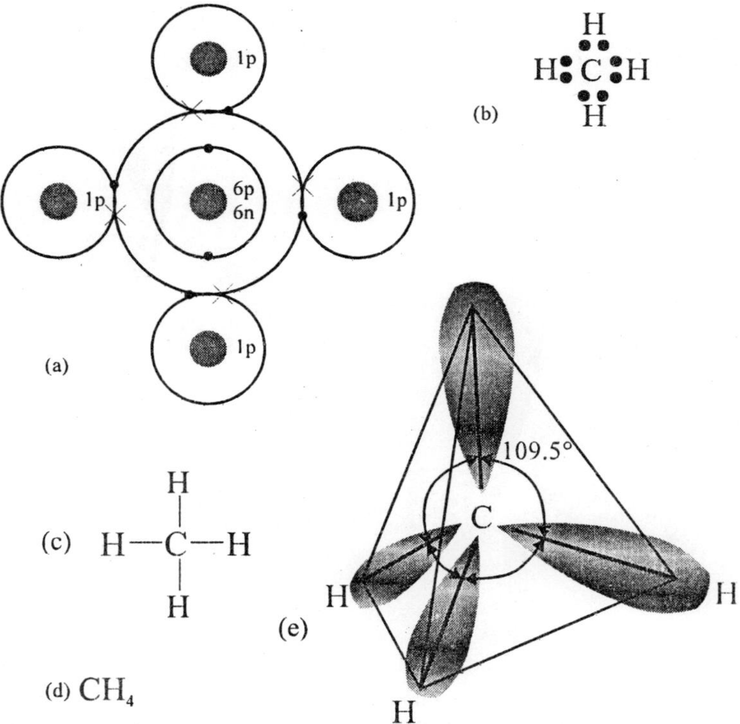

Fig. 2.12 : Five ways of depicting the methane molecule. (a) The complete atomic structure. (b) The electron-dot formula. (c) The structural formula. (d) The molecular formula. (e) A three-dimensional representation showing the tetrahedral orientation of the four covalent bonds. The regions shown in light gray represent the orbitals in which each shared pair of electrons is confined. The methane molecule is apolar.

formula (e.g., CH_4). This tells us only the number of each of the different atoms present in the molecule.

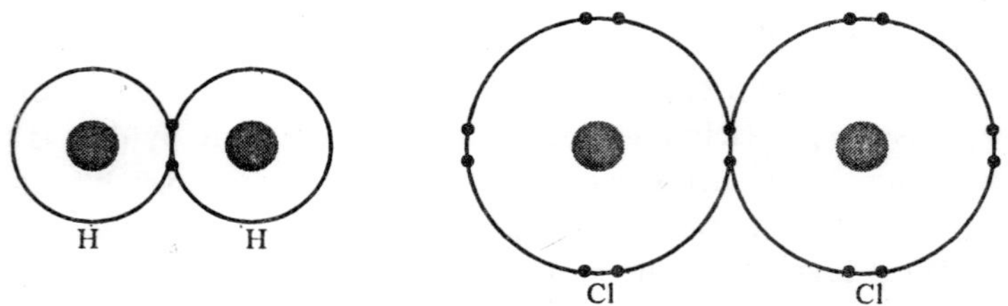

Fig. 2.13 : Structure of the molecules of hydrogen and chlorine. The bo bond between the two atoms of each of these diatomic molecules is covalent.

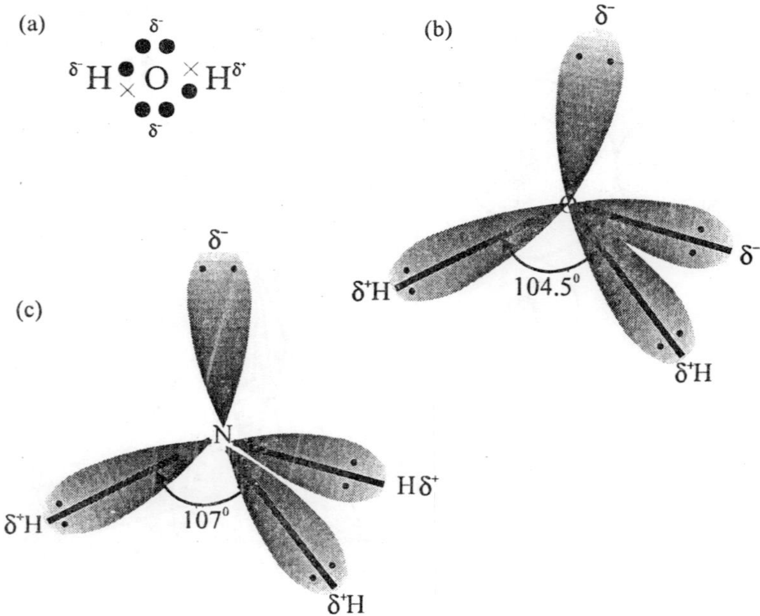

Fig. 2.14 : (a) A representation of the water molecule in two dimensions. The symbol δ represents a partial electrical charge. (b) A three-dimensional representation of the water molecule. Note that the partial positive charges are located at two corners of a tetrahedron, with the partial negative charges at the other two corners. Thus the molecules as a whole is polar. The areas in light gray are the orbitals in which both shared and unshared electron pairs are confined. (c) The geometry of the molecule of ammonia. Ammonia, like water, is strongly polar.

Although those of us who spend much time in the two-dimensional world of book pages and blackboards are apt to overlook it, atoms and molecules really exist in three dimensions. So even our electron-dot formula for methane is misleading. Actually, the four covalent bonds that are formed between the hydrogens and the carbon atom are directed as far apart from each other as is physically possible. This means that each is directed to one of the four corners of a regular tetrahedron. Thus each of the four covalent bonds forms an angle of 109.5° with *each* of its three neighbouring bonds. The result is a perfectly symmetrical distribution of all the electrons (and nuclei) of which this molecule is composed.

Two additional examples of covalent bond formation between atoms of similar electronegativity are of special interest. Many gaseous elements exist not as individual atoms but as diatomic molecules—that is, molecules consisting of two atoms. There is no electronegativity difference between identical atoms, so the bonding between them is covalent. Nitrogen and oxygen, which account for over 99% of the atmosphere, are both present as diatomic molecules, with the formulas N_2 and O_2, respectively. Chlorine and hydrogen gas also exist as the diatomic molecules Cl_2 and H_2.

Carbon atoms provide the backbone of living matter. In the next chapter we shall see several examples of giant molecules without which life as we know it could not exist. All of these giant molecules depend for their size on the ability of carbon atoms to bond together. There being no difference in electronegativity, the bonding in covalent. Thus chains of carbon atoms can be formed. These provide the structural basis for the elaboration of virtually all the molecules unique to living matter.

Polar Covalent Bonds

Oxygen is the second most electronegative of all the elements. Hydrogen is moderately electronegative. It is not as electronegative as oxygen or nitrogen, but is more electronegative than sodium or calcium. What happens when hydrogen and oxygen atoms are brought together? The difference in their electronegativities is not sufficient to yield ions. Consequently, they must share a pair of electrons between them-that is, x a covalent bond is formed. However, some kinds of sharing are more equal than others! In this case the greater electronegativity of the oxygen atom causes it to draw the electron pairs closer to its nucleus and, therefore, farther away from the nuclei of the hydrogen atoms. This result in a concentration of negative charges nearer the oxygen atom and thus farther from the positively charged protons that constitute the nuclei of the hydrogen atoms. The bond formed is thus intermediate in nature between a fully ionic bond on the one hand and a purely covalent one on the other. There is a separation of charges, but it is not complete as it is when ions are formed. We indicate the partial charges produced by the symbol δ.

If the orientation of the two hydrogen atoms about the oxygen atom were with the hydrogens opposite and in the same plane—these polar bonds would be of little significance. But, when we examine the situation as it really exists in three dimensions, we find that the distribution of partial charges around the water molecule is not symmetrical. The four pairs of electrons that surround the oxygen atom in a molecule of water are confined to four dumbbell-shaped regions in its *L* shell. These regions (technically kinown as orbitals) are directed at the four corners of a regular tetrahedron. Thus they take on a configuration similar to that which we found in methane. Two pairs of electrons represent polar covalent bonds to hydrogen nuclei and two are *unshared*. Because of the geometry of the regular tetrahedron, no matter which two corners we elect to place the hydrogen nuclei in, there will result an asymmetrical distribution of the partial charges. One side of the molecule (where the hydrogen atoms are) will be partially positively charged, the other side (with the unshared pairs of electrons) will be partially negatively charged. Thus the molecule *as a whole* is polar.

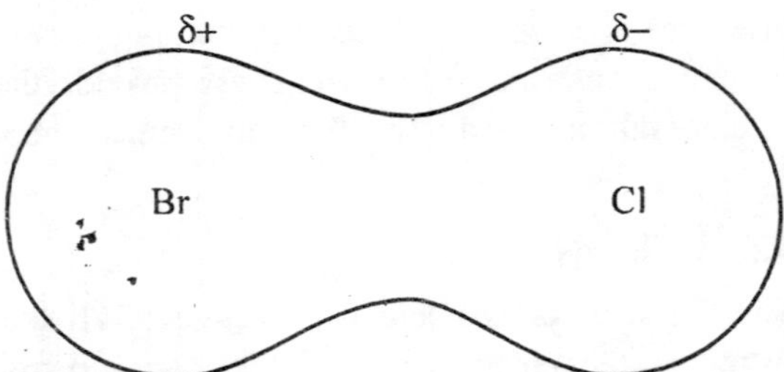

Fig. 2.15 : Electrical asymmetry in BrCl molecule

Actual measurement of the angle between the two polar covalent bonds of water shows that they are 104.5° apart, slightly less than the 109.5° that we would expect if the two hydrogen nuclei were as far apart as possible. What this probably indicates is that the unshared pairs of electrons occupy somewhat more space than the shared pairs, and thus force the two hydrogen atoms somewhat closer together. This would also explain the 107° (instead of the regular tetrahedral angle of 109.5°) found between the three polar covalent bonds in ammonia. And in this case, too, the overall molecule is polar, with a concentration of negativity at one corner of the tetrahedron and of positivity at the others.

Co-ordinate covalent bonds

Co-ordinate covalent bonds are also formed as a result of unequal sharing of electrons, as in polar covalent bonds. However, is this case the two electrons involved in the formation of a bond belong to the same atom. The atom which contributes the electrons is called 'donor' and the atom receiving them is called 'acceptor'. However, application of these terms (donor and acceptor) is rather erraneous because there is no complete transfer of electrons. Formation of ammonium boron trifluoride complex involves the co-ordinate covalent bond;

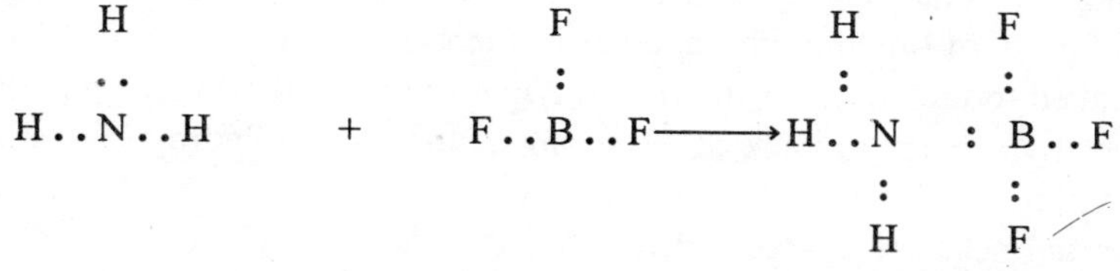

```
     H                    F            H      F
     ..                   :            :      :
H..N..H        +      F..B..F ——→ H..N   : B..F
                                       :      :
                                       H      F

   NH3                  BF3          NH3 ——→ BF3
```

In this case the two unpaired electrons of nitrogen are involved in binding with boron. The co-ordinate covalent bonds are generally indicated with an arrow and the molecule in this case is also electrically asymmetrical. Because of the shifting of the electrons towards one part, the other part of the molecule acquires a partial positive (δ^{+}) charge.

Co-ordinate covalent bonds are common in complex compounds. The two important biological molecules containing co-ordinate covalent bond are chlorophyll and heme in which either Mg of Fe atom is linked co-ordinately to nitrogen atoms. In vitamin B_{12} also, a cobalt atom is linked to nitrogen atoms through co-ordinate bonds. The compounds containing co-ordinate covalent bonds are called co-ordinate complexes.

Secondary Bonds

The primary ionic and covalent bonds are important in binding one atom to the other and thus forming small molecules. In biological systems, several smaller molecules formed as a result of these primary bonds exist together as coherent molecular aggregates. Further, some of the bigger molecules have secondary and tertiary foldings, so that they do not occupy too much of space. These molecular aggregates or secondary and tertiary foldings in bigger molecules are affected and stabilised through various weak secondary bonds. These bonds can be easily broken and formed, because they involve very little energy. This weakness of the bonds provides required dynamism to the biological molecules, which are rapidly synthesized and broken under physiological conditions.

Some important types of secondry bonds are described in the following paragraphs.

Hydrogen bonds

Hydrogen atom, covalently linked to one of the highly negative atoms of smaller sizes such as fluorine, oxygen or nitrogen, has unique property of binding with another of these atoms. The hydrogen atom linked to the atoms of fluorine, oxygen or nitrogen develops a partial positive charge because of the stronger attraction of hydrogen electron by those atoms. Consequently the other atoms (F, O or N) develops a partial negative charge. When the hydrogen atom of such a molecule comes in contact with F, O or N atom of any other molecule, there is attraction and a bond is formed. Such a bond is called hydrogen bond.

Hydrogen bond is formed in all states of matter; solid, liquid and gas. Molecules of sodium bi-carbonate are crystalline solids, having inter-molecular hydrogen bonds.

```
δ+          δ-            δ+        δ-
H ———————— O— — — — —H— — — —O
            |                  |
            C—O—Na             C—O—Na
          //                 //
         O                  O
```

Several molecules of $NaHCO_3$ may be joined together in this manner, to form a linear chain.

Water molecules are joined together through hydrogen bonds involving link between hydrogen and oxygen atoms.

Generally 5 molecules of water are grouped together in this manner. The solubility of a compound in water depends upon these bonds. When

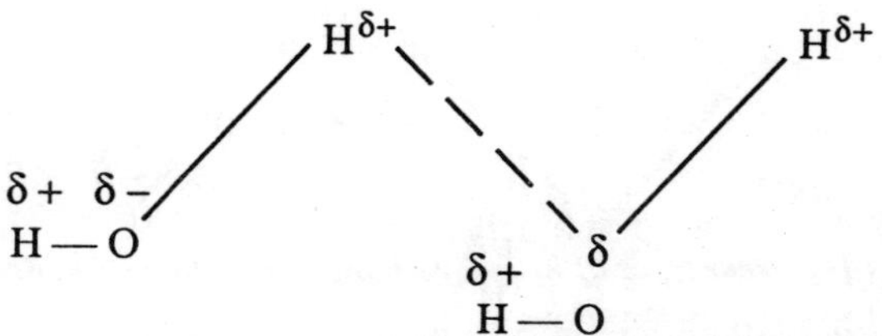

a compound is dissolved in water, some of the hydrogen bonds are broken and additional bonds between the compound and water molecules are formed. The compounds which are able to form such bonds with water are soluble. Glucose and pyruvate for example, have large number of groups that contain = O or – OH groups to form hydrogen bonds with water molecules and therefore they are water soluble.

δ+ H δ− F
δ− F
δ− F δ H δ+ δ+ H

Molecules of hydrogen fluoride in vapour form are joined together through hydrogen bonds.

1. Between peptide groups —

H
R — C(—C) C=O |||||||| H — N (C = O) (C—H, R)
H — N

2. Between two hydroxyl groups —

H
>C — O |||||||| H — O — C —
H H

3. Between a charged carboxyl group and a hydroxyl group —

–O |||||||| H — O — C —
— C H
‖
H

4. Between a charged amino group and a charged carboxyl group —

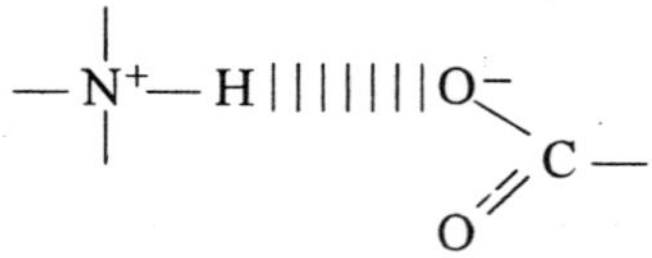

$$-\overset{|}{\underset{|}{N^{+}}}-H\,||||||\,O^{-}\!\!\diagdown C- \;\; (O{=}C)$$

Fig. : 2.16 : Different types of hydrogen bonds in biological molecules.

Because of these bonds, hydrogen fluoride in gaseous form, contains various polymeric species as H_2F_2, H_3F_3, and so on. It is apparent that the formation of hydorogen bonds is accompanied with a certain degreee of polymerization in the molecules. Because of this, these compounds have higher melting and boiling points than compounds which have similar structure but no hydrogen bonds. For example, melting and boiling points for water (a hydrogen bond compound) are 0° and 100°C respectively while the same constants for hydrogen sulfide (a non-hydrogen compound) are – 83° and – 60°C respectively. Further, the molecules with hydrogen bond have abnormally high heats of vaporization, heats of fusion viscosities and dielectric constants.

Vander Waals forces

Vander Waals forces (named after J.D. Vander Waals) are the weakest forces for binding atoms and molecules. This force is developed when a molecule or atom comes within a specified range of other molecule or atom. The existence of this force was realised during experiments with monitoring changes in volume (V) of some gases at different pressures (P), when temperature was kept constant. According to Boyle's law, the product PV for an ideal gas is constant and therefore a plot of PV and P will be straight line. However, during actual experimentation a deviation from linearity was observed. At certain pressure ranges the decrease in volume was more than expected. Increasing pressure over that range caused a increase in volume (consequently elevated PV level), indicating that the gas volume did not decrease as much as expected. Vander Waals explained this behaviour as the function of intra-atomic attraction at certain pressure. When atoms were within certain range (let us say at pressure P_1 in the graph) they attracted each other and therefore decrease in total volume of the gas was more than expected. This force of attraction which worked between the atoms at a precise distance was called Vander Waals forces. When the atoms came closer than this distance (let us say at P_2 in the graph), they repelled each other, consequently a lesser decrease in volume was observed.

The binding force in Vander Waals bond is also the opposite partial charge (as in polar covalent bond) between the atoms. But this partial charge develops as an induced dipole. Possibly there are two ways of developing such a dipole: (1) The distribution of electrons around the nucleus of an atom is on average symmetrical. However, at a given moment, the electron distribution may be non-spherical or asymmetrical. This creates a transient dipole on the atom. The neighbouring atom is affected by this dipole and as a result a complimentary dipole of opposite orientation is included on that atom. The complimentary dipole on two atoms results in an attraction between the two. The dipole may be induced from a polar molecule instead of from a asymmetrical atom. The polar compound will induce a complimentary dipole on adjacent molecule and both will be attracted. The particular distance between the two atoms or molecules, where Vander Waals force is operative is called *Vander Waals radii*. If the atoms come closer than this distance they are repulsed because of the electrons of their overlapping outer shells. It may be said that at a given distance of Vander Waals radii, these repulsive forces and Vander Waals attractive forces balance each other.

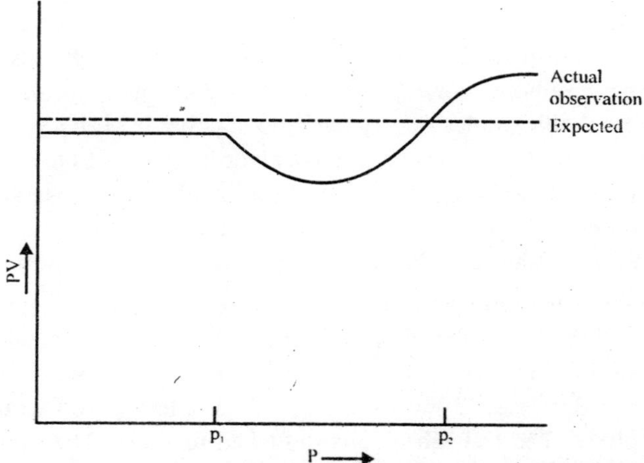

Fig. : 2.17 : A graph showing the changes in the product of pressure and volume of a gas at different pressure.

Since Vander Waals bonds are very weak, it is customary to call them as a force rather than a bond. The crystalline compounds possessing these bonds or liquified, are joined through these forces.

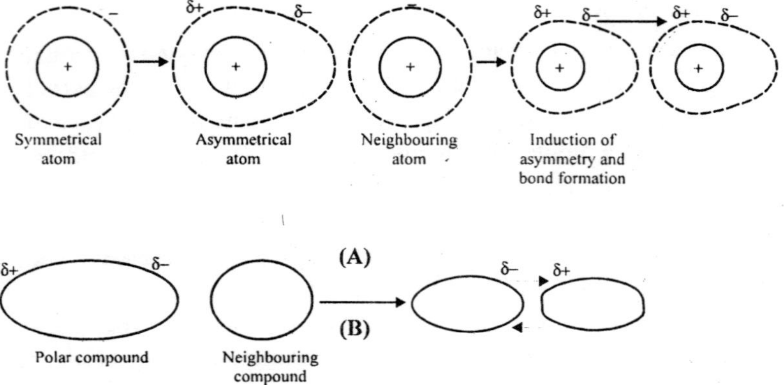

Fig. : 2.18 : Induction of dipole for bond formation (A) From an asymmetric carbon. (B) From a polar compound.

Vander Waals forces join many important biological molecules together. Such forces are however, stronger when the molecules to be joined have complimentary shapes such as enzyme-substrate, drug-receptor and antigen-antibody associations.

Hydrophobic bonds

The force of attraction for a hydrophobic bond are the same as Vander Waals forces. But these bonds are established exclusively between hgydrophobic compounds. In an aqueous medium, the non-polar (hydrophobic) molecules try to group together in such a way that they are not in contact with water. In other words, water tries to exclude non-polar compounds and in the process they are brought closer. Once the two non-polar compounds come within certain range, they may join together through Vander W0aals forces.

To illustrate the formation of hydrophobic bond, following analogy may be described. If a few drops of oil are dropped over water in a test tube, the oil drops from a layer over water. If the test tube is shaken, the oil layer disperses in the form of small droplets mixed with the aqueous phase. The test tube is then allowed to stand. The droplets coalesce together again to form bigger drops and eventually a layer of oil over water. The oil droplets in the aquoues phase, because of their hydrophobic nature, try to come together and form bigger drops and finally a continuous layer.

Hydrophobic bonds are common in the tertiary structure of proteins.

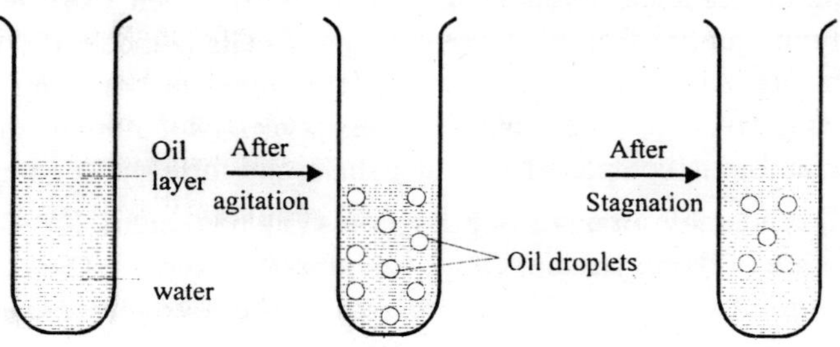

Fig. : 2.19 : Dispersion of oil in water.

Many amino acids of the proteins such as phenylalanine, leucine and isoleucine contain non-polar side chains. The side chains try to orient themselves together in the aqueous phase. This provides necessary stabilizing force for these proteins.

The constituents of living material

Elements Found in Living Material

One characteristic of living material is that it is selective with respect to its environment. The earth's crust is made mainly of oxygen, silicon, aluminium, sodium, calcium, iron, magnesium, and potassium; the

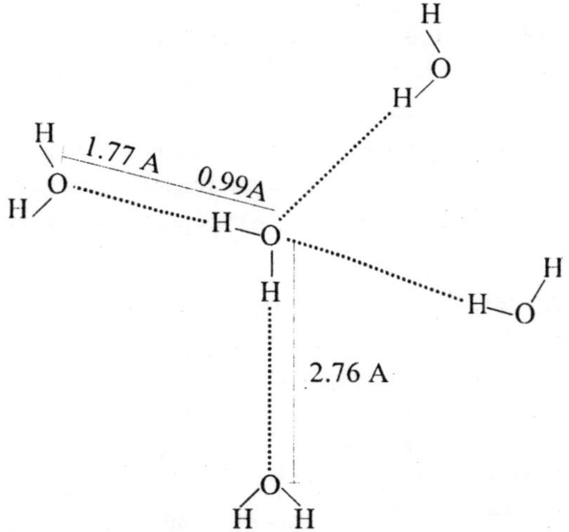

Fig. : 2.20 : Hydrogen bonding around a water molecule in ice.

remaining elements constitute less than one percent of the entire earth's crust. Living organisms are made mostly of hydrogen, oxygen, carbon, and nitrogen; in fact, 99 percent of many cells is made of these four elements. All of these four readily form covalent bonds by electron sharing. Hydrogen needs one electron, oxygen two, nitrogen three, and carbon four to complete their outer shells and form stable compounds.

Considerable amounts of phosphorus (P), potassium (K), sulfur (S), chlorine (Cl), magnesium (Mg), calcium (Ca), sodium (Na), and iron (Fe) are also found in living organisms. The following elements are found in smaller amounts in organisms: copper (Cu), cobalt (Co), manganese (Mn), zinc (Zn), boron (B), silicon (Si), iodine (I), selenium (Se), tin (Sn), chromium (Cr), and aluminium (Al). On the average only 25 of the elements found in the earth's crust are essential components in living organisms.

Molecules

A molecule consists of two or more atoms linked by chemical bonds. A substance whose molecules contain more than one kind of atom is a **compound**. Most biological substances are compounds. A substance (such as oxygen gas) that contains only one kind of atom is an **elemental substance**.

The **molecular formula** of a compound or an elemental substance shows how many atoms of each element are present in the molecule. This number is written to the lower right of the symbol. For example, the molecular formula for methane is CH_4 (each molecule contains one carbon atom and four hydrogen atoms), that for oxygen gas is O_2, that for nitric oxide is NO (Box 2.B), and that for sucrose (table sugar) is $C_{12}H_{22}O_{11}$. The hormone insulin is represented by the molecular formula $C_{254}H_{377}N_{65}O_{76}S_6$! Molecular formulas do not tell us anything about which atoms are linked to which. **Structural formulas**, give us this information.

Each compound has a **molecular weight**, which is simply the sum of the atomic weights of the atoms in the molecule. The atomic weights of hydrogen, carbon, and oxygen are, respectively, 1.008, 12.011, and 16.000. Thus the molecular weight of water (H_2O) is $(2 \times 1.008) + 16.000 = 18.016$, or about 18. What is the molecular weight of sucrose ($C_{12}H_{22}O_{11}$)? You can calculate this and find that the answer is approximately 342. If you remember the molecular weights of a few representative biological compounds, you will be able to picture the relative sizes of molecules that interact with one another (Fig. 2.22)

Suppose we want to compare how sodium chloride (NaCl), potassium chloride (KCl), and lithium chloride (LiCl) affect a biological process. You might at first think that we could simply give, say, 2 grams (g) of NaCl to one set of subjects, 2g of KCl to another, and 2 g of LiCl to the third. But because the molecular weights of NaCl, KCl and LiCl are different, 2-g samples of each of these substances contain different numbers of molecules. The comparison would thus not be legitimate. Instead, we want to give *equal numbers of molecules* of each substance so that we can compare the activity of one molecule of one substance with that of one molecule of another. But the weight of a single molecule of sodium chloride is 10^{-22} g—hardly a workable quantity.

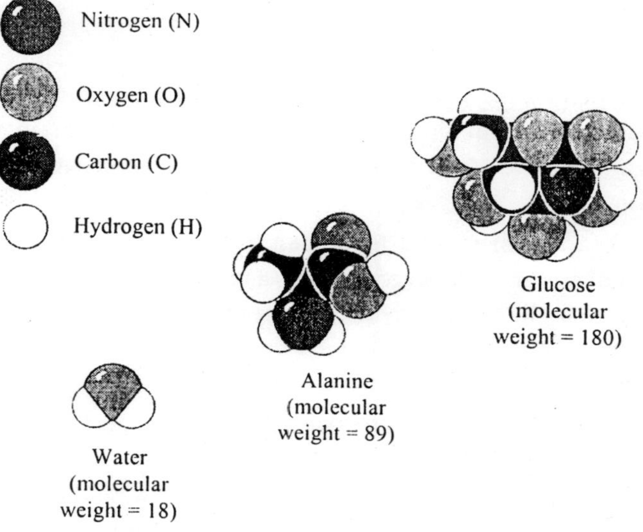

Since we can neither weigh nor count individual molecules, we work with **moles** (also known as gram molecular weights). *One mole of a substance is an amount whose weight in grams is numerically eaual to the molecular weight of the substance.* Potassium chloride (KCl) has a molecular weight of 74.55, so a mole of KCl weighs 74.55 g; a mole of NaCl weighs 58.45 g, and a mole of LiCl, 42.40 g. *A mole of one substance contains the same number of molecules as does a mole of any other substance.* This number, known as **Avogadro's number**, is 6.023×10^{23} molecules per mole. The concept of the mole is important for biology because it enables us to work easily with known numbers of molecules.

A solution containing one mole of solute per liter is a **molar** solution, 1 *M*. A solution containing one-half mole per liter is referred to as 0.5 M, or half-molar. How would you make 100 milliliters (ml) of a 0.02 M sucrose solution? The molecular weight of sucrose is 342, so one liter (1,000 ml) of a 1 M sucrose solution contains one mole, or 342 g, of sucrose. You were asked to make just 100 ml of 0.02 M sucrose solution you would use 0.02×34.2 g = 0.684 g of sucrose.

Chemical Reactions

When atoms combine or change bonding partners, a chemical reaction is occurring. Consider the flame of a propane kitchen stove or water heater. When propane (C_3H_8) reacts with oxygen gas (O_2), the carbon atoms become bonded to oxygen atoms instead of to hydrogen atoms , and the hydrogen atoms become bonded to oxygen instead of carbon. As the covalently bonded atoms change bonding partners, the composition of the matter changes, and propane and oxygen gas become carbon dioxide and water. Chemical reactions in which covalently bonded atoms change partners are common and important in organisms.

The heat of the stove's flame and its blue light reveal that the reaction of propane and oxygen releases a great deal of energy. Changes in energy ususally accompany chemical reactions: Energy may be given off to the environment, as in the reaction of propane with oxygen, or energy may be taken up from the environment.

We can measure the energy associated with chemical bonds. Work must be done to break a bond, and that work, or energy, can be expressed in calories. In most chemical reactions in which bond partners change, the total bond energies of the products differ from those of the reactants; these energies can also be expressed in calories. A calorie is the amount of heat energy needed to raise the temperature of 1 g of pure water (which contains no other substance) from 14.5°C to 15.5°C. (The nutritionist's Calorie, which biologists call a kilocalorie, is equal to 1,000 heat-energy calories.) Although defined in terms of heat, the calorie is a measure of any form of energy—mechanical, electric, or chemical.

Oxidation involves the loss of electrons; Reduction involves the gain of electrons

Rusting—the combination of iron (symbol Fe) with oxygen—is a familiar example of oxidation and reduction:

$$4\ Fe + 3\ O_2 \longrightarrow 2\ Fe_2O_3$$

Oxidation and reduction always occur together, but initially we will discuss them separately. **Oxidation** is a chemical process in which an atom, ion, or molecule loses electron. In rusting, each iron atom becomes oxidized as it loses three electrons.

$$4\ Fe \longrightarrow 4\ Fe^{3+} + 12e^-$$

The e^- is a symbol for an electron; the + superscript represents an electron deficit. (When an atom loses an electron, it acquires one unit of positive charge from the excess of one proton. In our example, each iron atom loses three electrons and acquires three units of positive charge.)

You will recall that the oxygen atom is very electronegative, able to remove electrons from other atoms. In this reaction, oxygen gains electrons from iron.

$$3\ O_2 + 12e^- \longrightarrow 6\ O^{2-}$$

Oxygen becomes reduced when it accepts electrons from the iron. **Reduction** is a chemical process in which an atom, ion, or molecule *gains* electrons. (The term reduction refers to the fact that the gain of an electron results in the *reduction* of any positive charge that might be present).

Oxidation and reduction reactions occur simultaneously because one substance must accept the electrons that are removed from the other. Oxidation-reduction reactions are sometimes referred to as **redox reactions**. In a redox reaction, one component, the *oxidizing agent*, accepts one or more electrons and becomes reduced. Oxidizing agents other than oxygen are known, but oxygen is such a common one that its name was given to the process. Another reaction component, the *reducing agent*, gives up one or more electrons and becomes oxidized.

In our example there was a complete transfer of electrons from iron to oxygen. Some redox reactions are not so obvious, however. In these, electrons simply move farther from the reducing agent and closer to the oxidizing agent.

Electrons are not easily removed from covalent compounds unless an entire atom is removed. In cells, oxidation often involves the removal of a hydrogen atom (an electron plus a proton that "goes along for the ride") from a compound; reduction often involves the addition of hydrogen.

Redox reactions are central to many of the energy conversions that go on in a cell. This is because the transfer of an electron also involves the transfer of the energy of that electron. Therefore, a reduced substance has gained not only one or more electrons, but also energy. Both cellular respiration and photosynthesis are essentially redox processes.

Buffers

The pH inside almost all living cells, and in the fluid surrounding cells in multicellular organisms, is fairly close to 7. Most of the biological catalysts (enzymes) in living systems are extremely sensitive to pH; often even a small change in pH will alter their shape, thereby disrupting their activities and rendering them useless. In a condition called *blood acidosis*, human blood, which normally has a pH of about 7.4, drops 0.2–0.4 points on the pH scale. This condition is fatal if not treated immediately. The reverse condition, *blood alkalosis*, involves an increase in blood pH of a similar magnitude and is just as serious.

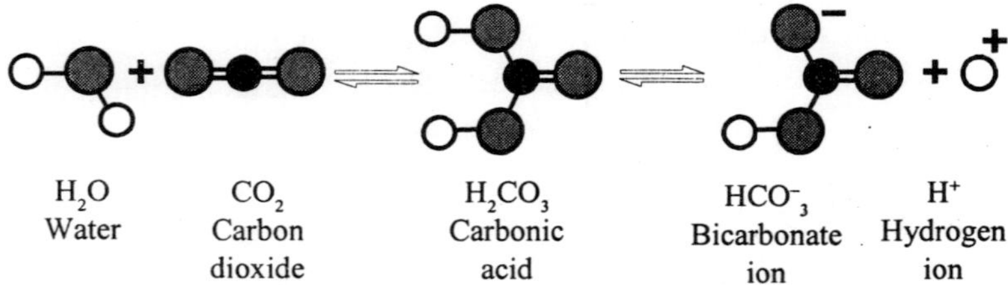

Fig. : 2.21 : Carbon dioxide and water combine chemically to form carbonic acid (H_2CO_3), which dissociates in water, freeing H^- ions. This reaction makes carbonated beverages acidic.

Yet, the chemical reactions of life constantly produce acids and bases within cells. Furthermore, many animals eat substances that are acidic or basic; Coca Cola, for example, is a strong (although dilute) acidic solution. Despite such variations in the concentrations of H^+ and OH^-, the pH of an organisms is kept at a relatively constant level by **buffers**.

A buffer is a substance that acts as a reservoir for hydrogen ions, donating them to the solutions when their concentration falls and taking them from the solution when their concentration rises. What sort of substance will act in this way? Within organisms, most buffers consist of pairs of substances, one an acid and the other a base. The key buffer in human blood is the acid-base pair, carbonic acid (acid)-bicarbonate (base). In a pair of reversible reactions, carbon dioxide (CO_2) and H_2O join to form carbonic acid (H_2CO_3), which dissociates to yield bicarbonate ion (HCO_3^-) and H^+. If some other substance adds H^+ ions to the blood, the HCO_3^- ions act as a base and remove the excess H^+ ions from solution by forming H_2CO_3. Similarly, if a substance removes H^+ ions from the blood, H_2CO_3 dissociates, releasing more H^+ ions into the blood. The blood's pH is thus stabilized by the equilibrium between the forward and reverse reactions that interconvert H_2CO_3 and HCO_3^-.

The reaction of carbon dioxide and water to form carbonic acid is important because it permits significant amounts of carbon to enter water from the air. As we will discuss in chapter 4, biologists believe that life first evolved in the early oceans, which were rich in carbon because of this reaction.

CHAPTER 3
WATER

Water: The Cradle Of Life

Of all the molecules that are common on earth, only water exists as a liquid at the relatively low temperatures that prevail on the earth's surface, three-fourths of which is covered by liquid water. When life was originating, water provide a medium in which other molecules could move around and interact without being held in place by strong covalent or ionic bonds. Life evolved as a result of these interactions, and it is still inextricably tied to water. About two-thirds of any

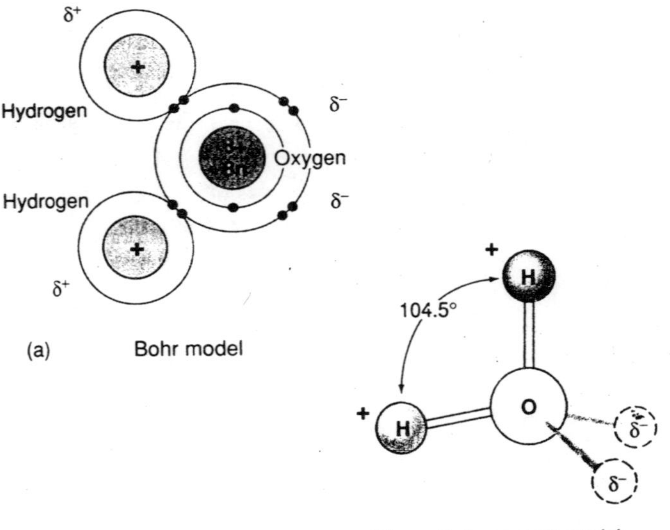

Figure 3.1 : **Water has a simple molecular structure.** ***(a) Each molecule is composed of one oxygen atom and two hydrogen atoms. The oxygen atom shares one electron with each hydrogen atom. (b) The greater electronegativity of the oxygen atom makes the water molecule polar: water carries a partial negative charge (δ^-) near the oxygen atom and a partial positive charge (δ^+) near the hydrogen atoms.***

organism's body is composed of water, and no organism can grow or reproduce in any but a water-rich environment. It is no accident that tropical rain forests are bursting with life, while dry deserts are almost lifeless except when water becomes temporarily plentiful, such as after a rainstorm. Farming is possible only in those areas of the earth where water is plentiful.

The chemistry of life, then, is water chemistry. The way in which life first evolved was determined in large part by the chemical properties of the liquid water in which that evolution occurred. The single most outstanding chemical property of water is its ability to form weak chemical associations with only 5–10% of the strength of covalent bonds. This property, which as you will see derives directly from the structure of water, is responsible for much of the organization of living chemistry.

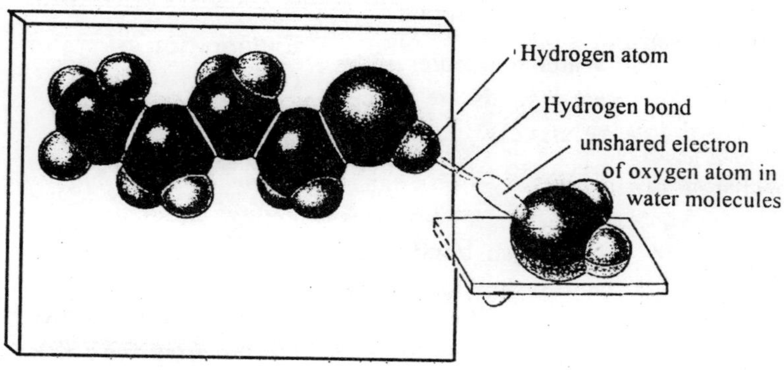

Fig. 3.2. : The structure of a hydrogen bond.

Water has a simple atomic structure, consisting of an oxygen atom bound to two hydrogen atoms by two single covalent bonds. The resulting molecule is stable: it satisfies the octet rule, has no unpaired electrons, and does not carry a net electrical charge.

Water Acts Like a Magnet

Both the oxygen and the hydrogen atoms attract the electrons they share in the covalent bonds of a water molecule; this attraction is called electronegativity. However, the oxygen atom is more electronegative than the hydrogen atoms, so it attracts the electrons more strongly than do the hydrogen atoms. As a result, the shared electrons in a water molecule are far more likely to be found near the oxygen nucleus than

Table 3.1 : The Property of Water

Property	Explanation	Example of Benefit to Life
High polarity	Polar water molecules are attracted to ions and polar compounds, making then soluble	Many kinds of molecules can move freely in cells, permitting a diverse array of chemical reactions
High specific heat	Hydrogen bonds absorb heat when they break, and release heat when they form, minimizing temperature changes	Water stabilizes the temperature of organisms and the environment
High heat of vaporization	Many hydrogen bonds must be broken for water to evaporate	Evaporation of water cools body surfaces
Lower density of ice	Water molecules in an ice crystals are spaced relatively far apart because of hydrogen bonding	Because ice is less dense than water, lakes do not freeze solid, and they overturn in spring
Cohesion	Hydrogen bonds hold water molecules together	Leaves pull water upward from the roots; seeds swell and germinate

near the hydrogen nuclei. This stronger attraction for electrons gives the oxygen atom a partial negative charge (δ^-), as though the electron cloud were denser near the oxygen atom than around the hydrogen atoms. Because the water molecule as a whole is electrically neutral, each hydrogen atom carries a partial positive charge (δ^+).

What would you expect the shape of a water molecule to be? Each of water's two covalent bonds has a partial charge at each end, (δ^-) at the oxygen end and (δ^+) at the hydrogen end. The most stable arrangement of these charges is a *tetrahedron*, in which the two negative and two positive charges are approximately equidistant from one another . The oxygen atom lies at the center of the tetrahedron, the hydrogen atoms occupy two of the apexes, and the partial negative charges occupy the other two apexes. This results in a bond angle of 104.5° between the two covalent oxygen-hydrogen bonds.

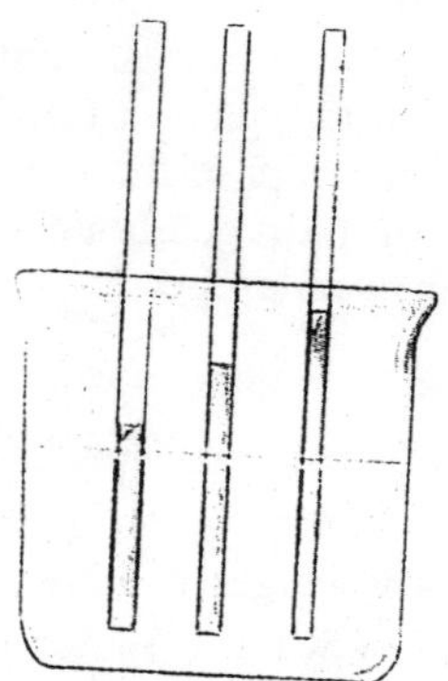

Fig. 3.3. : Capillary action.

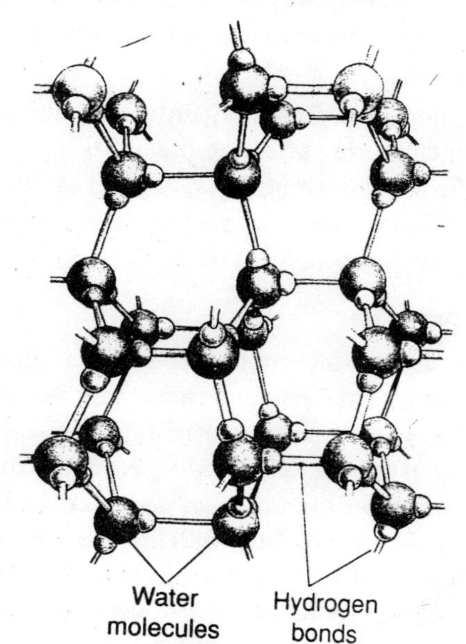

Fig. 3.4. : The role of hydrogen bonds in an ice crystals.

The water molecule, thus, has distinct "ends," each with a partial charges, like the two poles of a magnet. Molecules that exhibit charge separation are called **polar molecules** because of their magnet-like poles, and water is one of the most polar molecules known. *The polarity of water underlies its chemistry and the chemistry of life.*

Polar molecules interact with one another, as the δ^- of one molecule is attracted to the δ^+ of another. Since many of these interactions involve hydrogen atoms, they are called **hydrogen bonds**. Each hydrogen bond is individually very weak and transient, lasting on average only 1/100,000,000,000 second (10^{-11} sec). However, the cumulative effects of large numbers of these bonds can be enormous. Water forms an abundance of hydrogen bonds, which are responsible for many of its important physical properties

Water is Essential To Life

A large part of the mass of most organisms is water. In human tissues the percentage of water ranges from 20% in bones to 85% in brain cells. About 70% of our total body weight is water; as much as 95% of a jellyfish and certain plants is water . Water is the source, through photosythesis, of the oxygen in the air we breathe, and its hydrogen atoms become incorported into many organic compounds. Water is also the solvent for most biological reactions and a reactant or product in many chemical reactions.

Water is not only important inside organisms but also is one of the principal environmental factors affecting them. Many organisms live within the sea or in freshwater rivers, lakes, or puddles. Water's unique combination of physical and chemical properties has permitted living things to originate, survive, and evolve on earth.

Water molecules are polar

As discussed previously, water molecules are polar, that is, one end of each molecule bears a partial positive charge and the other a partial negative charge. The water molecules in liquid water and in ice associate by hydrogen bonds. The hydrogen atom of one water molecule, with its partial positive charge, is attracted to the oxygen atom of a neighbouring water molecule, with its partial negative charge, forming a hydrogen bond. Each water molecule can form hydrogen bonds with a maximum of four neighbouring water molecules.

Water Stores Heat

The temperature of any substance is a measure of how rapidly its individual molecules are moving. Because of the many hydrogen bonds that water molecules form with one another, a large input of thermal energy is required to break these bonds before the individual water molecules can begin moving about more freely and so have a higher temperature. Therefore, water is said is to have a high **specific heat**, which is defined as the amount of heat that must be absorbed or lost by one gram of a substance to change its temperature by one degree Celsius (°C). Specific heat measures the extent to which a substance resists changing its temperature when it absorbs or loses heat. The more polar a substance is, the higher is its specific heat. The specific heat of water (1 calorie/gram/°C) is twice that of most carbon compounds and nine times that of iron. Only ammonia, which is more polar than water and forms very strong hydrogen bonds, has a higher specific heat than water (1.23 calories/gram/°C). Still, only 20% of the hydrogen bonds are broken as water heats from 0 to 100°C.

With its high specific heat, water heats up more slowly than almost any other compound and holds its temperature longer when heat is no longer applied. This characteristic enables organisms, which have a high water content, to maintain a relatively constant internal temperature. The heat generated by the chemical reactions inside cells would destroy the cells, if it were not for the high specific heat of the water within them.

At low temperature, the molecules of water are locked into a crystal-like lattice of hydrogen bonds, forming the solid we call ice. Interestingly, ice is less dense than liquid water because the hydrogen bonds in ice space the water molecules relatively far apart. Were it otherwise, ice would be the majority of all surface water, with only shallow surface melting annually.

A considerable amount of heat energy (586 calories) is required to change one gram of liquid water into a gas. Hence, water also has a high **heat of vapourization**. Since the transition of water from a liquid to a gas requires the input of energy, the evaporation of water from a surface causes cooling of that surface. Many organisms dispose of excess body heat by evaporative cooling; for example, humans and many other vertebrates sweat.

Water Is a Powerful Solvent

Water molecules gather closely around any substance that bears an electrical charge, whether that substance carries a full charge (ion) or a charge separation (polar molecule). For example, sucrose (table sugar) is composed of molecules that contain slightly hydroxyl (– OH) groups. A sugar crystal dissolves rapidly in water because water molecules can form hydrogen bonds with the hydroxyl groups of the sucrose molecules. Therefore, sucrose and other polar molecules are said to be *soluble* in water. Every time a sucrose molecule dissociates or breaks away from the crystal, water molecules surround it in a cloud, forming a **hydration shell** and preventing it from associating with other sucrose molecules. Hydration shells also form around ions. Life originated in water not only because water is a liquid at moderate temperatures, but also because so many other substances are polar or ionized and, thus, are water-soluble.

Water Organizes Nonpolar Molecules

Water molecules always tend to form the maximum possible number of hydrogen bonds. When nonpolar molecules such as oils, which do not form hydrogen bonds, are placed in water, the water molecules are forced into association with one another, thus minimizing their disruption of the hydrogen bonding of water. In effect, they shrink from contact with water, and for this reason they are referred to as **hydrophobic** (Greek *hydros*, "*water*" + *phobos*, "fearing"). In contrast, polar molecules, which readily form hydrogen bonds with water, are said to be **hydrophilic** ("water-loving").

The tendency for nonpolar molecules to aggregate in water is known as **hydrophobic exclusion**. Hydrophobic forces induced by the hydrogen bonding of water were important in the evolution of life, because some of the exterior portions of many of the molecules on which life came to be based are nonpolar. By forcing these hydrophobic portions of molecules into proximity to one another, water causes such molecules to assume particular shapes. Different molecular shapes have evolved by alteration of the location and strength of nonpolar regions. As you will see, much of the evolution of life reflects changes in molecular shape that can be induced in just this way.

Ionization of water

Because of the small mass of the hydrogen atom and the fact that its single electron is tightly held by the oxygen atom, there is a finite tendency for the hydrogen atom to dissociate from the oxygen atom to which it is covalently bound in one water molecule and "jump" to the oxygen atom of the adjacent water molecule to which it is hydrogen-bonded, provided that the internal energy of each molecule is favourable (see below). In this reaction, two ions are produced, the hydronium (H_3O^+) ion and the hydroxyl (OH^-) ion. In a liter of pure water at 25°C, there are at any given time only 1.0×10^{-7} mole of H_3O^+ ions and an equal amount of OH^- ions, as shown by electrical-conductivity measurements.

Although the ionization of water is by convention written simply as

$$H_2O \rightleftharpoons H^+ + OH^-$$

"bare" protons do not exist in water. Actually, the H_3O^+ ion is itself hydrated through further hydrogen bonding to form the ion $H_3O_4^+$.

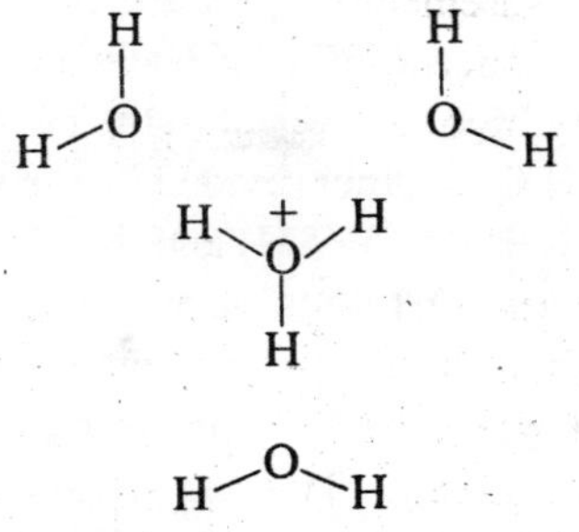

Fig. 3.5. : Hydrated form of hydronium ion ($H_2O_4^+$). The hydration shell is stable to 100°.

In table 3.2, it can be seen that the rate of migration of H^+ ions in an electrical field is many times greater than that of Na^+ or K^+ ions. The high electrical mobility of the H_3O^+ ion is due to the fact that a proton may "jump" from a hydronium ion to a neighbouring water molecule, a reaction that occurs with an extremely high frequency. A series of proton jumps has the effect of translating protons at a rate that is much higher than the rate of diffusive or bulk movement of H_3O^+ ions per se. Such proton jumps also are responsible for the exceptionally high electrical conductivity of ice. Conduction of protons through hydrogen-bonded water molecules may be an important phenomenon in biological systems.

Table 3.2. Electrical mobility of some cations at infinite dilution (25°C)

Ion	Mobility cm² volt⁻¹ sec⁻¹ (× 10⁴)
H^+	36.3
Na^+	5.2
K^+	7.62
NH_4^+	7.60
Mg^{2+}	5.4
Li^+	4.0

The Ion Product of Water; the pH Scale

The dissociation of water is an equilibrium process:

$$H_2O \rightleftharpoons H^+ + OH^-$$

for which we can write the equilibrium constant

$$K_{eq} = \frac{[H^+][OH^-]}{[H_2O]}$$

where the brackets indicate concentrations in moles per liter. The magnitude of this equilibrium constant at any given temperature can be calculated from conductivity measurement on pure distilled water. Since the concentration of water in pure water is very high (it is equal to the number of grams of H_2O in a liter divided by the gram mol wt, or 1000/18=55.5M) and the concentration of H^+ and OH^- ions very low (1×10^{-7} M at 25°C), the molar concentration of water is not significantly changed by its very slight ionization. The equilibrium-constant expression may thus be simplified to:

$$55.5 \times K_{eq} = [H^+][OH^-]$$

and the term $55.5 \times K_{eq}$ can be replaced by a "lumped" constant Kw, called the *ion product* of water,:

$$K_w = [H^+][OH^-]$$

The value of K_w at 25°C is 1.0×10^{-14}. In an acid solution, the H^+ concentration is relatively high and the OH^- concentration correspondingly low; in a basic solution, the situation is reversed.

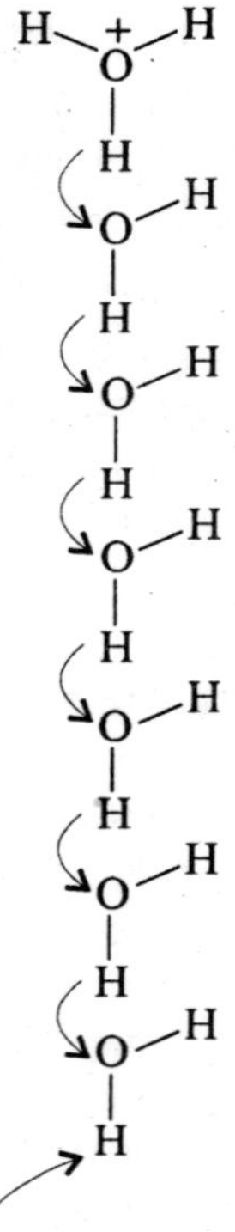

This H_2O molecule becomes a hydronium ion at the end of the series of proton jumps

Fig. 3.6. : Proton jumps

K_w, the ion product of water, is the basis for the pH scale (table 3.3), a means of designating the actual concentration of H^+ (and thus of OH^-) ions in any aqueous solution in the acidity range between 1.0 M H^+ and 1.0 *M* OH^-. The term pH is defined as

$$pH = \log_{10} \frac{1}{[H^+]} = -\log_{10}[H^+]$$

In a precisely neutral solution at 25°C,

$$[H^+] = [OH^-] = 1.0 \times 10^{-7} M$$

The pH of such a solution is

$$pH = \log_{10} \frac{1}{1 \times 10^{-7}} = 7.0$$

The value of 7.0 for the pH of a precisely neutral solution is thus not an arbitrarily chosen figure; it is derived from the absolute value of the ion product of water at 25°C. It is especially important to note that the pH scale is logarithmic, not arithmetic. To say that two solutions differ in pH by 1 pH unit means that one solution has ten times the hydrogen-ion concentration of the other. Table 3.4 lists the pH of some fluids.

Table 3.3. The pH scale

$[H^+]$ [M]	pH	$[OH^-]$ [M]
1.0	0	10^{-14}
0.1	1	10^{-13}
0.01	2	10^{-12}
0.001	3	10^{-11}
0.0001	4	10^{-10}
0.00001	5	10^{-9}
10^{-6}	6	10^{-8}
10^{-7}	7	10^{-7}
10^{-8}	8	10^{-6}
10^{-9}	9	10^{-5}
10^{-10}	10	10^{-4}
10^{-11}	11	0.001
10^{-12}	12	0.01
10^{-13}	13	0.1
10^{-14}	14	1.0

Measurement of pH

Measurement of pH is one of the most important and frequently used analytical procedures in biochemistry since the pH determines many important features of the structure and activity of biological macromolecules, and thus of the behaviour of cells and organisms. The primary standard for measurement of H^+-ion concentration (and thus of pH) is the *hydrogen electrode.* This is a specially treated platinum electrode that is immersed in the solution whose pH is to be measured. The solution is in equilibrium with gaseous hydrogen at a known pressure and temperature. The electromotive force at the electrode responds to the equilibrium.

$$H_2 \rightleftharpoons 2H^+ + e^-$$

The potential difference between the hydrogen electrode and a reference electrode of known emf (for example, a calomel electrode) is measured and used to calculate the H^+-ion

Table 3.4. pH of some fluids

	pH
Seawater	7.0 – 7.5
Blood plasma	7.4
Interstitial fluid	7.4
Intracellular fluids	
Muscle	6.1
Liver	6.9
Gastric juice	1.2 – 3.0
Pancreatic juice	7.8 – 8.0
Saliva	6.35 – 6.85
Cow's milk	6.6
Urine	5 – 8
Tomato juice	4.3
Grapefruit juice	3.2
Soft drink (cola)	2.8
Lemon juice	2.3

concentration.

The hydrogen electrode is too cumbersome for general use and has been replaced by the glass electrode, which is directly sensitive to H+-ion concentration in the absence of hydrogen gas. The response of the glass electrode must be calibrated against buffers of precisely known pH. Another way of measuring pH is by the use of acid-base indicators.

Acids and Bases

The most general and comprehensive definitions of acids and bases, applicable to nonaqueous as well as aqueous systems, are those of G.N. Lewis. A Lewis acid is a potential *electron-pair acceptor,* and a Lewis base a potential *electron-pair donor*. The Lewis definitions are very useful in the analysis of reaction mechanisms, as we shall see later. However, the formalism, introduced by J.N. Bronsted and T.M. Lowry is more useful than that of Lewis in describing acid-base reactions in dilute aqueous systems. According to the Bronsted-Lowry concepts, an acid is a proton donor and a base is a proton acceptor (margin). An acid-base reaction always involves a conjugate acid-base pair, made up of a proton donor and the corresponding proton acceptor. Foe example, acetic acid (CH_3COOH) and the acetate anion (CH_2COO^-) form a conjugate acid-base pair. Each acid has a characteristic affinity for its proton. Those with high affinity for protons are weak acids and dissociate only slightly; those with low affinity are strong acids and lose H- ions readily. The tendency of any given acid to dissociate is given by its dissociation constant. For the acid HA, the dissociation constant at a given temperature is

Conjugate acid-base pairs of acetic acid, ammonium ion, and water.

Proton donor		Proton acceptor
CH_3COOH	$\rightleftharpoons$	$H^+ + CH_3COO$
NH_4^+	$\rightleftharpoons$	$H^+ + NH_3$
HOH	$\rightleftharpoons$	$H^+ + OH$

$$K' = \frac{[H^+][A^-]}{[HA]}$$

where the brackets indicate concentrations in moles per liter. It is the convention in biochemistry to employ disconcentrations of reactants and products under a given set of experimental conditions, i.e., at a given total concentration and ionic strength and with other solutes specified. Such a constant is called an *apparent* dissociation constant and is designated K' to distinguish it from the true, or *thermodynamic*, dissociation constant K employed by the physical chemist, which is corrected for deviation of the system from ideal behaviours caused by such factors as concentrations and ionic strength.

In the Brönsted-Lowry formalism, acids and bases are treated alike, i.e., solely in terms of the tendency of protons to dissociate from the proton-donor species. (So called "basic dissociation constants," such as K_b for the dissociation $NH_4OH \rightleftharpoons NH_4^+ + OH^-$, are not employed.

Table 3.5. Apparent dissociation constant and pK' of some acid (25°C)

Acid *(Proton donor)*	K' *(M)*	*pK'*
HCOOH	1.78×10^{-4}	3.75
CH_3COOH	1.74×10^{-3}	4.76
CH_3CH_2COOH (propionic acid)	1.35×10^{-3}	4.87
$CH_3CHOHCOOH$ (lactic acid)	1.38×10^{-4}	3.86
$COOHCH_2CH_2COOH$ (succinic acid)	6.16×10^{-5}	4.21
$COOHCH_2CH_2COO^-$	2.34×10^{-6}	5.63
H_3PO_4	7.25×10^{-3}	2.14
$H_2PO_4^+$	6.31×10^{-8}	7.20
HPO_4^{2-}	3.98×10^{-13}	12.4
H_2CO_3	1.70×10^{-4}	3.77
HCO_3^-	6.31×10^{-11}	10.2
NH_4^+	5.62×10^{-10}	9.25
$CH_3NH_3^+$	2.46×10^{-11}	10.61

In fact, in the Bronsted-Lowry formalism, NH_4OH is *neither* an acid nor base). Table 3.5 also gives values for the expression pK', which is a logarithmic transformation of K', just as the term pH is a logarithmic transformation of $[H^+]$:

$$pK' = \log_{10} \frac{1}{K'} = -\log_{10} K'$$

Strong acids have low pK' values, and strong bases have high pk' values.

The titration curves of some conjugate acid-base pairs have been shown in figure. The pH after each increment of NaOH is added during the titration of an acid is plotted vs. the equivalents of OH^- added. The shapes of such titration curves are very similar from one acid to another; the important difference is that the curves are displaced vertically along the pH scale. The pH intercept at the midpoint of the titration is numerically equal to the pK' of the acid titrated. At the midpoint, equimolar concentrations of proton-donor (HA) and proton-acceptor species (A^-) of the acid are present. In fact, the p*K*' of an acid can be calculated from the pH at any point on the titration curve of an acid if the concentrations of the proton-donor and proton-acceptor species at this point are known. The shape of the titration curve can be expressed by the *Henderson-Hasselbalch equation*, which is a logarithmic transformation of the expression for the dissociation constant. It is derived as follows:

$$K' = \frac{[H^+][A^-]}{[HA]}$$

Solve for $[H^+]$:

$$H^+ = K' \frac{[HA]}{[A^-]}$$

Take the negative logarithm of both sides :

$$-\log[H^+] = -\log K' - \log\frac{[HA]}{[A^-]}$$

Substitute pH for $-\log[H^+]$ and pK' for $-\log K'$:

$$pH = pK' - \log\frac{[HA]}{[A^-]}$$

If we now change signs, we obtain the Henderson-Hasselbalch equation :

$$pH = pK' + \log\frac{[A^-]}{[HA]}$$

which, in more general form, is

$$pH = pK' + \log\frac{[\text{proton acceptor}]}{[\text{proton donor}]}$$

This equation makes it possible to calculate the pK' of any acid from the molar ratio of proton-donor and proton-acceptor species at a given pH, calculate the pH of a conjugate acid-base pair of a given pK' and a given molar ratio, and to calculate the molar ratio of proton donor and proton acceptor given the pH and pK'. The Henderson-Hasselbalch equation is fundamental for quantitative treatment of all acid-base equilibria in biological systems.

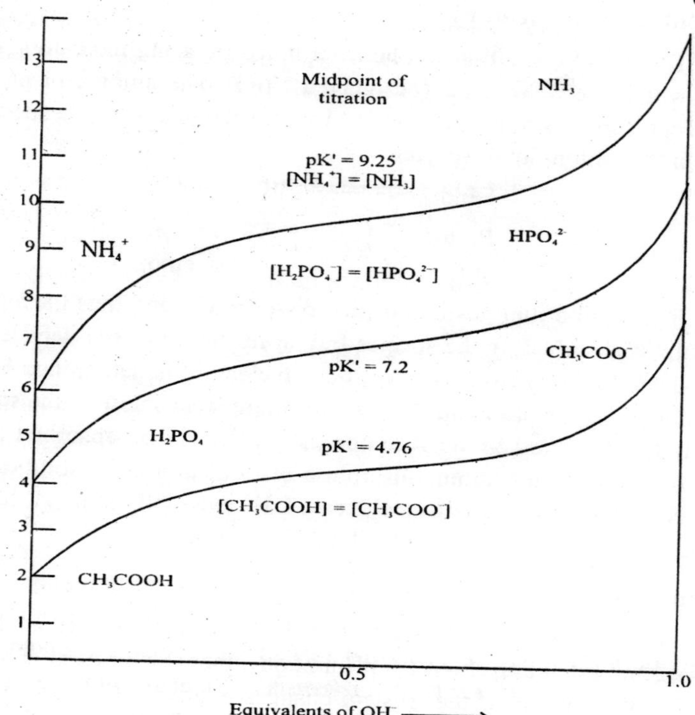

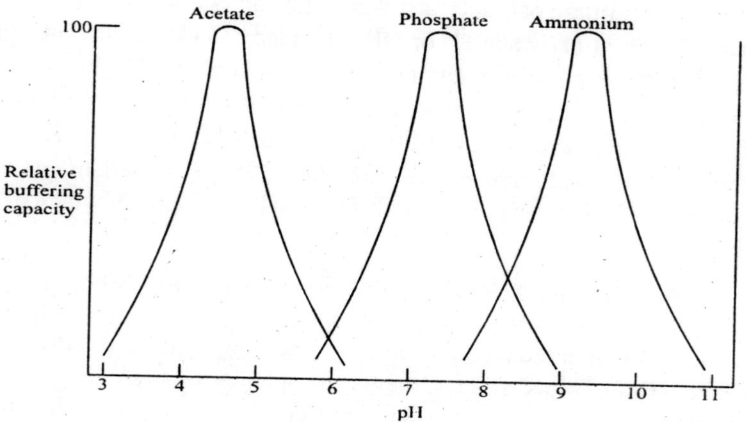

Fig. 3.7. : Acid-base titraton curves of some acids, showing the major ionic species at beginning, midpoint, and end of titration. (Below) The relative buffering power plotted against pH. Maximum buffering power is given at pH = pK'.

Acid–Base Indicators

The pH of a solution can be determined by using indicators, most of which are weak acids (designated HInd). The anionic or proton-acceptor species of indicator (Ind-) has a different absorption spectrum than the proton-donor species:

$$\underset{\text{Proton donor}}{\text{HInd}} \rightleftharpoons \text{H}^+ + \underset{\text{Proton acceptor}}{\text{Ind}^-}$$

The equilibrium position of this dissociation, and thus the amount of light absorbed by the species Ind^- at its characteristic wavelength, is determined by the characteristic wavelength, is determined by the ambient H+-ion concentration. In a strongly acid solution, the species HInd is favoured by mass action and little will be absorbed at the wavelength of maximum absorption of Ind^-. In a basic solution, the species Ind^- is favoured since excess OH^- ions will "trap" H^+ ions as H_2O; in this case, much light will be absorbed.

Buffers

It has already been shown that the titration curve of each acid has a flat zone extending about 1.0 pH unit on either side of its midpoint. In this zone, the pH of the system changes relatively little when small increments of H^+ or OH^- are added.This is the zone in which the conjugate acid-base pair acts as a *buffer*. i.e.. a system which tends to resist change in pH when H^+ or OH^- is added. At pH values outside this zone, there is less capacity to resist changes in pH. The buffering power is maximum at the exact midpoint of the titration curve, i.e., when the ratio [proton acceptor]/[proton donor]=1.0, at which pH=pk' (figure 2-8). Each conjugate acid- base pair has a characteristic pH at which its buffering capacity is greatest, namely, the point at which pH = pK'.

Incrácellular and extracellular fluids of living organisms contain conjugate acid-base pairs which act as buffers at the normal pH of these fluids. The major intracellular buffer is the conjugate acid-base pair $H_2PO_4^-$-HPO_4^{2-} (pK'=7.2). Organic phosphates such as glucose 6-phosphate and ATP also contribute buffering power in the cell. The major extracellular buffer in the blood and interstitial fluid of vertebrates is the bicarbonate buffer system. The extraordinary buffering power of blood plasma may be shown by the following comparision. If 1 ml of 10 *N* HCI is added to 1.0 liter of neutral physiological saline (i.e., about 0.15 *M* NaCI), the pH of the saline will fall to pH 2.0, since NaCl solutions have no buffering power. However, if 1 ml of 10 *N* HCI is added to 1 liter of blood plasma, the pH will decline only slightly, from pH 7.4 to about pH 7.2.

The bicarbonate buffer system (H_2CO_3-HCO_3^-) has some distinctive features. While it functions as a buffer in the same way as other conjugate acid-base pairs, the pK' of H_2CO_3 a relatively strong acid, is about 3.8 which is far lower than the normal range of blood pH. In a bicaronate buffer system, the proton-donor species carbonic acid is in reversible equilibrium with dissolved CO_2:

$$H_2CO_3 \rightleftharpoons CO_2\ (\text{diss.}) + H_2O$$

If such an aqueous system is in contact with a gas phase, then the dissolved CO_2 will, in turn, equilibrate bertween the gaseous and aqueous phases:

$$CO_2\ (\text{diss.}) \rightleftharpoons CO_2\ (\text{gas})$$

Since by Henry's law the solubility of a gas in water is proportional to its partial pressure, the pH of the bicarbonate buffer system is a function of the partial pressure, of CO_2 in the gas phase over the buffer solution. If the CO_2 pressure is increased, all other variables remaining constant, the pH of the bicarbonate buffer declines, and vice versa. The bicarbonate system can buffer effectively near pH 7.0, at which the proton acceptor/proton donor ratio is very high, because a small amount of proton donor H_2CO_3 is in labile equililbrium with a relatively large reserve capacity of gaseous CO_2 in the lungs. Under any conditions in which the blood must absorb excess OH^-, the H_2CO_3 which is used up and converted to HCO_3^- is quickly replaced from the large pool of gaseous CO_2 in the lungs.

There is another distinctive feature of the bicarbonate buffer system. CO_2 is the end product of the aerobic combustion of fuel molecules in the vertebrate and is ultimately expired from the lungs. The steady-state ratio of $[HCO_3^-]/[H_2CO_3]$ in the blood is a reflection of the rate of loss of CO_2 production during tissue oxidation and of the rate of loss of CO_2 by expiration.

The pH of blood plasma in vertebrates is held at remarkable constant values. The blood plasma of man normally has a pH of 7.40. Should the pH-regulating mechanisms fail, as may happen in disease, and the pH of the blood fall below 7.0 or rise above 7.8, irreparable damage may occur. We may ask: What molecular mechanisms in cells are so

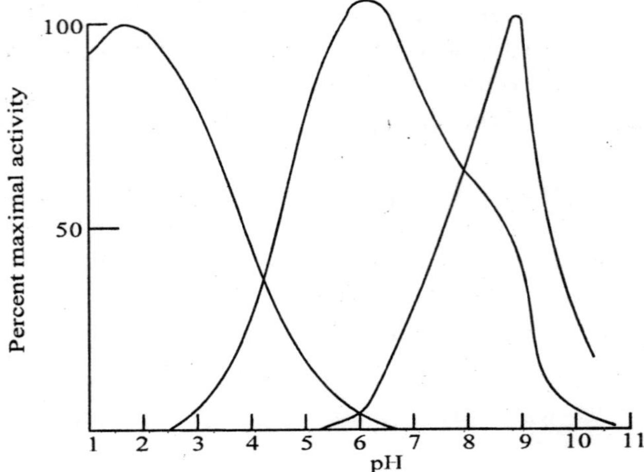

Fig. 3.8. : The effect of pH on the activity of some enzymes. Each enzyme has a characteristic pH-activity profile. The pH may influence the degree of ionization not only of enzymes but also of coenzyme and substrate molecules.

extraordinarily sensitive that a change in H^+ concentration of as little as 3×10^{-8} M (approximately the difference between blood at pH 7.4 and blood at pH 7.0) can be lethal? Although many aspects of cell structure and fuction are influenced by pH, it is the catalytic activity of enzymes that is especially sensitive. The typical curves in Figure show that enzymes have maximal activity at a characteristic pH, called the *optimum pH*, and that their activity declines sharply on either side of the optimum. Thus biological control of the pH of cells and body fluids is of central importance in all aspects of intermediary metabolism and cellular function.

CHAPTER 4
THE MAGIC OF CARBON

It is sometimes said that life on our planet is based on carbon. Carbon is an element that is found in all organic molecules. The term organic, meaning living, was used originally because it was thought that only living things could make organic compounds. This myth was dispelled in 1828 when the German chemist Wöhler synthesised the organic molecule urea from inorganic starting materials. This made scientists realise that there was no 'magic' or special life force needed to make the biochemicals of life, and today we have even reached the point where we could in theory synthesise DNA, the genetic material, from inorganic starting materials, and hence 'create' life.

But why is carbon so important? Carbon forms strong covalent bonds, in other words it shares electrons, with other elements. It forms four such bonds, that is it has a valency of four. A simple example is methane, whose **molecular formula** is CH_4 and whose structural formula is shown in fig 4.1.

Table 4.1: The elements found in living organisms.

Chief elements of organic molecules	*Ions*	*Trace elements*	
H hydrogen	Na^+ sodium	Mn manganese	B boron
C carbon	Mg^{2+} magnesium	Fe iron	Al aluminium
N nitrogen	Cl^- chlorine	Co cobalt	Si silicon
O oxygen	K^+ potassium	Cu copper	V vanadium
P phosphorus	Ca^{2+} calcium	Zn zinc	Mo molybdenum
S sulphur			I iodine

Carbon has been described as the element most intimately associated with life. Carbon has six protons and six electrons, and thus it presents a special case. Since two electrons occupy its inner shell, it can only have four in its outer shell. In order to satisfy its shell requirements then, it must either give up these four electrons or gain four more. Either of these changes, however, would throw its electric charge too far out of balance. The only way carbon can resolve this dilemma is to share its four outer electrons with other atoms through covalent bonding.

The problem becomes interesting at this point because carbon can combine with either negatively charged atoms. For example, it can share the electrons of oxygen, nitrogen, sulfur, or phosphorus atoms, or it can share electrons with hydrogen atoms. If a carbon atom simply add four hydrogens, the result is CH_4 the **swamp gas** called methane.

It should be pointed out that the four hydrogens do not simply stick out at right angles, as the diagram seems to indicate. Instead, they protrude three-dimensionally in such a way as to be as far from each other as possible. Thus, the structure would be more accurately represented as:

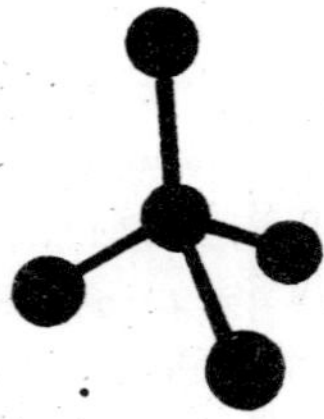

This configuration is called a **caltrops**. In ancient times, spikes were arranged in such a way because when they were thrown in the path of an advancing army, one spike always pointed up.

In addition, because of their peculiar bonding properties, carbon atoms can interact with each other as well. In this way, carbon atoms can join together to form complex molecules, as we shall see shortly shows some of the simpler carbon chains. It is interesting that of the some 2.2 million chemical compounds known, about 2 million contain carbon.

Ethan-e Propane Butane

Figure 4.1: Some of the simpler carbon chains.

Molecules produced by any combination of hydrogen and carbon atoms are called **hydrocarbons**. There are some very long hydrocarbon chains. You can see that in these kinds of molecules the electron-shell requirements of every atom are satisfied; each carbon atom has four other atoms with which to share electrons, and each hydrogen atom is satisfied through a single covalent bond with a carbon atom.

Because of the properties of carbon some very complex configurations can be formed. For example, carbon chains can branch. A five-carbon branch can be written as:

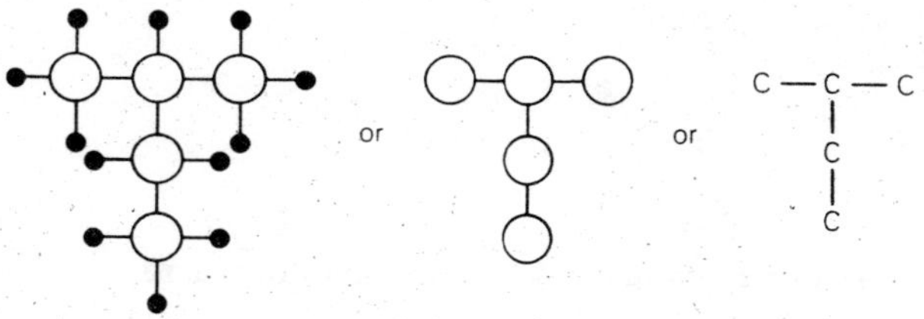

or

or

C — C — C
|
C
|
C

Carbon can also join ends to form rings. Five- or six-carbon rings are especially common:

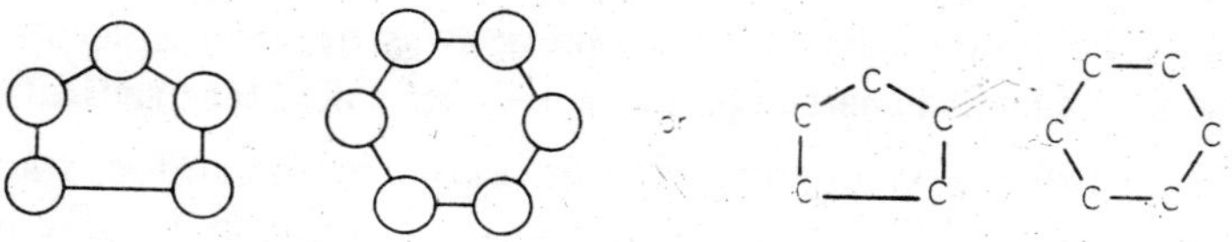

or

The notation of such molecules can be further simplified and written as:

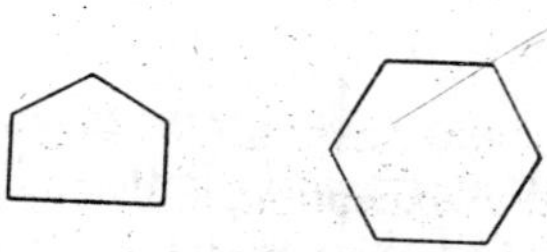

You should be aware that if no atom is shown where one is expected along a carbon backbone, that atom exists and is hydrogen. In such a case it is assumed to be present. Other atoms attached to carbon are always noted.

Carbon tends to be associated with certain kinds of molecules. Among these are those with **functional groups**. Functional groups are molecular side groups with specific chemical responses. Among them are the **hydroxyl groups**, —OH, which combine with carbon with

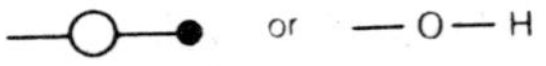

or —O—H

carbon to form alcohols, the **amino groups**, —NH_2. Which can form substances called amines.

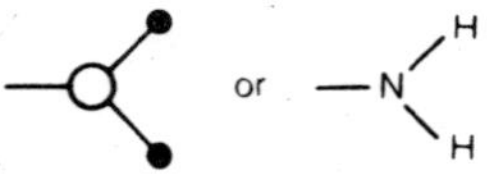

or —N H H

and the **carboxyl groups**, —COOH which form acids:

or —C OH O

Functional groups tend to behave as a single atom. That is, they move together as a group when the larger molecule of which they were a part becomes dissociated. If they break away from the molecule and exist as unattached charged particles, they are called **free radicals**.

The peculiar bonding of carbon not only permits the formation of large and complex molecules, but those sorts of molecules have precisely the traits that permit the complex processes of life. In fact, it can be said that because carbon is so intimatey associated with the processes of living things, the chemistry of carbon is, in a sense, the chemistry of life.

The Importance of Carbon

To study cellular molecules really means to study carbon-containing compounds. Almost without exception, molecules of importance to the cell biologist have a backbone,or skeleton, of carbon atoms linked together covalently. Actually, the study of carbon-containing compounds is the domain of **organic chemistry**. In its early days, organic chemistry was almost synominous with biological chemistry because most of the carbon-containing compounds that were first investigated were obtained from biological sources (hence the word *organic*, acknowledging the organismal origins of the compounds). The terms have long since gone their separate ways, however, because organic chemists have now synthesized a bewildering variety of carbon-containing compounds that do not occur naturally (that is, in the biological world). Organic chemistry therefore includes all classes of carbon-containing compounds, whereas **biological chemistry** (**biochemistry** for short) deals specifically with the chemistry of living systems and is, as we have already seen, one of the several historical strands that form an integral part of modern cell biology.

The **carbon atom** (c) is the most important atom in biological molecules. The diversity and stability of carbon–containing compounds are due to specific properties of the carbon atom and especially to the nature of the interactions of carbon atoms with one another as well as with a limited number of other elements found in molecules of biological importance.

The single most fundamental property of the carbon atom is its **valence** of four, which means that the outermost electron orbital of the atom lacks four of the eight electrons needed to fill it completely. Since a complete outer orbital is required for the most stable chemical state of an atom, carbon atoms tend to associate with one another or with other electron-deficient atoms, allowing adjacent atoms to share a pair of electrons. For each such pair, one electron comes from each of the atoms. Atoms that share each other's elelctrons in this way are said to be joined together by a **covalent bond**. Carbon atoms are most likely to form covalent bonds with one another and with atoms of oxygen (O), hydrogen(H), nitrogen (N), and sulfur (S).

The electronic configurations of several of these atoms are shown in Figure. Notice that, in each case, one or more electrons are required to complete the outer orbital.

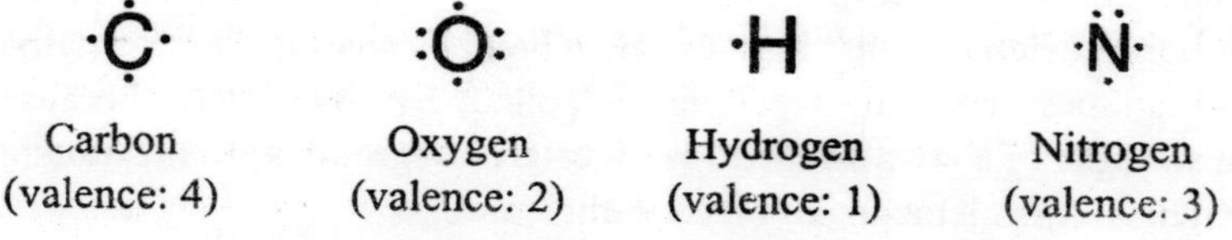

(a) Some biologically important atoms and their valences

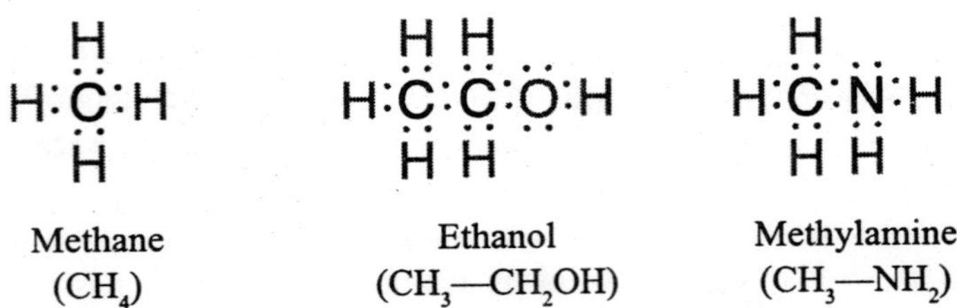

Methane (CH_4) | Ethanol (CH_3—CH_2OH) | Methylamine (CH_3—NH_2)

(b) Some simple organic molecules with single bonds

Ethylene ($CH_2 = CH_2$) | Carbon dioxide (CO_2)

(c) Some simple molecules with double bonds.

Molecular nitrogen (N_2) | Hydrogen cyanide (HCN)

(d) Some simple molecules with triple bonds.

Fig. 4.2 : Electron configurations of some biologically important atoms and molecules. Electronic configurations are shown for **(a)** atoms of carbon, oxygen, hydrogen and nitrogen and for simple organic molecules with **(b)** single bonds, **(c)** double bonds and **(d)** triple bonds. Only electrons in the outermost electron orbital are shown. In each case, the two electrons positioned bytween adjacent atoms represent a shared electron pair, with one electron provided by each of the two atoms. Electrons are color-coded; those from carbon, oxygen, hydrogen, and nitrogen are black, light pink, dark pink, and grey, respectively.

The number of "missing" electrons corresponds in each case to the valence of the atom, which indicates, in turn, the number of covalent bonds the atom can form. Carbon oxygen, hydrogen, and nitrogen are the lightest elements that form covalent bond by sharing electron pairs. This lightness makes the resulting compounds especially stable because the strength of a covalent bond is inversely proportional to the atomic weights of the elements involved in the bond.

Because four electron are required to fill the outer orbital of carbon, stable organic compounds have four covalent bonds for every carbon atom. Methane, ethanol, and methylamine are simple examples of such compounds, containing only **single bonds** between atoms. Sometimes, two or even three pairs of electrons can be shared by two atoms, giving rise to **double bonds** or even **triple bonds**. Ethylene and carbon dioxide are examples of double-bonded compounds. Triple bonds are found in molecular nitrogen (N_2) and hydrogen cyanide

Carbon-Containing Molecules Are Stable

As already implied, the stability of organic molecules is a property of the favourable electronic configuration of each carbon atom in the molecule. This stability is expressed in terms of **bond energy**—the amount of energy required to break 1 mole (6×10^{23}) of such bonds. Bond energies are usually expressed in *calories per mole* (*cal/mol*), where a **calorie** is the amount of energy needed to raise the temperature of one gram of water by one degree centigrade.

It takes a great deal of energy to break a covalent bond. For example, the carbon-carbon -(C—C) bond has a bond energy of 83 Kilocalories per mole (kcal/mol). The bond energies for carbon-nitrogen (C—N), carbon-oxygen (C—O), and carbon-hydrogen (C—H) bonds are all in the same range: 70,84, and 99 kcal/mol, respectively. Even more energy is required to break a carbon-carbon double bond (C ═ C; 146 kcal/mol) or a carbon-carbon triple bond (C ≡ C; 212 kcal/mol), so these compounds are even more stable.

Most noncovalent bonds in biologically important molecules have energies of only a few kilocalories per mole, and the energy of thermal vibration is even lower—about 0.6 kcal/mol. Covalent bonds are much higher in energy and therefore very stable. In fact, specific enzyme-catalyzed reactions capable of quite high energy input are required to break such bonds during chemical reactions in cells.

The "fitness" of the carbon-carbon bond for biological chemistry on Earth is especially clear when its energy is compared with that of solar radiation. As shown in Figure 4.3, there is an inverse relationship between the wavelength of electromagnetic radiation and its energy content. Specifically, the energy of electromagnetic radiation is related to the wavelength by the equation $E = 28,60 / \lambda$, where λ is the wavelength in nm, E is the energy in kilocalories per einstein, and 28,600 is a constant with the units kcal-nm/einstein.(An **einstein** is equal to 1 mole of photons.) Using this equation, you can readily

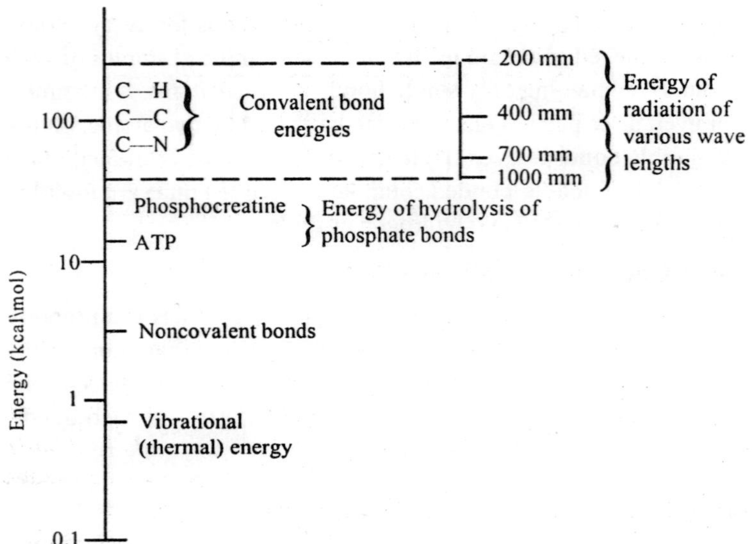

Fig. 4.3: Energies of bilogically important transitions, Bonds, and wavelengths of Electromagnetic Radiation.

calculate that the visible portion of sunlight (wavelengths of 400–700 nm) is lower in energy than the carbon-carbon bond. For example, green light with a wavelength of 500 nm has an energy content of about 57.2 kcal/einstein. The energy of green light is therefore well below the energies of covalent bonds. If this were not the case, visible light would break covalent bonds spontaneously, and life as we know it would not exist.

At a wavelength of 300 nm, for example, ultraviolet light has an energy content of about 95.3kcal/einstein, clearly enough to break carbon-carbon bonds spontaneously. It is this threat that has led to the current concern about pollutants that destroy the ozone layer in the upper atmosphere because the ozone layer filters out much of the ultraviolet radition that would otherwise reach the Earth's surface.

Carbon -Containing Molecules Are Diverse

In addition to their stability, carbon-containing compounds are characterized by the great diversity of molecules that can be generated from relatively few different kinds of atoms. Again, this diversity is due to the tetravalent nature of the carbon atom and the resulting

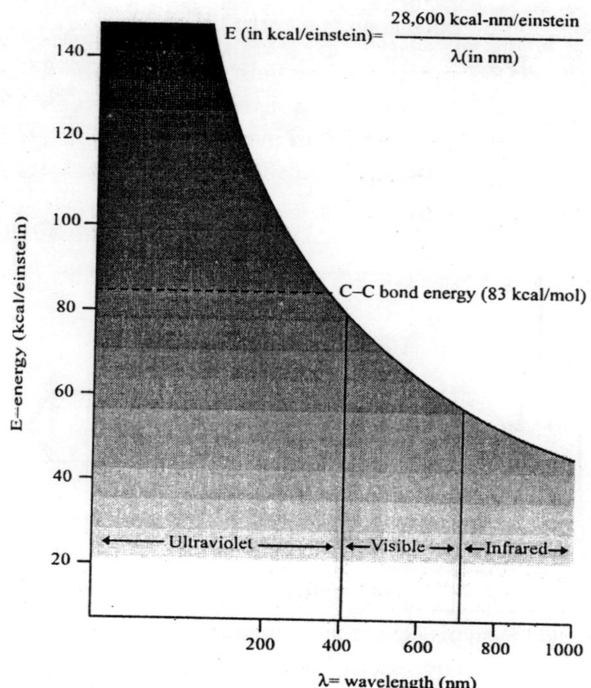

Fig. 4.4: The relationship between energy (E) and Wavelength (l) for electromagnetic Radiation. The dashed line marks the energy content of the C—C single bond (83 kcal/mol.)

propensity of each carbon atom to form covalent bonds to four other atoms. Because one or more of these bonds can be to other carbon atoms, molecules consisting of long chains of carbon atoms can be built up. Ring compounds are also common. Further variety is possible by the introduction of branching and of double and single bonds into the carbon-carbon chains.

When only hydrogen atoms are used to complete the valence requirements of such linear or circular molecules, the resulting compounds are called **hydrocarbons**. Hydrocarbons are very important economically because gasoline and other petroleum products are mixtures of short-chain hydrocarbons such as *octane*, an eight-carbon compound ($C_8 H_{18}$).

In biology, on the other hand, hydrocarbons play only a limited role because they are essentially insoluble in water, the universal solvent

CH_3—CH_3
Ethane

CH_3—CH_2—CH_3
Propane

CH_2=CH_2
Ethylene

CH_2≡CH_2
Acetylene

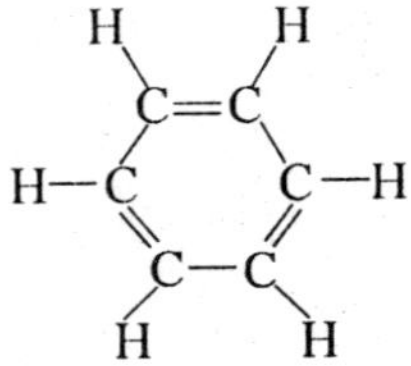

H H
C=C
H—C C—H OR
C—C
H H

Benzene

Fig. 4.5: Some simple hydrocarbon compounds.

in biological systems. There is an important exception to this general rule, however : The interior of every biological membrane is a nonaqueous environment from which water and water-soluble compounds are excluded by the long hydrocarbon "tails"of phospholipid molecules that project into the interior of the membranes has important implications for their role as permeability barriers, as we will see shortly.

Most biological compounds contain, in addition to carbon and hydrogen, one or more atoms of oxygen and often nitrogen, phosphorus, or sulfur as well. These atoms are usually part of various **functional groups** that confer both water solubility and chemical reactivity on the molecules of which they are a part.

Some of the more common functional groups present in biological molecules are shown in Figure ; Several of these groups are ionized or protonated at the near-neutral pH of most cells, including the negatively charged *carboxyl* and *phosphoryl* groups and the positively charged amino group. Other groups, such as the *hydroxyl*, *sulfhydryl*, *carbonyl*, and *aldehyde* groups, are uncharged at pH values near neutrality. However, they are much more polar than hydrocarbon and so cause a significant redistribution of electrons within the molecules to which they are attached, thereby conferring on these molecules greater water solubility and chemical reactivity.

Carbon-Containing Molecules

Can Form Stereoisomers

Carbon-containing molecules are capable of still greater diversity because the carbon atom is a **tetrahedral** structure with *geometric symmetry.* When four different atoms or groups of atoms are bonded to the four corners of such a tetrahedral structure, two different spatial configurations are possible. Although both forms have the same structural formula, they are not superimposable but are, in fact, mirror images of each other. Such mirror-image forms of the same compound are called **stereoisomers**.

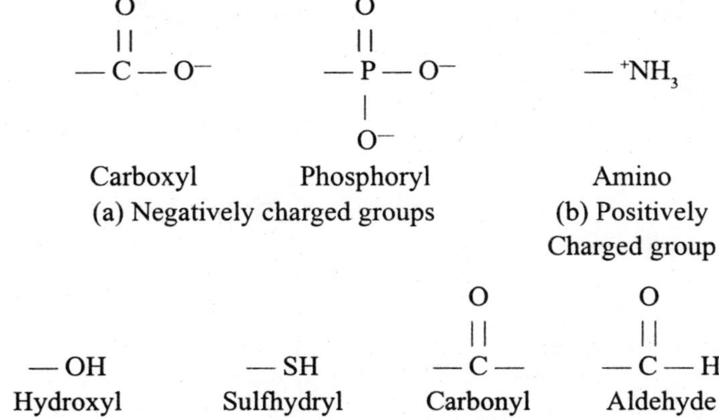

Fig. 4.6: Some common functional groups found in biological molecules.

A Carbon atom that has four different substituents is called an **asymmetric carbon atom**. Because two stereoisomers are possible for each asymmetric carbon atom, a compound with *n* asymmetric carbon atoms will have 2^n possible stereoisomers. The three-carbon amino acid *alanine* has a single asymmertric carbon atom (in the centre) and thus has two stereoisomers, called L-alanine and D-alanine. (Neither of the other two carbon atoms of alanine is an asymmetric carbon atom because one has two bonds to a single oxygen atom.) Both stereoisomers of alanine occur in nature, but only L-alanine is present as a component of proteins.

As an example of a compound with multiple asymmetric carbon

atoms, consider the six-carbon sugar *glucose*. Of the six carbon atoms of glucose, the four shown in pink are asymmetric. (Again, you should be able to figure out why the other two carbon atoms are not asymmetric.) With four asymmetric carbon atoms, the structure shown, D-glucose, is only one of 2^4, or 16, possible stereoisomers of the $C_6H_{12}O_6$ molecule. In this case, however, not all of the other possible stereoisomers exist in nature, mainly because some are energetically much less favourable than others.

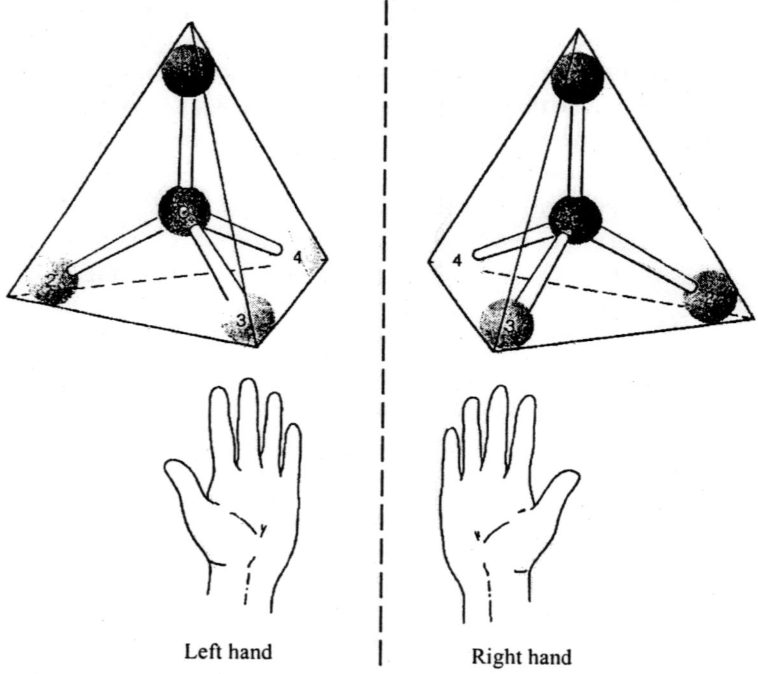

Figure 4.7: **Stereoisomers.** ***Stereoisomers of organic compound occur when four different groups are attached to a tetrahedral carbon atom. Stereoisomers, like left and right hands, are mirror images of each other and cannot be superimposed on one another.(The dashed line down the center of the figure is the plane of the mirror.)***

Carbon atoms can form an enormous variety of structures

Although carbon can exist in simple inorganic form, it has unique properties that permit formation of the large, complex molecules essential to life. A carbon atom has a total of six electrons—two in its lowest energy level and four valence electrons in its highest energy level. This means that a carbon atom can complete its outer shell by forming a total of four covalent bonds. Each bond can link it to another carbon atom or to an atom of a different element. Carbon is particularly well suited to serve as the backbone of a large molecule because carbon-to-carbon bonds are strong and not easily broken.

However, they are not so strong that an unreasonable amount of energy is required to break them. Carbon-to-carbon bonds are not limited to single bonds (based on sharing of one electron pair). Two carbon atoms can share two electron pairs with each other, foroming double bonds ($>C=C<$). In some compounds, triple carbon-to-carbon ($—C\equiv C—$) are formed.Carbon chains can be unbranched or branched, and carbon atoms can also be joined in some compounds.

The shape of a molecule is important in determining its biological properties and function. A carbon-containing molecule has a three-dimensional structure due to the terahedral nature of its bond angles.When a carbon atom forms four covalent single bonds with other atoms, the electron orbitals in its outer energy level become elongated and project from the carbon atoms toward the corners of tetrahedron. The angle between any two of these bonds is about 109.5 degrees. This bond angle is similar in diverse organic compounds. Keep in mind that, for simplicity, many of the figures in this book are drawn as two-dimensional graphic representations of three-dimensional molecules. For example, hydrocarbon chain, are not actually straight, but have a three-dimensional zigzag structure.

Generally, There is freedom of rotation around each carbon-to-carbon single bond. This property permits organic molecules to be flexible and assume a variety of shapes, depending on the extent to which each single bond is rotated. Double and triple bonds do not permit rotation, so regions of a molecule with such bonds tend to be inflexible.

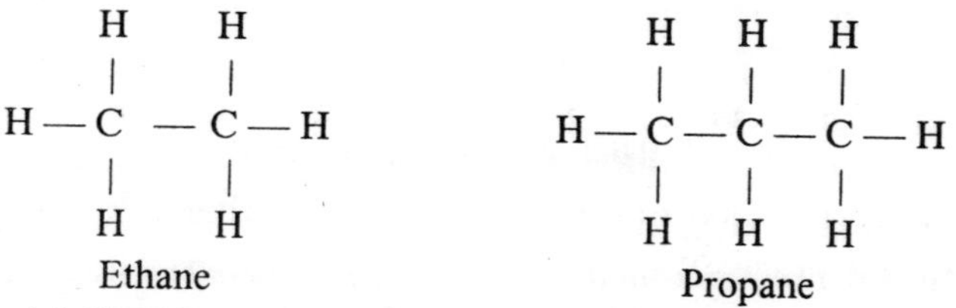

(*a*) Carbon atoms can form chains of varying length

1-Butene 2-Butene

(*b*) Carbon atoms may form double bonds with one another

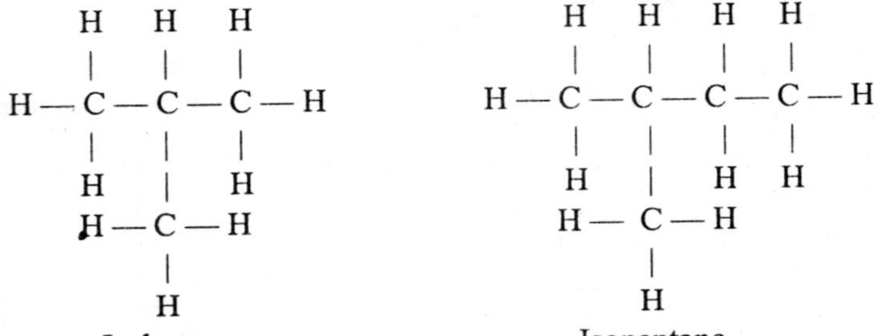

Isobutane Isopentane

(*c*) Carbon atoms can form branched chains

Cyclopentane Benzene

(*d*) Carbon atoms can form rings

Histidine (an amino acid)

(*e*) Rings and chains may be joined

Fig. 4.8: These structural formulas illustrate common variations in the architecture of organic molecules.

Isomers have the same molecular formula, but differ in structure

Compounds with the same molecular formulas but different structures and thus different properties are called **isomers**. Isomers do not have identical physical or chemical properties and may have different common names. Cells can distinguish between isomers. Usually, one isomer is biologically active and the other is not. Three types of isomers are structural isomers, geometric isomers, and enantiomers.

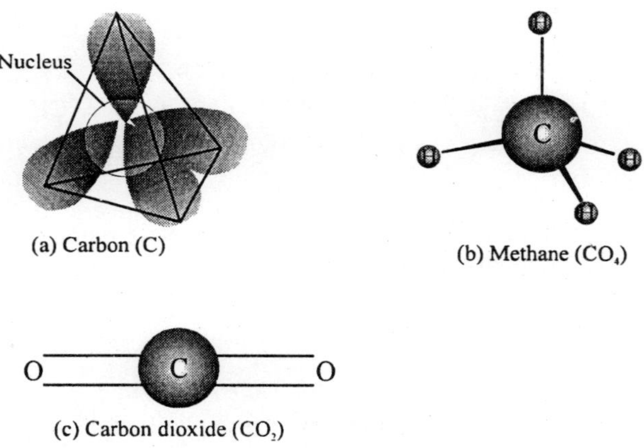

(a) Carbon (C)

(b) Methane (CO_4)

(c) Carbon dioxide (CO_2)

Fig.4.9: A carbon atom can form four covalent bonds in three-dimensional space.

Structural isomers are compounds that differ in the covalent arrangements of their atoms. For example, there are two structural isomers of the four-carbon hydrocarbon butane, one with a straight chain and the other with a branched chain (isobutane). Large compounds have more possible structural isomers. There are only two structural isomers of butane, but there can be up to 366,319 isomers of $C_{20}H_{42}$.

Geometric isomers are compounds that are identical in the arrangement of their covalent bonds, but different in the spatial arrangement of groups of atoms. Geometric isomers, also called *cis-trans* isomers, are present in some compounds with carbon-to-carbon double bonds. Because double bonds are not flexible like single bonds, atoms joined to the carbons of a double bond cannot rotate freely about the axis of the bonds The *cis-trans* isomers may be drawn. The

H
|
H—C—C—OH
|
H

Ethanol (C_2H_6O)

(a) Structural isomers

H H
| |
H—C—O—C—H
| |
H H

Dimethyl ether (C_2H_6O)

H_3C H
C=C
H CH_3

trans-2-butene

(b) Geometric isomers

H_3C CH_3
C=C
H H

cis-2-butene

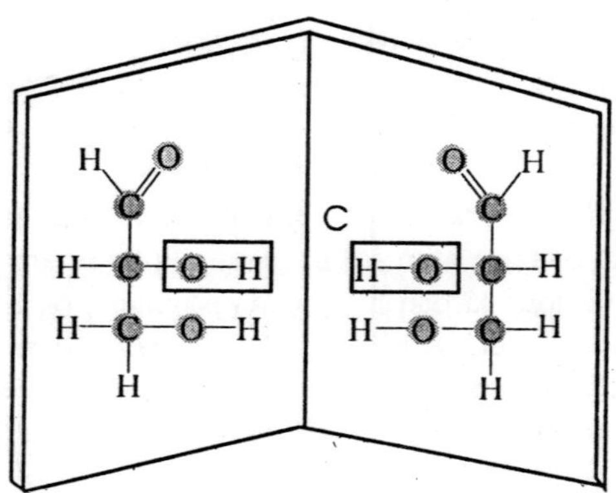

(c) Enantiomers of glyceraldehyde

Fig.4.10 : Isomers have the same molecular formula, but their atoms are arranged differently.(a)Structural isomers differ in the covalent arrangement of their atoms. (b) Geometric, or Cis-trans, isomers have identical covalent bonds, but differ in the order in which groups of atoms are arranged in space. Enantiomers are isomers that are mirror image of one another.

designation *cis* indicates that the two larger components are on the same side of the double bond. If they are on opposite sides of the double bond, the compound is designated a *trans* isomer.

Enantiomers are molecules that are mirror images of one another. Recall that the four groups bonded to a single carbon atom are arranged at the vertices of a tetrahedron. If the four bonded groups are all different, the central carbon is described as asymmetric. Figure 3–4*d* illustrates that the four groups can be arranged about the asymmetric carbon in two different ways that are mirror images of each other. The two molecules are enantiomers if they cannot be superimposed on one another no matter how they are rotated in space.Although enantiomers have similar chemical properties and identical physical properties, cells distinguish between them, and usually only one form is found in organisms.

Functional groups change the properties of organic molecules

Covalent bonds between hydrogen and carbon are nonpolar. Hydrocarbons therefore lack distinct charged regions. They are quite hydrophobic and do not interact readily with other compounds. However, the characteristics of a molecule can be changed dramatically by replacing one of the hydrogens with a group of atoms known as a **functional group**. Functional groups help determine the types of chemical reactions in which the compound participates. Most functional groups readily form associations, such as ionic and hydrogen bonds,with other molecules. Polar and ionic functional groups are hydrophilic because they associate strongly with polar water molecules

Each class of organic compounds is characterized by the presence of one or more specific functional groups. Note that the symbol R is used to represent the *remainder* of the molecule of which the functional group is a part. Most compounds present in cells contain two or more different functional groups. When we know what kinds of functional groups are present in an organic compound, we can predict its chemical behaviour.

The **hydroxyl group** (abbreviated R—OH) must not be confused with the hydroxide ion (OH^-). The hydroxyl group is polar due to the presence of a strongly electronegative oxygen atom. If a hydroxyl group replaces one of the hydrogens of a hydrocarbon, the resulting molecule can have significantly altered properties. For example, ethane is a hydrocarbon that takes the form of a gas at room temperature. If a hydrogen is replaced by a hydroxyl group, the resulting molecule is ethyl alcohol, or ethanol. Ethanol is somewhat cohesive because the polar hydroxyl groups of adjacent molecules interact ; it is therefore liquid at room temperature. Unlike ethane, ethyl alcohol can dissolve in water because the polar hydroxyl groups interact with the polar water molecules.

The **carbonyl group** consists of a carbon atom that has a double covalent bond with an oxygen atom. This double bond is polar because of the electronegativity of the oxygen; thus the group is hydrophilic. The position of the carbonyl group in the molecule determines the class to which the molecule belongs. An **aldehyde** has a carbonyl group positioned at the end of the carbon skeleton (abbreviated R—CO); a **ketone** has an internal carbonyl group (abbreviated R—CO—R).

The **carboxyl group** (abbreviated R—COOH) consists of a carbon atom joined by a double covalent bond to an oxygen atom, and by a single covalent bond to another oxygen, which is in turn bonded to a hydrogen atom. Two electronegative oxygen atoms in such close proximity establish an extremely polarized condition, which can cause the hydrogen atom to be stripped of its electron and released as a hydrogen ion (H^+). The carboxyl group then has one unit of negative charge ($R—COO^-$).

$$-C(=O)-O-H \longrightarrow -C(=O)-O^- + H^+$$

Carboxyl groups are weakly acidic; only a fraction of the molecules ionize in this way. This group can therefore exist in one of two hydrophilic states: ionic or polar. Carboxyl groups are essential constituents of amino acids.

An **amino group** (abbreviated $R—NH_2$) includes a nitrogen atom covalently bonded to two hydrogen atoms. Amino groups are weakly basic; some of the molecules accept a hydrogen ion (proton), thus acquiring a unit of positive charge. Amino groups are components of amino acids and of nucleic acids.

A **phosphate group** (abbreviated $R—PO_4H_2$) is weakly acidic. It can release one or two hydrogen ions, resulting in ionized forms with one or two units of negative charge. Phosphates are constituents of nucleic acids and certain lipids. They are very significant in many energy transfer reactions that take place in the cell.

The **sulfhydryl group** (abbreviated R—SH), consisting of an atom of sulfur covalently bonded to a hydrogen atom, is found in molecules called *thiols*. As we shall see, amino acids that contain a sulfhydryl group can make important contributions to the structure of proteins.

Many biological molecules are polymers

Many biological molecules such as proteins and nucleic acids are very large, consisting of thousands of atoms. Such gaint molecules are known as **macromolecules**. Most macromolecules are **polymers**, produced by linking small organic compounds called **monomers**. Just as all the words in this book have been written by arranging the 26 letters of the alphabet in various combinations, monomers can be grouped together to form an almost infinite variety of larger molecules. Just as we use different words to convey information, cells use different molecules to convey information. The thousands of different complex organic compounds present in organisms are constructed from about 40 small, simple monomers. For example, the 20 common types of amino acid monomers can be linked end-to-end in countless ways to form the polymers we know as proteins.

Each organism is unique because of differences in the monomer sequence within its DNA, the polymer that constitutes the information in the genes. Cells and tissues within the same organism are also different, due to variations in their component polymers. Muscle tissue and brain tissues differ in large part because of differences in the types and sequences of amino acids in proteins. Ultimately this protein structure is dictated by the sequence of monomers within the DNA of the organism, which is expressed somewhat differently in each cell type.

Table 4.2: Some biologically important functional groups

Functional group	Structural formula	Class of compounds characterized by group	Description
Hydroxyl	R — OH	Alcohols H H | | H — C — C — OH | | H H Ethanol	Polar because electronegative oxygen attracts covalent electrons

Functional group	Structural formula	Class of compounds characterized by group	Description
Amino	$R-NH_2$ (R—N with two H) Nonionized $R-N^{+}H_3$ (R—N+ with three H) Ionized	Amines $R-CH(NH_2)-C(=O)-OH$ Amino acid	Weakly basic; can accept an H^{+} ion
Carbonyl	$R-C(=O)-R$	Aldehydes $H-C(=O)-H$ Formaldehyde	Carbonyl group carbon is bonded to at least one H atom; polar because electrone-gative oxygen attracts covalent electrons
	$R-C(=O)-R$	Ketones $H-CH_2-C(=O)-CH_2-H$ Acetone	Carbonyl group carbon is bonded to two other carbons; polar because electronegative oxygen attracts covalent electrons.
Carboxyl	$R-C(=O)-OH$ Nonionized $R-C(=O)-O^{-} + H^{+}$ Ionized	Carboxylic acids (organic acid) $R-CH(NH_2)-C(=O)-OH$ Amino acid	Weakly acidic; can release an H^{+} ion

Polymers can be degraded to their component monomers by **hydrolysis** (which means "to break with water"). In a reaction regulated by a specific enzyme, a hydrogen from a water molecule attaches to one monomer, and a hydroxyl from water attaches to the adjacent monomers.

Functional group	Structural formula	Class of compounds characterized by group	Description
Methyl	$R—CH_3$	Component of many organic compounds $H-\overset{\overset{H}{\vert}}{\underset{\underset{H}{\vert}}{C}}-H$ Methane	Hydrocarbon; nonpolar
Phosphate	$R—O—\overset{\overset{O}{\Vert}}{\underset{\underset{OH}{\vert}}{P}}—OH$ Noninized $R—O—\overset{\overset{O}{\Vert}}{\underset{\underset{O^-}{\vert}}{P}}—O^- + 2H^+$ Ionized	Organic Phosphates $OP—\overset{\overset{O}{\Vert}}{\underset{\underset{OH}{\vert}}{P}}—O—R$ Phosphate ester (as found in ATP)	Weakly acidic; one or two H^+ ions can be released.
Sulfhydryl	$R—SH$	Thiols $\underset{\underset{SH}{\vert}}{CH_2}—\underset{\underset{NH_2}{\vert}}{CH}—\overset{\overset{O}{\Vert}}{C}—R$ Cysteine	Help stabilize internal structure of proteins.

The synthetic process by which monomers are covalently linked is called **condensation**. Because the *equivalent* of a molecule of water is removed during the reactions that combine monomers, the term *dehydration synthesis* is sometimes used to describe the process. However, in biological systems the synthesis of a polymer is not simply the reverse of hydrolysis, even though the net effect is the opposite of hydrolysis. Synthetic processes require energy and are regulated by different enzymes.

Carbohydrates include sugars, starches, and cellulose

Sugars, starches, and cellulose are **carbohydrates**. Sugars and **starches** serve as energy sources for cells; **cellulose** is the main structural component of the walls that surround plant cells. Carbohydrates contain carbon, hydrogen, and oxygen atoms in a ratio of approximately one carbon to two hydrogens to one oxygen $(CH_2O)_n$. The term *carbohydrate*, meaning "hydrate (water) of carbon," reflects the 2:1 ratio of hydrogen to oxygen, the same ratio found in water(H_2O). Carbohydrates contain one sugar unit (*mono*saccharides), two units (*dis*accharides), or many sugar units (*poly*saccharides).

CHAPTER 5
THE CELL

All organisms are composed of cells, and all cells come from preexisting cells. These two statements constitute the **cell theory**. Cells show the characteristics by which living systems are recognized: They use DNA as hereditary material and proteins as catalysts. In addition, most cells reproduce, transform matter and energy, and respond to their enviornment.

Viruses, on the other hand, show only a few of these characteristics and thus are not usually considered alive. Viruses depend entirely on living cells to reproduce, and they neither transform matter and energy nor respond to the environment. Some cells can develop into entire multicellular organisms, but no part of a cell can do that—nor can viruses.

Most cells are tiny. The volume of cells ranges from 1 to 1,000 μm^3. The eggs of some birds are enormous exceptions, to be sure, and individual cells of several types of algae are large enough to be viewed with the unaided eye. And athough neurons (nerve cells) have a volume that is within the "normal cell" range, they often have fine projections that may extend for meters, carrying signals from one part of a large animal to another. But by and large, cells are minuscule. Why?

Properties and Strategies of Cells

As we begin to consider what cells are and how they function, several general characteristics of cells quickly emerge. These include the sizes and shapes of cells, the classification of cells on the basis of their organizational complexity, and the specializations that cells undergo.

Cell Sizes and shapes

Cells come in a variety of sizes, shapes and forms. Some of the smallest bacterial cells, for example, are only about 0.2-0.3 μm in diameter—so small that about 30,000 such cells could fit side by side

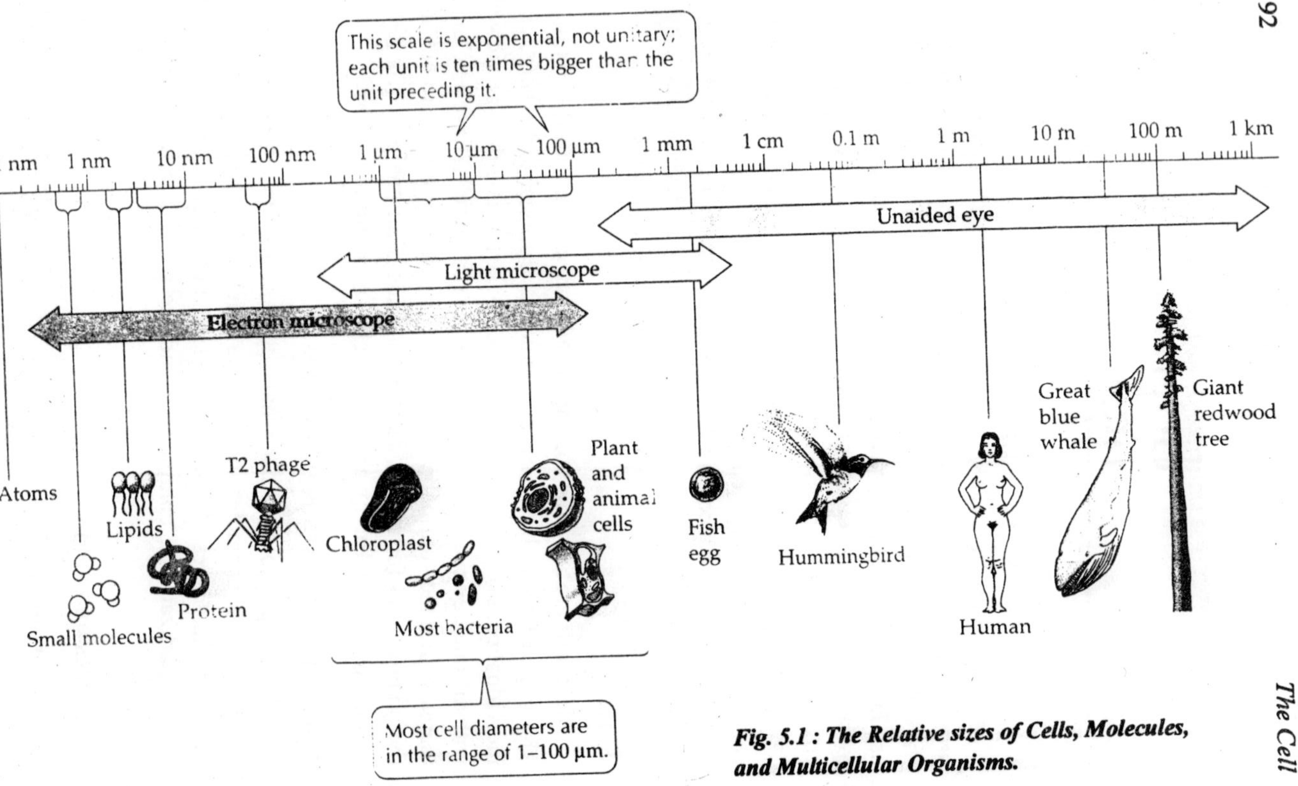

Fig. 5.1 : The Relative sizes of Cells, Molecules, and Multicellular Organisms.

across the head of a thumbtack! At the other extreme are highly elongated nerve cells, which may extend one or more meters. Those in the neck and legs of a giraffe are dramatic examples. On the other hand, the oft-cited examples of bird eggs, especially ostrich eggs, are rather misleading because they are indeed single cells, but with most of their internal volume occupied by yolk—large deposits of stored food intended as nourishment for the developing embryo.

Most cells fall into a rather narrow and predictable range of sizes, however. Bacterial cells, for example, are usually about 1–5 μm in diameter, and most cells of higher plants and animals have dimensions in the range of 10–50 μm. The two most important factors that limit cell size are the requirement for an adequate surface area/volume ratio and the need to maintain adequate concentrations of the substances and enzymes involved in various cellular processes.

The surface Area/Volume Ratio. The main constraint on cell size is that set by the need to maintain an adequate **surface area/volume ratio**. Surface area is important because it is here that the exchange between the cell and its environment takes place. The internal volume of the cell determines the amount of nutrients that will have to be imported and the quantity of waste products that must be excreted, but the surface area effectively measures the amount of membrane available for such uptake and excretion.

The problem of maintaining adequate surface area arises because the volume of a cell increases with the cube of the cell's length or diameter, whereas its surface area only increases with the square. The cell on the left is 20 μm on a side and has a volume of $8000\ \mu m^3$. ($V = s^3$, where $s = 20\ \mu m$) and a surface area of $2400\ \mu m^2$ ($A = 6s^2$). The surface area/volume ratio is therefore $2400\ \mu m^2 / 8000\ \mu m^3$, or $0.3\ \mu m^{-1}$. When this single large cell is divided into smaller cells, the total volume remains the same but the surface area increases. Thus, the surface area/volume ratio increases as the linear dimension of the cell decreases. The 1000 cells on the right, for example, still have a total volume of $8000\ \mu m^3 (1000 \times 2^3)$ but the total surface area is $24{,}000\ \mu m^2$ $(1000 \times 6 \times 2^2)$, so the surface area/volume ratio is $24{,}000\ \mu m^2$ / $8000\ \mu m^3$, or $3.0\ \mu m^{-1}$.

This comparison illustrates a major constraint on cell size. As a cell increases in size, its surface area does not keep pace with its volume, and the necessary exchange of substances between the cell and its

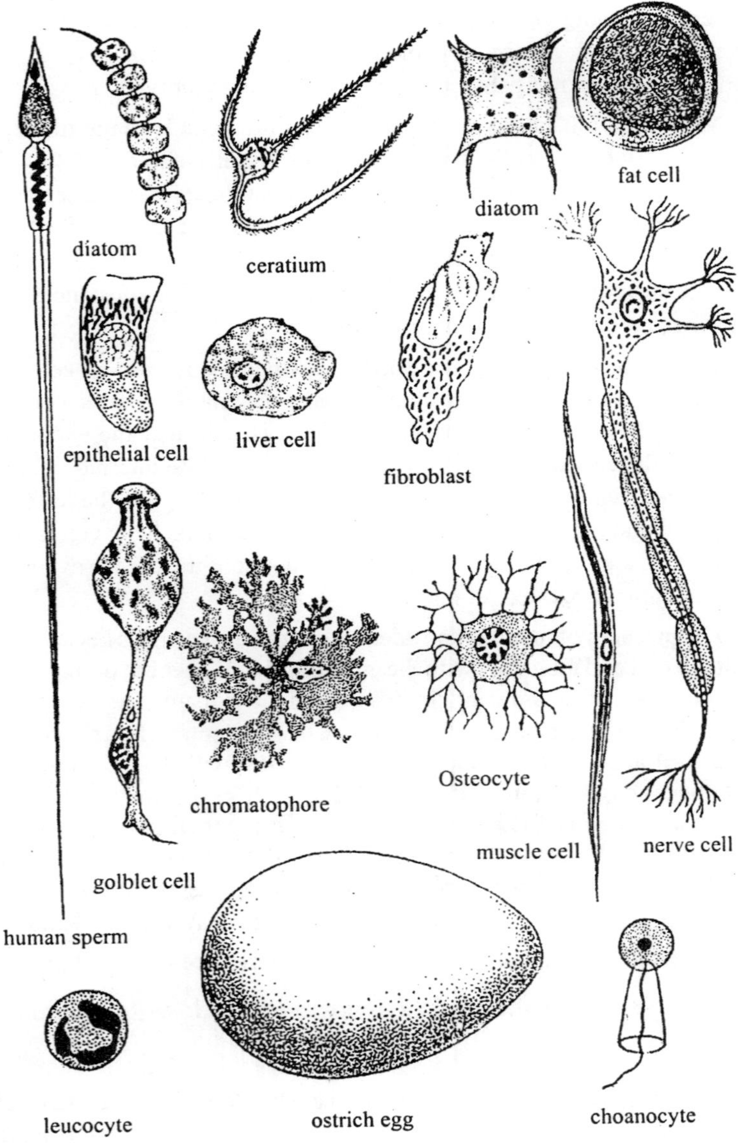

Fig. 5.2 : Various types of eukaryotic cells showing different shapes.

surroundings becomes more and more problematic. Cell size can therefore increase only over the range of values for which the membrane surface area is still adequate for the passage of materials into and out of the cell. Once the limiting surface area/volume ratio is reached, further increases in cell size would generate more cytoplasmic volume and therefore greater exchange requirements than could be met by the more modest increases in membrane surface area.

Some cells, particularly those that play a role in absorption, have additional characteristics that maximize their surface area. Effective surface area is most commonly increased by the inward folding or outward protrusion of the cell membrane. The cells that line your intestine, for example, contain many fingerlike projections called *microvilli* that greatly increase the effective membrane surface area and therefore the absorbing capacity of these cells.

Compartmentalization. Another limit on cell size is that imposed by the need to maintain adequate concentrations of the essential compounds and catalysts (enzymes) for the various processes cells must carry out. For a chemical reaction to occur in a cell, the appropriate reactants must collide with and bind to the surface of a particular enzyme. The frequency of such collisions will be greatly influenced by the concentrations of the reactants and of the enzyme itself. To maintain appropriate levels of reactants and enzymes as the size of a cell increases, the number of all such molecules must increase eight fold every time the three dimensions of the cell double. This increase taxes the synthetic capabilities of the cell.

The **compartmentalization** of activities within the cell is one solution to the problem of concentration. If all the enzymes and compounds involved in a specific process are localized within a specific region of the cell, a locally high concentration of those enzymes and compounds can be maintained in that region, rather than throughout the whole cell. This is what happens in plant and animal cells and is presumably the main reason they can be so much larger than bacterial cells and still function efficiently.

Plant and animal cells have a variety of **organelles**, internal compartments that are delineated by membranes and are highly specialized for specific functions. For example, the cells in a plant leaf have most of the enzymes, compounds, and pigments needed for photosynthesis compartmentalized together into structures called *chloroplasts*. Such cells can therefore maintain appropriately high concentrations of everything that is essential for photosynthesis within the chloroplasts without having to have similarly high levels of these substances elsewhere in the cell. In a similar way, other functions are localized within other compartments. This internal compartmentalization of specific functions makes it possible for the large cells of plants and animals to maintain locally high concentrations of the specific enzymes and compounds involved in particular cellular processes. Such processes can therefore proceed efficiently, even though as a whole such cells are orders of magnitude larger than bacterial cells.

Prokaryotes and Eukaryotes: An Organizational Dichotomy

With the advent of electron microscopy, biologists came to recognize two fundamentally different plans of cellular organization, the simpler one found in bacteria and the more complex one found in all other kinds of cells. Based on these structural differences, organisms can be divided into two broad groups, the prokaryotes (bacteria) and the eukaryotes (all other forms of life). The most fundamental distinction is that eukaryotic cells have a true, membrane-bounded nucleus (*eu-* is Greek for "true" or "genuine"; *karyon* means "nucleus"), whereas prokaryotic cells do not (*pro-* means "before", suggesting an evolutionarily earlier form of life).

Prokaryotes, in turn, are either *eubacteria* or *archebacteria*. The eubacteria ("true bacteria") include most present-day *bacteria and the cyanobacteria* (previously called blue green algae). Although similar to eubacteria in cellular structure, archebacteria are as different from eubacteria as they are from eukaryotes in terms of molecular structure and biochemistry. They are regarded as modern descendants of an evolutionarily ancient form of prokaryote that differed fundamentally from the ancestors of present eubacteria. (Arche- is a Greek prefix meaning "ancient" or "original"). Present-day archebacteria include the *methanogens*, which produce methane gas (CH_4) from carbon dioxide; the *halophiles*, which can grow in the presence of high salt concentrations (up to 5.5 *M* NaCl); and the *thermacidophiles*, which thrive in acidic hot springs (pH as low as 2, temperatures as high as 80°C)

Eukaryotes represent a much broader spectrum of organisms. Whereas all prokaryotes are single-celled organisms, eukaryotes may be either unicellular or multicellular organisms. Included in this group are all plants, animals, *fungi*, and *protists* (single-celled eukaryotes such as *algae* and *protozoa*).

Prokaryotes and eukaryotes differ at the cellular level in important structural, biochemical, and genetic features. Some of these differences are summarized in Table and discussed briefly here.

Table 5.1: A Comparison of some properties of prokaryotic and eukaryotic cells

Property	Prokaryotic Cells	Eukaryotic Cells
Size*	Small (a few micrometers in length or diameter, in most cases)	Large (10-50) times the length or diameter of prokaryotes, in most case)
Membrane-bounded nucleus	No	Yes
Organelles	No	Yes
Microtubules	No	Yes
Microfilaments	No	Yes
Intermediate filaments	No	Yes
Exocytosis and endocytosis	No	Yes
Mode of cell division	Cell fission	Mitosis and meiosis
Genetic information	DNA molecule complex--ed with relatively few proteins	DNA complexed with proteins (notably histones) to form chromosomes
Processing of RNA	Little	Much
Ribosomes**	Small (70S): 3 RNA molecules and 55 proteins	Large (80S): 4 RNA molecules and about 78 proteins

The disparity in size between prokaryotic and eukaryotic cells indicated here is generally valid, but cells of both types vary greatly in size, with some overlap in size ranges.

*****Ribosomes are characterized in terms of their sedimentation coefficients or S values, a measure of their sedimentation rate based on size and shape. Sedimentation coefficients are normally expressed in Svedberg units (S), where*** $1S = 1 \times 10^{-13}$ sec.

Some Organelles Contain DNA

Mitochondria: The Cell's Chemical Furnaces

Mitochondria (singular, *mitochondrion*) are typically tubular or sausage-shaped organelles 1 to 3 micrometers long found in all types of eukaryotic cells. Mitochondria are bounded by two membranes. The outer membrane is smooth, while the inner one is folded into numerous contiguous layers called **cristae**. The cristae partition the mitochondrion into two compartments: **a matrix**, lying inside the inner membrane; and an **outer compartment** or **intermembrane space**, lying between the two mitochondrial membranes. On the surfaces of the inner

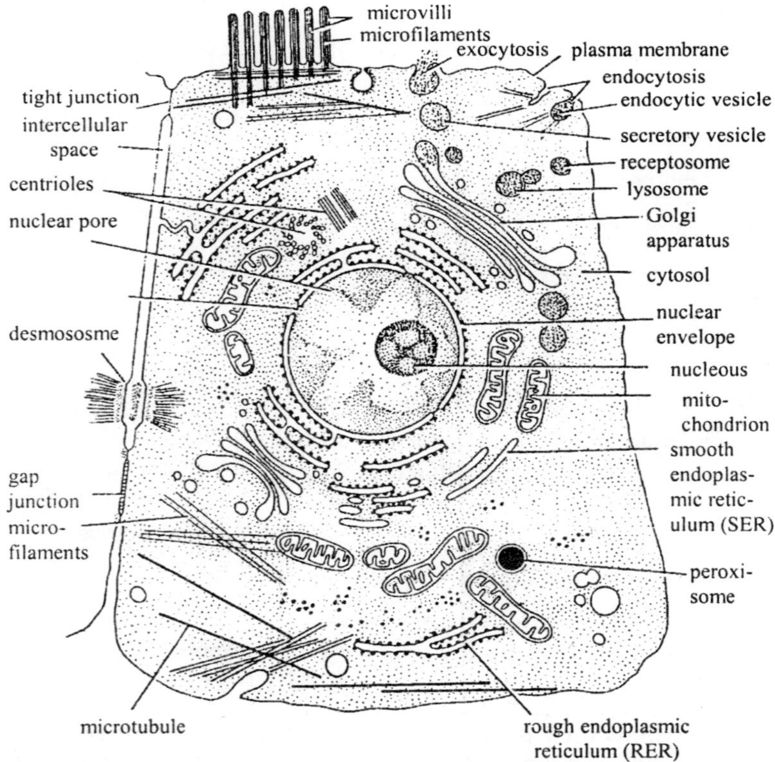

Fig. 5.3 : Ultrastrucutre of a typical animal cell as seen in the electron microscope.

membrane, and also submerged within it, are the proteins that carry out oxidative metabolism, the oxygen-requiring process by which the energy in macromolecules is stored in ATP.

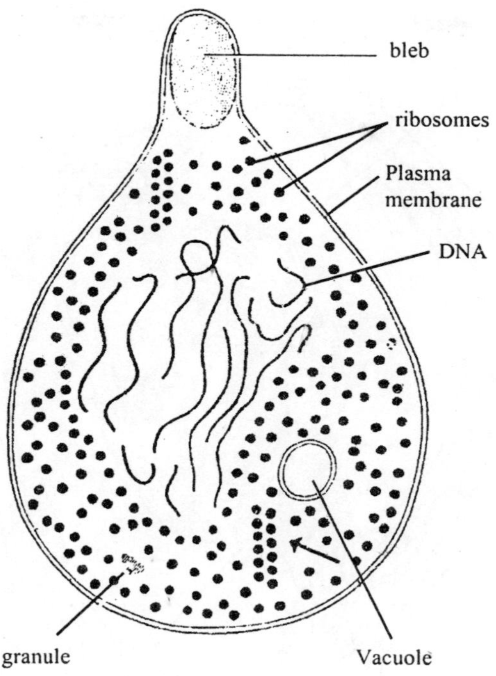

Fig. 5.4: A schematic diagram of typical PPLO cell.

Mitochondria contain their own DNA, on which are located several genes that produce some of the proteins essential for the role of mitochondria in oxidative metabolism. All of these genes are copied into RNA within the mitochondrion and used there to make proteins. In this process, the mitochondria use small RNA molecules and ribosomal components that are also encoded by the mitochondrial DNA. However, most of the genes that produce the enzymes used in oxidative metabolism are located in the nucleus.

A eukaryotic cell does not produce brand new mitochondria each time the cell itself divides. Instead, the mitochondria it has divide in two, doubling the number, and these are partitioned between the new cells. Most of the components required for mitochondrial division are encoded by genes in the nucleus and translated into proteins by cytoplasmic ribosomes. Mitochondrial replication is, therefore, impossible without nuclear participation, and mitochondria cannot be grown in a cell-free culture.

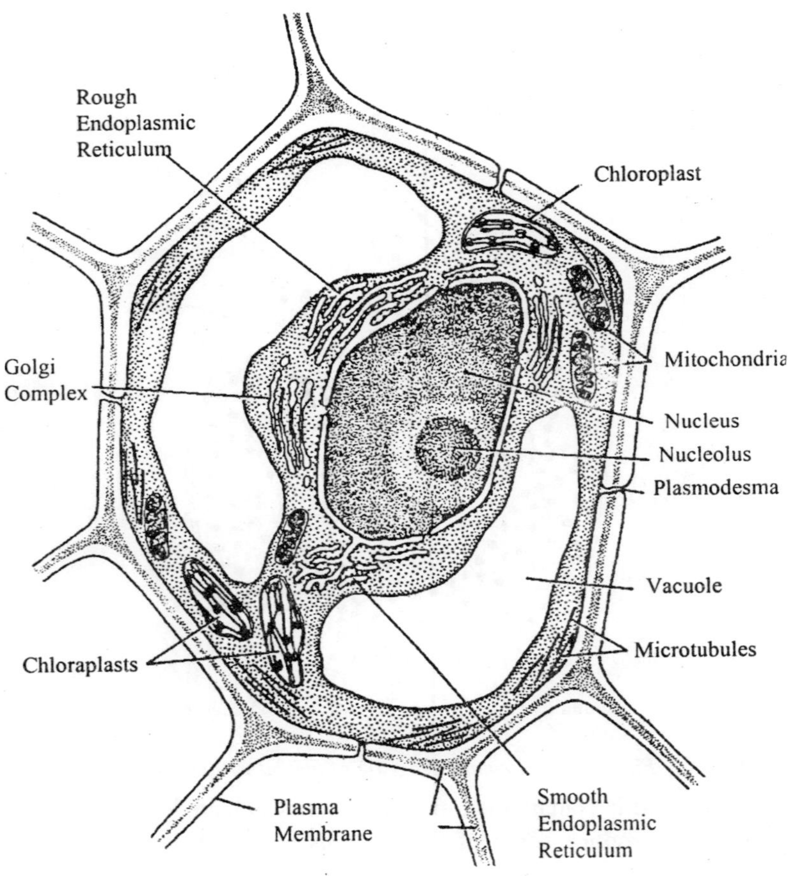

Fig. 5.5: Electron microscopic structure of a plant cell.

Chloroplasts: Where Photosynthesis Takes Place

The photosynthetic cells of algae and plants typically contain from one to several hundred chloroplasts, which give these eukaryotes the ability to perform photosynthesis. The advantage that chloroplasts bestow on the organisms that possess them is obvious: they can manufacture their own food. The number of chloroplasts contained in a cell depends on the organism involved or, in the case of multicellular photosynthetic organisms, the kind of cell.

The chloroplast body in bounded, like the mitochondrion, by two membranes that resemble those of mitochondria. However, chloroplasts are larger than mitochondria, and they have a more complex organization. In addition to the outer and inner membranes, which lie in close association with each other, chloroplasts have a closed compartment of stacked membranes called **grana** (singular, *granum*), which lie internal to the inner membrane. There may be a hundred or more grana in each chloroplast, and each granum may contain from a few to several dozen disk-shaped structures called **thylakoids**. On the surface of the thylakoids are the light-capturing photosynthetic pigments.

Like mitochondria, chloroplasts contain DNA, but many of the genes that specify chloroplast components are located in the nucleus. Some of the elements used in the photosynthetic process, including the specific RNA and protein components necessary to accomplish the reaction, are synthesized entirely within the chloroplast.

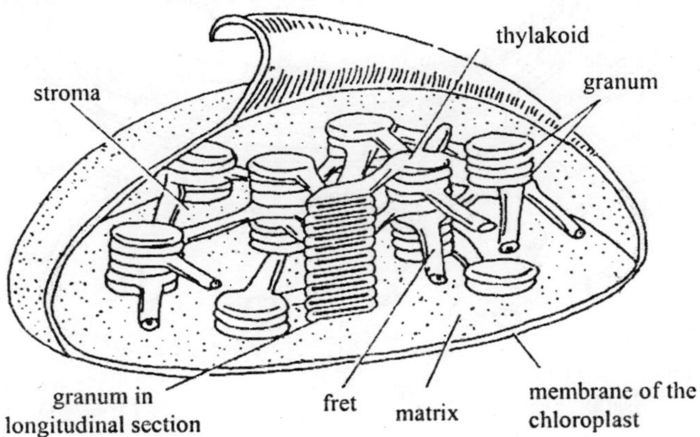

Fig. 5.6: Submicroscopic structure of the chloroplast.

If deprived of light for long periods, many chloroplasts lose much of their internal structure and are called **leucoplasts**. In the root cells and some other cells of plants, leucoplasts may serve as sites of storage for starch. A leucoplast that stores starch is sometimes termed an **amyloplast**. A collective term for these different kinds of organelles, all derived from **proplastids**, is **plastid**. All plastids come from the division of existing plastids.

The Golgi Complex: The Delivery System of the Cell

At various locations within the cytoplasm, flattened stacks of membranes called **Golgi bodies** occur. These structures are named for Camillo Golgi, the nineteenth-century Italian physician who first called attention to them. The numbers of Golgi bodies range from 1 or a few in the cells of protists, to 20 or more in animal cells and several hundred in plant cells. They are especially abundant in glandular cells, which manufacture and secrete substances. Collectively the Golgi bodies are referred to as **Golgi Complex**.

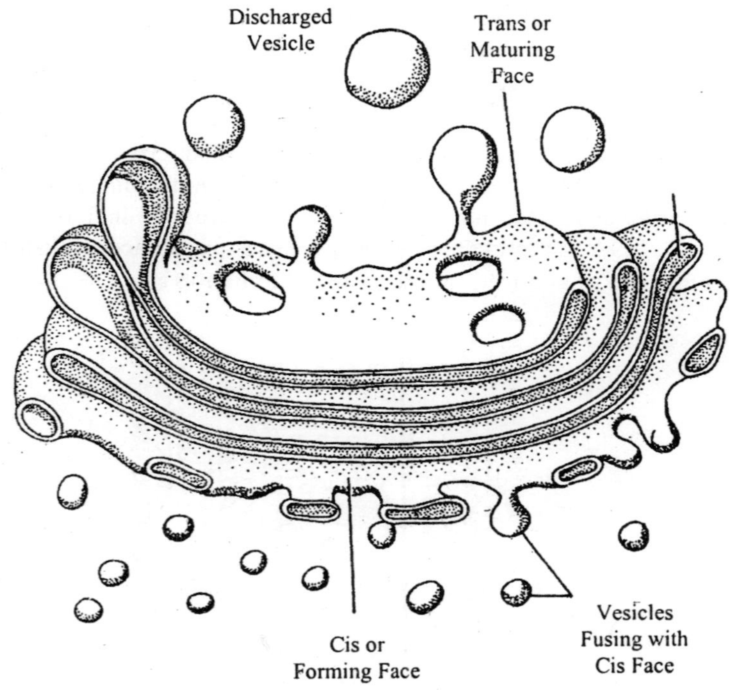

Fig. 5.7: Structure of Golgi complex.

The Golgi complex functions in the collection, packaging, and distribution of molecules synthesized in the cell. The proteins and lipids manufactured on the rough and smooth ER membranes are transported through the channels of the ER, or as vesicles budded off from it, into the Golgi complex. Within the Golgi complex, many of these molecules are modified by having short sugar chains attached to them, forming **glycoproteins**, which consist of a polysaccharide complexed to a protein, and **glycolipids**, consisting of a polysaccharide bound to a lipid. The newly formed **glycoproteins** and **glycolipids** collect at the ends of the Golgi bodies, in flattened stacked membrane folds called **cisternae** (Latin, "collecting vessels"). Periodically, the membranes of the cisternae push together, pinching off small membrane-bounded transport vesicles containing the glycoprotein and glycolipid molecules. These vesicles, which are often referred to as **liposomes**, then move to other locations in the cell, distributing the newly synthesized molecules within them to their appropriate destinations. Liposomes can be manufactured synthetically to contain any variety of desirable substances (such as drugs), and can then be injected into the body. Since the membrane of liposomes is similar to plasma and organellar membranes, these liposomes serve as an effective and natural delivery system to cells, and may prove to be of great therapeutic value.

Lysosomes: Producers of Digestive Enzymes for the Cell

Lysosomes, another class of membrane-bounded organelles, provide an impressive example of the metabolic compartmentalization achieved by the activity of the Golgi complex. They contain in a concentrated mix the digestive enzymes of the cell, which catalyze the rapid breakdown of proteins, nucleic acids, lipids, and carbohydrates. Throughout the lives of eukaryotic cells, lysosomal enzymes break down old organelles, recycling their component molecules and making room for newly formed organelles. As an example, mitochondria are replaced in some tissues every 10 days.

Lysosomes that are actively engaged in digestive activities keep their battery of hydrolytic enzymes (those that catalyze the hydrolysis of molecules) fully active by pumping protons into their interiors and thereby maintaining a low internal pH. Only at such acidic pH values are the hydrolytic enzymes maximally active. Lysosomes that are not functioning actively do not maintain an acidic internal pH are called **primary lysosomes.** When a primary lysosome fuses with a food vacuole or other organelle, its pH falls and its arsenal of hydrolytic enzymes is activated; it is then called a **secondary lysosome**.

What prevents lysosomes from digesting *themselves?* Although the answer to this question is not entirely clear, the process must require energy, because eukaryotic cells that are metabolically inactive die as the hydrolytic enzymes of primary lysosomes digest the lysosomal membranes from within. When these membranes disintegrate, the digestive enzymes of the lysosomes pour out into the cytoplasm of the cell and destroy it. Thus, the very process that repairs the ravages of time in eukaryotic cells may also lead to their destruction. Bacteria, in contrast, do not possess lysosomes and do not die when they are metabolically inactive. Instead, they are able to remain quiescent until altered conditions restore their metabolic activity, a property that greatly heightens their ability to persist under unfavorable environmental conditions.

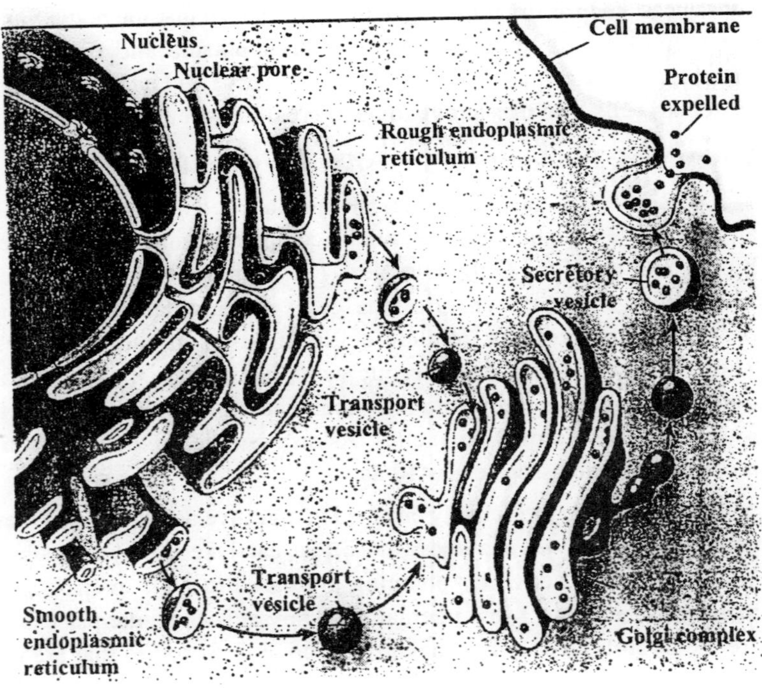

Fig. 5.8: How proteins are secreted across membranes.

In addition to their role in eliminating organelles and other structure within cells, lysosomes also eliminate substances (including other cells) taken internally via phagocytosis. When a white blood cell, for example, phagocytizes a passing pathogen, lysosomes fuse with the resulting "food vesicle," releasing their enzymes into the vesicle and degrading the material contained within it.

Lysosomal enzymes participate in the phenomenon of selective cell death, a characteristic feature in the development of many multicellular organisms. When a tadpole develops into a frog, the cells of the tail are destroyed by the enzymes from lysosomes. Many cells in your brain die during development, as do the webs between your fingers you had as an embryo. This directed cellular suicide is one of the principal mechanisms used by multicellular organisms in their achievement of complex patterns during development.

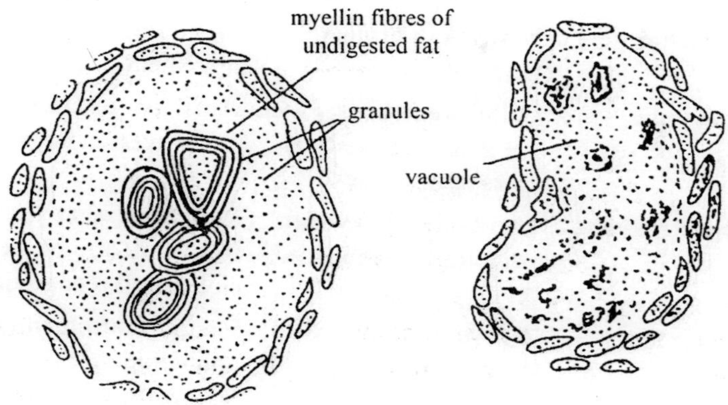

Fig. 5.9: Two types of Lysosomes from the kidney cells of rat.

Peroxisomes: Detoxifiers of Hydrogen Peroxide

Eukaryotic cells contain a variety of enzyme-bearing membrane-bounded vesicles called **microbodies**. Originally believed to be derived from the ER, microbodies are now known to arise only from preexisting microbodies through growth and fission. Microbodies are found in the cells of plants, animals, fungi, and protists, although traditionally they have been called **peroxisomes** in animal cells and **glyoxysomes** in plant cells.

The distribution of enzymes into microbodies is one of the principal ways in which eukaryotic cells organize their metabolism. Some of these enzymes convert fats into carbohydrates. Others, referred to as oxidative enzymes, catalyze the removal of electrons and associated hydrogen atoms. If oxidative enzymes were not isolated within microbodies, they would tend to short-circuit the metabolism of the cytoplasm, much of which involves the addition of hydrogen atoms to oxygen. The name given to animal microbodies, peroxisomes, refers to the hydrogen peroxide that is produced as a by-product of the activities of many of the oxidative enzymes within the microbody. Hydrogen peroxide is dangerous to cells because of its violent chemical reactivity. However, microbodies contain the enzyme catalase, which breaks down hydrogen peroxide into harmless water and oxygen.

Table 5.2: Eukaryotic cell structures and their functions

Structure	Description	Function
Cell wall	Outer layer of cellulose or chitin, or absent	Protection; support
Cytoskeleton	Network of protein filaments	Structural support; cell movement
Flagella (cilia)	Cellular extensions with 9 + 2 arrangements of pairs of microtubules	Motility or moving fluids over surfaces
Plasma membrane	Lipid bilayer in which proteins are embedded	Regulates what passes into and out of cell; cell-to-cell recognition
Endoplasmic reticulum	Network of internal membranes	Forms compartments and vesicles
Nucleus	Spherical structure bounded by double membrane; contains chromosomes	Control center of cell; directs protein synthesis and cell reproduction
Golgi complex	Stacks of flattened vesicles	Package proteins for export from cell; forms secretory vesicles
Lysosomes	Vesicles derived from Golgi complex that contain hydrolytic digestive enzymes	Digest worn-out mitoch-ondria and cell debris; play role in cell death
Microbodies	Vesicles derived from endoplasmic reticulum, containing oxidative and other enzymes	Isolate particular che-mical activities from rest of cell
Mitochondria	Bacteria-like elements with inner membrane	"Power plant" of the cell; site of oxidative metabolism
Chloroplasts	Bacteria-like elements with vesicles containing chlorophyll	Site of photosynthesis in plant cells

Chromosomes	Long threads of DNA that form a complex with protein	Contain hereditary information
Nucleolus	Site of chromosomes of rRNA synthesis	Assembles ribosomes
Ribosomes	Small, complex assemblies of protein and RNA, often bound to endoplasmic reticulum	Sites of protein synthesis

The Endoplasmic Reticulum: Compartmentalization of the Cell

As seen through a light microscope, the interiors of eukaryotic cells exhibit a relatively featureless matrix, within which various organelles are embedded. With an electron microscope, however, a striking difference becomes evident—the cell interiors appear to be packed with membranes. So thin that they are invisible with the relatively low resolving power of light microscopes, these membranes. So thin that they are invisible with the relatively low resolving power of light microscopes, these membranes fill the cell, dividing it into compartments, channeling the transport of molecules through the interior of the cell, and providing the surfaces on which enzymes act. The presence of these membranes in eukaryotic cells constitutes the most fundamental distinction between eukaryotes and prokaryotes.

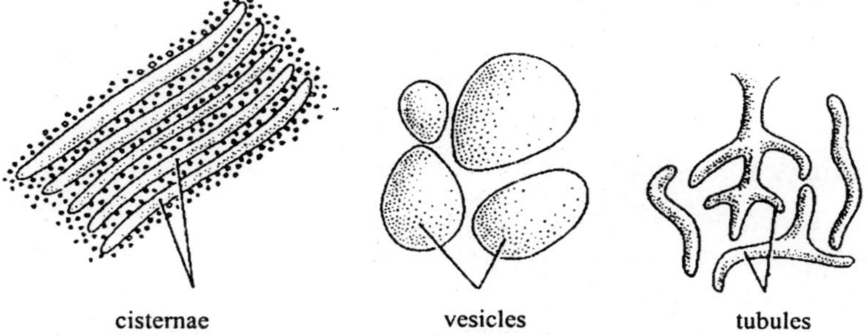

Fig. 5.10 : Various components of the endoplasmic reticulum.

This internal membrane system is called the **endoplasmic reticulum (ER)**. The term *endoplasmic* means "within the cytoplasm," and the term reticulum comes from a Latin word that means "a little net". Like the plasma membrane, the ER is composed of a lipid bilayer containing embedded proteins. It weaves in sheets through the interior of the cell, creating a series of channels and interconnections between its membranes.

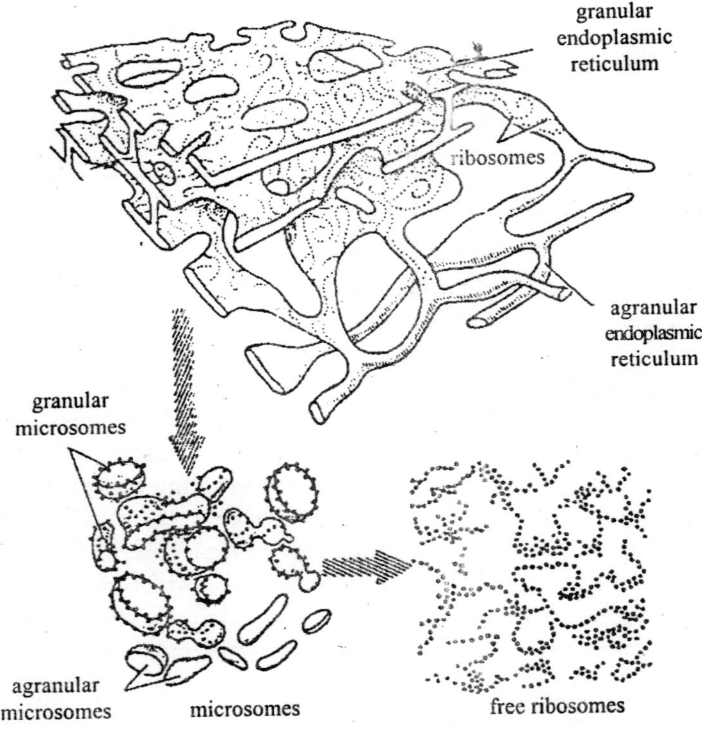

Fig. 5.11: Three-dimensional structure of endoplasmic reticulum showing microcomes and ribosomes.

Rough ER: Manufacturer of Proteins for Export

The surface of those regions of the ER that are devoted to the synthesis of proteins are heavily studded with ribosomes, large molecular aggregates of protein and ribonucleic acid (RNA) that translate RNA copies of genes into protein. Through the electron microscope, these regions of the ER appear pebbly and are therefore called **rough ER**.

The proteins synthesized on the surface of the rough ER are exported from the cell. These proteins contain special amino acid sequences called *signal sequences*. As a new protein is made by a free ribosome (one not attached to a membrane), the signal sequence of the growing polypeptide attaches to a recognition factor that carries the ribosome and its partially completed protein to a "docking site" on the surface of the ER. As the protein is made it passes through the ER membrane into the interior ER compartment and moves to the vesicle-forming system called the Golgi complex. It then travels within vesicles to the inner surface of the plasma membrane, where it is released to the outside.

Smooth ER: Organizer of Internal Activities

Regions of the ER with relatively few bound ribosomes are referred to as smooth ER. The membranes of the smooth ER contain many embedded enzymes, most of which are active only when associated with a membrane. Enzymes anchored with the ER, for example, catalyze the synthesis of a variety of carbohydrates and lipids. In cells that carry out extensive lipid synthesis, such as those in the testes, intestine, and brain, smooth ER is particularly abundant. In the liver, the enzymes of the smooth ER are involved in the detoxification of drugs including amphetamines, morphine, codeine, and phenobarbital.

Ribosomes. The last portrait in our gallery of organelles is the **ribosomes**, which serves as the site of protein synthesis. Strictly speaking, a ribosome should not be considered an organelle because it is not bounded by a membrane. But it is convenient to include ribosomes at this time because ribosomes, like organelles, are the focal point for a specific cellular activity—in this case, protein synthesis. Unlike true organelles, ribosomes are found in both eukaryotic and prokaryotic cells. Even here, however, the dichotomy between the two basic cell types manifests itself in that prokaryotic and eukaryotic ribosomes differ characteristically in size and in the numnber and kinds of protein and RNA molecules they contain.

Compared to membrane-bounded ogranelles, ribosomes are tiny structures. The ribosomes of eukaryotic cells have a diameter of about 30 nm, and those of prokaryotic cells are slightly smaller. An electron microscope is therefore required to visualize ribosomes. To appreciate how small ribosomes are, consider that more than 350,000 ribosomes could fit inside a typical bacterial cell, with room to spare!

Another way to express the size of such a small particle is to refer to its **sedimentation coefficient**. The sedimentation coefficient of a particle or macromolecule is a measure of how rapidly the particle sediments in an ultracentrifuge and is expressed in *Svedberg units* (*S*).

Sedimentation coefficients are widely used to indicate relative size, especially for large macromolecules such as proteins and nucleic acids and small particles such as ribosomes. Ribosomes from eukaryotic cells have sedimentation coefficients of about 80S; those from prokaryotic cells are about 70S.

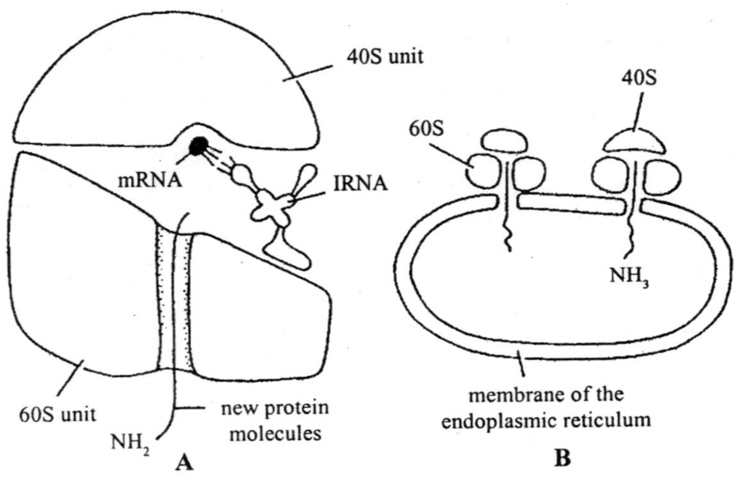

Fig. 5.12 : Functional 80S ribosomes: A—Diagram of ribosome showing the two subunits and the probable position of mRNA.and tRNA. The nascent polypeptide chain passes through a kind of tunnel with the large subunit; B—Diagram showing the relationship between the ribosome and the membrane of the endoplasmic reticulum and the entrance of the polypeptide chain into the centre of endoplasmic reticulum during the process of protein synthesis.

A ribosome consists of two subunits differing in size, shape and composition. In eukaryotic cells, the **large** and **small ribosomal subunits** have sedimentation coefficients of about 60S and 40S, respectively. For prokaryotic ribosomes, the corresponding values are about 50S and 30S. In both eukaryotic and prokaryotic cells, ribosomal subunits are synthesized and assembled separately in the cell but come together for the purpose of making proteins.

Ribosomes are far more numerous than most other cellular structures. Prokaryotic cells usually contain thousands of ribosomes, and eukaryotic cells may have hundreds of thousands or even millions of them. Ribosomes are also found in both chloroplasts and mitochondria, where they function in organelle-specific protein synthesis. Although the ribosomes of these eukaryotic organelles differ in size and composition from the ribosomes found in the cytoplasm of the same cell, they are strikingly similar to those found in bacteria and cyanobacteria.

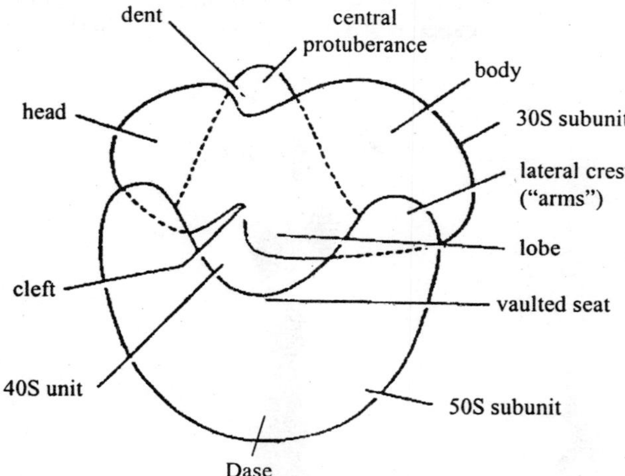

Fig. 5.13 : Stoffler and Wittman's model of 70S ribosome.

This similarly is particularly striking when the nucleotide sequences of ribosomal RNA (rRNA) from mitochondria and chloroplasts are compared with those of prokaryotic rRNAs. These similarities support the **endosymbiont** theory for the evolutionary origins of mitochondria and chloroplasts. This theory, developed most fully by Lynn Margulis, proposes that both of these organelles originated from prokaryotes that gained entry to, and established a symbiotic relationship within, the cytoplasm of ancestral cells (sometimes called *protoeukaryotic* cells) that eventually gave rise to present-day eukaryotic cells. (In biology, *symbiosis* involves the intimate living together of two organisms or cells for the mutual benefit of both). Specifically, the endosymbiont theory postulates that both mitochondria and chloroplasts arose from specific prokaryotic ancestors—bacteria and cyanobacteria, respectively. Although still speculative, the endosymbiont theory is the most widely accepted explanation of the origin of eukaryotic organelles.

Vacuoles

Cells also contain a variety of other membrane-bounded organelles called **vacuoles**. In animal cells, vacuoles are frequently used for temporary storage or transport. Some protozoa, for example, take up food particles or other materials from their environment by a process called *phagocytosis* ("cell eating"). Phagocytosis is a form of endocytosis that involves an inpocketing of the plasma membrane around the desired substance, followed by a pinching-off process that internalizes the membrane-bounded particles as a vacuole.

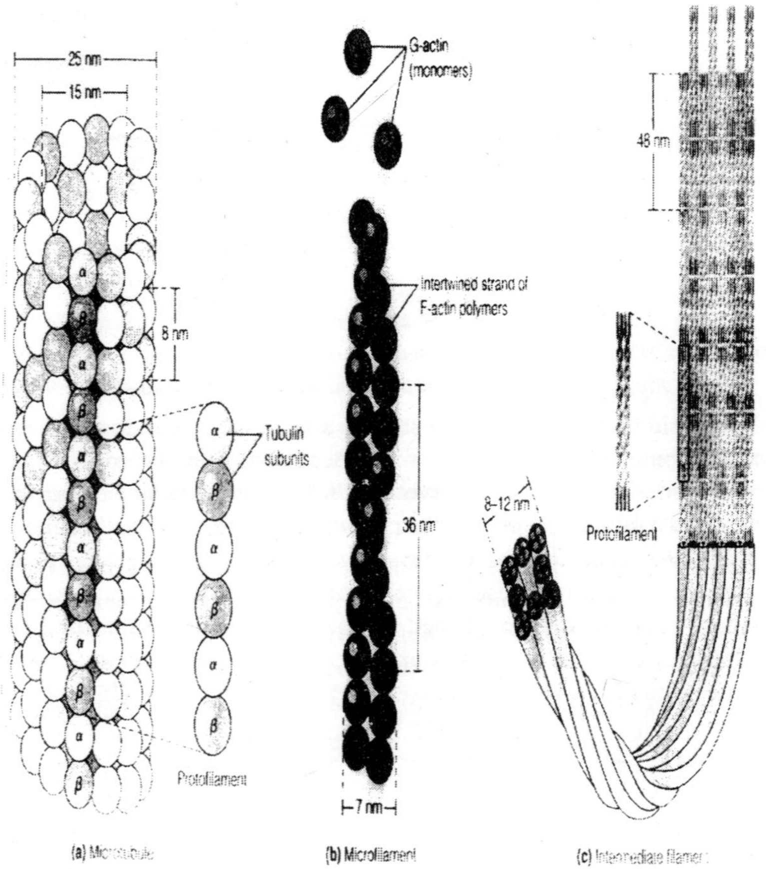

Fig. 5.14 : Structure of Microtubules, Microfilaments, and Intermediate Filaments.

The term *vacuole* is also used with plant cells, but usually in reference to the large **central vacuole** that characterizes most mature plant cells. Although the central vacuole may play a limited role in storage and appears also to be capable of a lysosomelike function in intracellular digestion, its real importance lies in the maintenance of the *turgor pressure* of the plant cell.

As you already know, a plant cell is surrounded by a rigid cell wall. The central vacuole is like an inflatable sphere in the center of the cell that is "pumped up" with liquid, pressing the rest of the cellular constituents out against the cell wall and thereby maintaining the turgor pressure characteristic of nonwilted plant tissue. The limp, flaccid appearance associated with wilting comes about when the central vacuole does not provide adequate pressure, allowing the cell (and hence the tissue) to go limp. We can easily demonstrate this by placing a piece of crisp celery in salt water. The high concentration of salt on the outside of the cells will cause water to move out of the cells; the turgor pressure will then decrease, and the tissue will quickly become flaccid and lose its crispness.

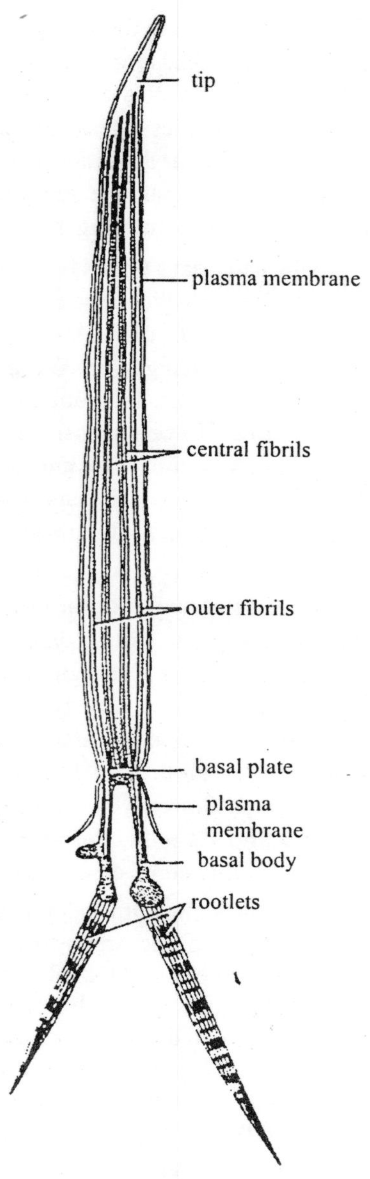

Fig. 5.15 : A ciliary apparatus in L.S. showing fundamental structure of cilium or shaft, basal bodies and ciliary rootlets

The Cytoplasm, Cytosol, and Cytoskeleton

The **cytoplasm** of a eukaryotic cell consists of that portion of the interior of the cell not occupied by the nucleus. Thus, the cytoplasm includes organelles such as the mitochondria; it also includes the **cytosol**, the semifluid substance in which the organelles are suspended. In a typical animal cell, the cytoplasm occupies more than half of the total internal volume of the cell. Many cellular activities take place in the cytoplasm, including the synthesis of proteins, the synthesis of fats, and the initial steps in the release of energy from sugars.

In the early days of cell biology, the cytosol was regarded as a rather amorphous, gel-like substance, and its proteins were thought to be soluble and freely diffusible. However, several new techniques have done much to change this view greatly. We now know that the cytosol of eukaryotic cells, far from being a structureless fluid, is permeated by an intricate three-dimensional array of interconnected filaments and tubules called the **cytoskeleton**, and the **cytoskeletal network**. Specifically, the cytoskeleton consists of a network of microtubules, microfilaments, and intermediate filaments, all of which are unique to eukaryotes.

As the name suggests, the cytoskeleton is an internal framework that gives a eukaryotic cell its distinctive shape and high level of internal organization. This elaborate array of filaments and tubules forms a highly structured yet very dynamic matrix that not only helps establish and maintain shape but also plays important roles in cell movement and cell division. The filaments and tubules that make up the cytoskeleton function in the contraction of muscle cells, the beating of cilia and flagella, the movement of chromosomes during cell division, and in some cases the locomotion of the cell itself.

In addition, the cytoskeleton serves as a framework for positioning and actively moving organelles within the cytosol. The same may be true of ribosomes and enzymes. Some researchers estimate that up to 80% of the proteins of the cytosol are not freely diffusible but are instead associated with cytoskeleton. Even water, which accounts for about 70% of the cell volume, may be influenced by the cytoskeleton. It has been estimated that as much as 20–40% of the water in the cytosol may be bound to the filaments and tubules of the cytoskeleton.

The three major structural elements of the cytoskeleton are *microtubules*, *microfilaments*, and *intermediate filaments*. These structures occur only in eukaryotic cells. They can be visualized by phase-contrast and immunofluorescence microscopy and by electron microscopy. Some of the structures in which they are found (such as cilia, flagella, or muscle fibrils) can even be seen by ordinary light microsocpy. Microfilaments and microtubules are best known for their roles in contraction and motility. In fact, these roles were appreciated well before it became clear that the same structural elements are also integral parts of the pervasive network of filaments and tubules that gives cells their characteristic shape and structure.

Microtubules. Of the structural elements found in the cytoskeleton, **microtubules (MTs)** are the largest. A well known MT-based cellular structure is the *axoneme* of cilia and flagella, the appendages responsible for motility of eukaryotic cells. The axoneme of the sperm tail consists of microtubules. MTs also form the *spindle fibers* that separate chromosomes prior to cell division.

In addition to their involvement in motility and chromosome movement, MTs also play an important role in the organization of the cytoplasm. They contribute to the polarity and overall shape of the cell, the satial disposition of its organelles, and the distribution of microfilaments and intermediate filaments. Examples of the diverse phenomena that are governed by MTs include the asymmetric shapes of animal cells, the plane of cell division in plant cells, the ordering of filaments during muscle development, and the positioning of mitichondria around the axoneme of motile appendages.

MTs are straight, hollow cylinders with an outer diameter of about 25 nm and an inner diameter of about 15 nm. The wall of the microtubule consists of longitudinal arrays of protofilaments, usually 13 arranged side by side around the hollow center, or lumen. Each protofilament is a linear polymer of *tubulin* molecules. Tubulin is a dimeric protein, consisting of two similar but distinct polypeptide subunits, α *-tubulin* and β *-tubulin*. All of the tubulin dimers in each of the protofilaments are oriented in the same direction, such that all of the subunits face the same end of the microtubule. This uniform orientation gives the MT an inherent polarity. The polarity of microtubules has important implications for their assembly.

Microfilaments. Microfilaments (MFs) are much thinner than microtubules. They have a diameter of about 7 nm, which makes them the smallest of the major cytoskeletal components. MFs are best known of the major cytoskeletal components. MFs are best known for their role in the contractile fibrils of muscle cells.

However, microfilaments are involved in a variety of other cellular phenomena as well. They can form connections with the plasma membrane and therby influence *locomotion, amoeboid movement* and *cytoplasmic streaming,* a cyclic or back-and-forth flow of cytoplasm seen in a variety of algal, plan, and animal cells. MFs also produce the *cleavage furrows* that divide the cytoplasm of animal cells after chromosomes have been separated by the spindle fibres. In addition, MFs contribute importantly to the development and maintenance of cell shape.

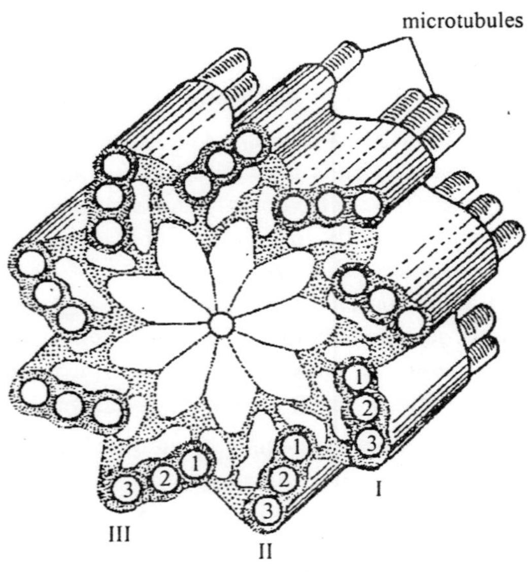

Fig. 5.16 : Cross-sectional structure of a centriole. There are nine groups of three microtubules (after De With, 1977)

MFs are polymers of the protein *actin*. Actin is synthesized as a monomer called G-actin (G for globular). *G-actin* monomers polymerize reversibly into long, double-helical strands of *F-actin* (F for filamentous), each strand about 4 nm wide. A microfilament therefore consists of a stand of F-actin wrapped around itself to form a right-handed double helix with a diameter of about 7 nm. Like microtubules, microfilaments are polar structures; all of the subunits are oriented in the same direction. This polarity influences the direction of MF elongation; assembly usually proceeds more readily at one end of the growing microfilament, where disassembly is favoured at the other end.

Intermediate Filaments. Intermediate filaments (IFs) comprise the third structural element of the cytoskeleton. As their name suggests, IFs have a diameter of about 8–12 nm, larger than the diameter of microfilaments but smaller than that of microtubules. Intermediate filaments are the most stable and the least soluble constituents of the cytoskeleton. Because of this stability, some researchers regard IFs as a scaffold that supports the entire cytoskeletal framework. Intermediate filaments are also thought to have a tension bearing role in some cells because they often occur in areas that are subject to mechanical stress.

In contrast to microtubules and microfilaments, intermediate filaments differ in their composition from tissue to tissue. Based on biochemcial and immunological criteria, IFs from animal cells can be grouped into six classes. A specific cell type usually contains only one or sometimes two classes of IF proteins. Because of this tissue specificity, animal cells from different tissues can be distinguished on the basis of the IF protein present. This *intermediate filament typing* serves as a diagnostic tool in medicine.

Despite their heterogeneity of size and chemical properties, all IF proteins share common structural features. They all have a central rodlike segment that is remarkably similar from one IF protein to the other. Flanking the central region of the protein are N-terminal and C-terminal segments that differ greatly in size and sequence, presumably accounting for the functional diversity of these proteins.

Figure 4-24C shows a possible model for IF structure. The basic structural unit is a dimer of two intertwined IF polypeptides. Two such dimers align laterally to form a tetrameric protofilament. Protofilaments then interact with each other to form an intermediate filament that is thought to be eight protofilaments thick at any point, with protofilaments probably joined end to end in an overlapping manner.

Information-processing Organelles

Living things depend on accurate, appropriate information. Information is *stored* as the sequence of bases in DNA molecules. The bulk of the DNA in eukaryotic cells resides in the nucleus. Information is translated, from the language of DNA into the language of proteins, on the surfaces of the ribosomes.

The Nucleus

Typically, the nucleus is the largest organelle in the eukaryotic cell. Most animals cells have a nucleus that is approximately 5 μm in diameter. The possession of a membrane-bounded nucleus is the

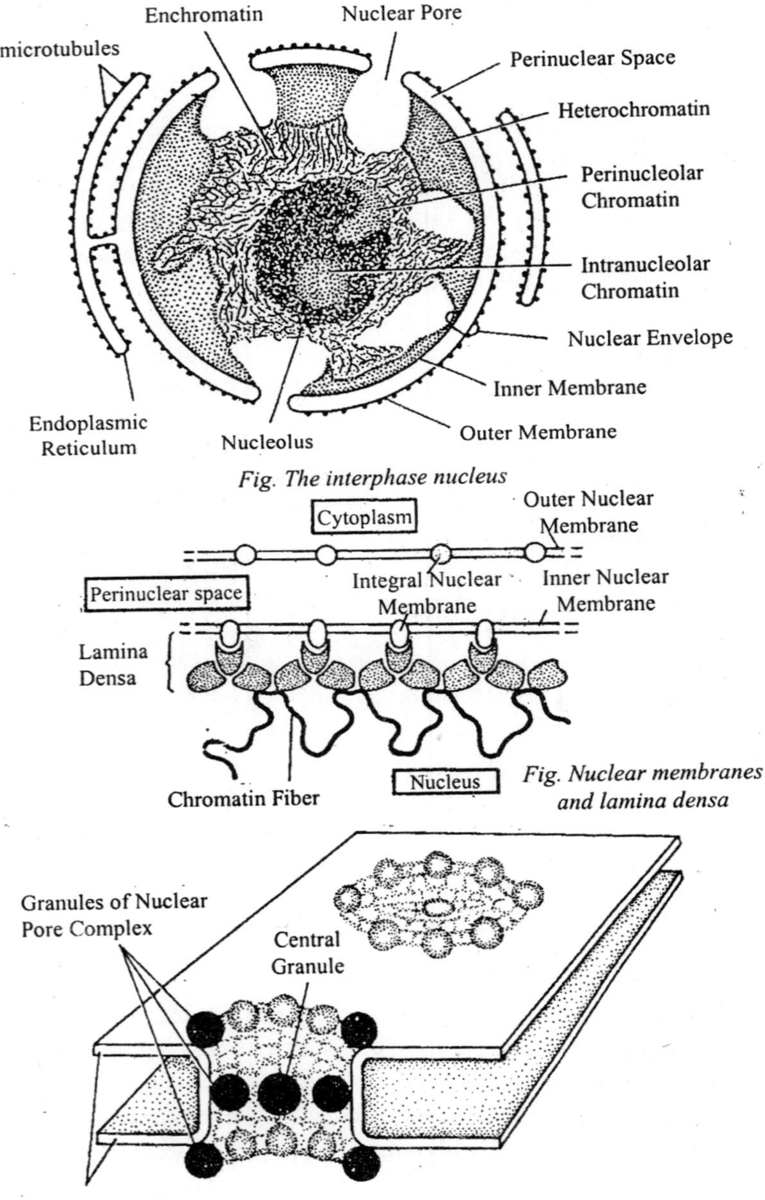

Fig. The interphase nucleus

Fig. Nuclear membranes and lamina densa

Fig. 5.17 : Structure of nuclear pore complex.

defining property of the eukaryotic cell. (Remember that in prokaryotes there is no membrane separating the nucleoid from the surrounding cytoplasm.) As viewed under the electron microscope, a nucleus is surrounded by *two* membranes separated by a few tens of nanometers. The **nuclear envelope**, as this pair of membranes is called, is perforated by **nuclear pores** approximately 9 nm in diameter. Each pore is surrounded by eight large protein granules arranged in an octagon where the inner and outer membranes merge. RNA and water-soluble molecules pass through the pores to enter or leave the nucleus.

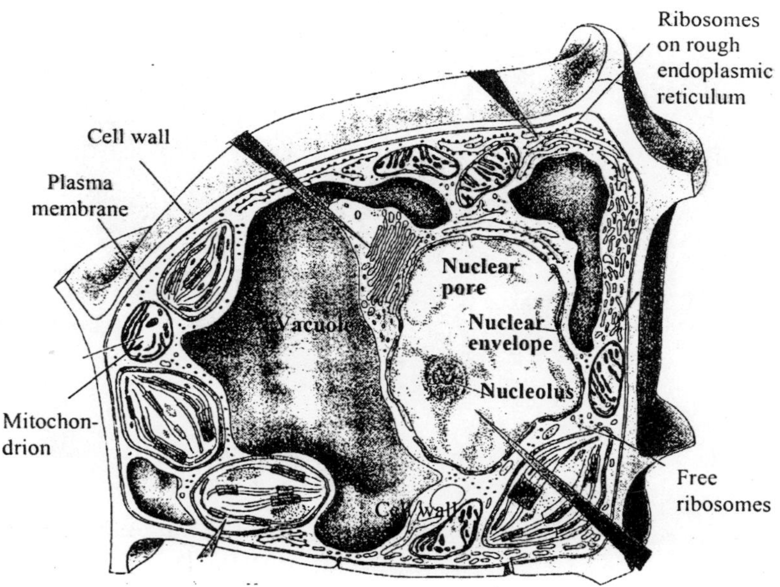

Fig. 5.18 : A Plant Cell : This drawing is based on electron micrograph of a photosynthetic cell from a leaf. Several of the major structures are shown in detail in the electron micrographs.

The outer membrane of the nuclear envelope sometimes folds outward into the cytoplasm and is continuous with the network called the endoplasmic reticulum. The endoplasmic reticulum and, to a lesser extent, the outer surface of the outer membrane of the nuclear envelope often carry great numbers of ribosomes. There are no ribosomes on the inner surface of the outer membrane, or on either surface of the nuclear envelope's inner membrane.

In the nucleus, DNA combines with proteins in a fibrous complex called **chromatin**. Throughout most of the life cycle of the cell, the chromatin exists as exceedingly long, fine threads that are so tangled that they cannot be seen clearly with any microscope. When the nucleus is about to divide the chromatin condenses and coils tightly to form a

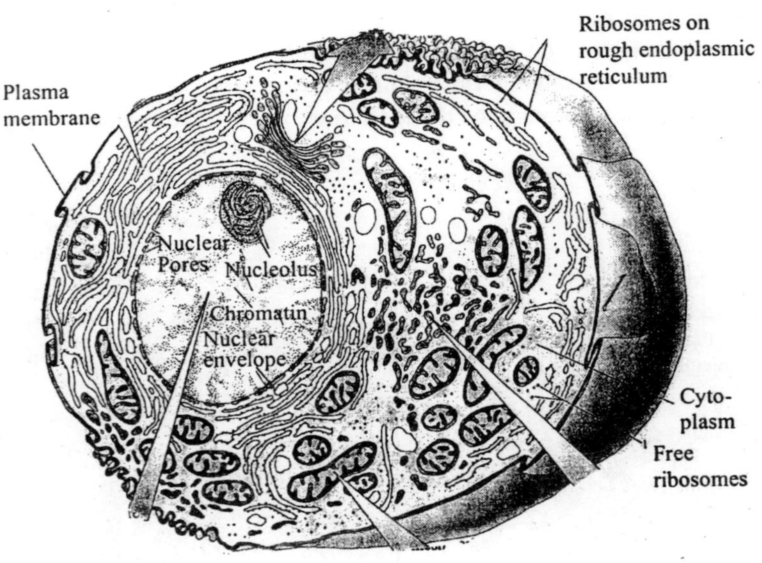

Fig. 5.19 : An Animal Cell : A typical animal cell. Details of individual structures are shown in the electron micrographs.

precise number of readily visible objects called **chromosomes**. Each chromosome contains one long molecule of DNA. The chromosomes are the bearers of hereditary instructions; their DNA carries the information required to carry out the synthetic functions of the cell and to endow the cell's descendants with the same instructions. Between nuclear divisions, the chromatin attaches to a protein meshwork, the nuclear lamina, on the inside of the nuclear envelop. The chromatin detaches when nuclear division commences, and the envelope breaks up into vesicles.

During most of the nuclear cycle, dense, roughly spherical bodies called **nucleoli** are visible in the nucleus. Taken together, the nucleoli contain from 10 to 20 percent of a cell's RNA. Ribosomes are assembled in the nucleolus. Protein molecules move into the nucleus and then into the nucleoli, where they combine with RNA molecules to form the cell's ribosomes. The ribosomes then move out of the nucleus. Each nucleus must have at least one nucleolus, and those of some species have several. The exact number of nucleoli in its cells is characteristic of a species.

Chromosomes and nucleoli float in a fluid called **nucleoplasm**. This fluid is a suspension of various particles, fibers, proteins, and other compounds. (Whereas in a solution solid particles dissolve, in a suspension they disperse but remain solid). Recall that the fluid portion of the cytoplasm, in which the various organelles, particles, and fibres are suspended, is called the cytosol.

Introducing Acetabularia

In the imaginations of the more romantically inclined biologists, the little seaweed Acetabularia resembles a mermaid's wineglass. Less imaginatively, it has been described as a little green toadstool measuring, at most, 2 or 3 inches in length. Although it a typical alga of tropical seas, it also occurs in some subtropical waters that are both shallow and somewhat rocky.

Nineteenth-century biologists discovered that this insignificant underwater plant consists of a single giant cell. Small for a seaweed, *Acetabularia* is gigantic for a cell. It consists of (1) a rootlike holdfast, (2) a long cylindrical stalk, and (3) at sexual maturity, a cuplike cap. The nucleus is found in the holdfast, about as far away from the cap as it can be. In due course, the nucleus divides by meiosis, and its progeny of pronuclei swim up the stalk into the cap, where they become the pronuclei of sex cells. These will be released upon maturity to swim away in search of partners. Although there are several species of Acetabularia, with caps of different shapes, all species function similarly.

Hammerling's and Brachet's Experiments

If the cap of *Acetabularia* is removed experimentally just before reproduction, another one will grow after a few weeks. Such behaviour, common among lower organisms, is called **regeneration**. This fact attracted the attention of investigators, especially A. H_mmerling and Jean Brachet, who in the early decades of this century became interested in the relationship that might exist between the nucleus and the physical characteristics of the plant. Because of its great size, Acetabularia could be subjected to surgery impossible with smaller cells. These investigators and their colleagues performed a brilliant series of experiments that in many ways laid the foundation for much of our modern knowledge of the nucleus. In most of these experiments they

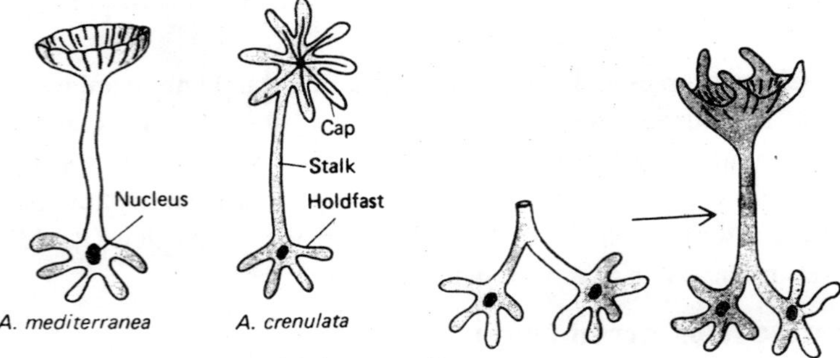

employed two species of *Acetabularia*,

A. *mediterranea*, which has a smooth cap, and

A. *crenulata*, with a cap broken up into a series of finger-like projections.

The kind of cap that is regenerated depends upon the species of *Acetabularia* used in the experiment. As you might expect, *A. crenulata* will ordinarily regenerate a *crenulata* cap. And *A. mediterranea* will regenerate a mediterranea cap. But it is possible to graft two capless plants of different species together by telescoping their stalks into one another—not as easy as it may sound, incidentally. After this union, they will regenerate a common cap that has characteristics intermediate between those of the two species involved. Thus, there is evidently something about the lower part of the cell that controls cap shape.

Stalk Exchange

It is also possible to attach a section of *Acetabularia* to a holdfast that is not its own by telescoping the cell walls of the two into one another. In this way the stalks and holdfasts of different species may be intermixed.

First, we take A. *mediterranea* and A. *crenulata* and remove their caps. Now we sever the stalks from the holdfasts. Finally, we exchange the parts.

What happens? Not, perhaps, what you would expect. The caps that regenerate are characteristic not of the species that donated the holdfasts but of those that donated the stalks.

However, if the same caps are removed still again, this time the caps that regenerate will be characteristic of the species that donated the holdfasts. That will continue to be the case no matter how many times

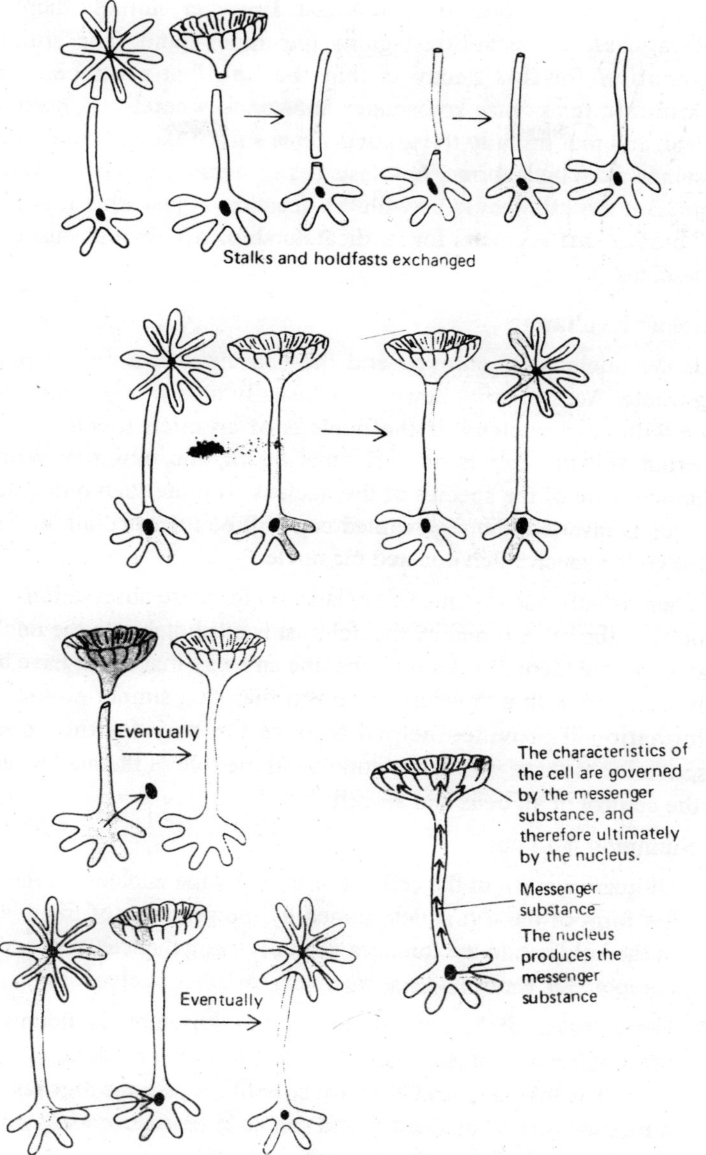

the regenerated caps are removed.

From all this we may deduce that the ultimate control of the cell is vested in the holdfast, for from now on, no matter how often the caps of these grafted plants are removed, they are always regenerated according to the species of the holdfast. However, initially there is a time lag before the holdfast gains the upper hand. The simplest explanation for this decay is that the holdfast produces some cytoplasmic temporary messenger substance whereby it exerts its control, and that initially the grafted stems still contain enough of that substance from their former holdfasts to regenerate a cap of the former shape. But this still leaves us with the question of just what it is about the holdfast that accounts for its dicatatorship. An obvious suspect is the nucleus.

Nuclear Exchange

If the nucleus is removed and the cap cut off, a new cap will regenerate. Acetabularia, however, is usually able to regenerate only once without a nucleus. If the nucleus of an alien species is now inserted and the cap is cut off once again, the new cap will be caharacteristic of the species of the nucleus. If more than one kind of nucleus is inserted, the regenerated cap will be intermediate in shape between the species that donated the nuclei.

There is only one reasonable explanation for these observations: The control of the cell exerted by the holdfast is attributable to the nucleus that is located there. It is hard to imagine any way that could have been demonstrated with a "higher" organism than this simple protist. The information it provided helped to pave the way for the epochal discoveries of those who have worked out the role of the nucleic acids in the control of all cells and all cellular life.

Summing it all up:

1. Ultimate control of the cell is exercised by the nucleus. In the end, the form of the cap is determined by the presence of the nucleus in the holdfast. In the long run, the only thing that can successfully compete for control with a nucleus is another nucleus.
2. Some control of the form of the cap is exercised by the nonnuclear parts of the cell, presumably the cytoplams or something in it. But since that substance can exercise control for just one regeneration, it must be limited in quantity and unable to reproduce itself without the nucleus. Perhaps it is perishable as well.
3. The source of the messenger substance must be the nucleus.

CHAPTER 6
THE CATALYSTS OF LIFE

Enzymes are the biological catalysts and are proteinaceous in nature. They are able to accelerate the rate of chemical reactions in a living cell. Like catalysts, they are not used up in the reaction, but unlike catalysts they are produced by living cells only. These are the enzymes which run myriads of biochemical reactions under normal cellular conditions. For hydrolysing starch in the food for example, high temperature or acidity is required. In the alimentary canal however, it is digested to glucose under relatively normal conditions of temperature and pH by the enzymes. Similarly the enzymes are able to hydrolyse other macromolecules such as the protein and lipid also under normal physiological conditions.

Historical

Earlier the agents causing fermentation of sugars were named as ferments. Pasture working on microorganisms concluded that fermentation and similar processes could be performed only by living cells. This concept was later disputed by J. Liebig who proposed that fermentation could take place even in the absence of living cells. Therefore, a distinction was made between organized ferments present in cells and unorganized ferments that were not associated with microorganism. In 1836, J. J. Berzelius named the biological agent as 'diastase' which is known today as amylase capable of converting starch from malt extract into sugar. The term enzyme ('in yeast') was proposed by Kühne in 1877 to distinguish enzyme from organized and unorganized ferments.

Occurrence and distribution

Enzymes occur in all living cells but not all enzymes are found in all the cells. The enzymes catalyze a wide variety of biochemical reactions many of which are localized in specific organs or are peculiar to certain species of plant or animal life. Thus, for example, pepsin is produced only in the cells of gastric mucosa and trypsin only in the pancreas. In the plant world lipases are not generally distributed but are found chiefly in plants that produce oilseeds.

A few of the enzymes are present in most forms of life. For example, catalases and peroxidases are widely distributed in all higher plants and animals.

The amount of enzymes may be different in different tissues. For example, resting seeds have low activity of analyses and proteinanses while the germinated seedlings have more activity. In some fruits, enzymes are concentrated near pit. Some enzymes are also organelle-specific.

Nomenclature

Naming of enzymes

In the past, enzymes were named in a haphazard manner as and when they were discovered. The ways in which enzymes are or have been named are listed below:

1. The first enzymes studied have been named for their colour, their localization within the body or after the person who discovered them. However, this nomenclature had not been agreeable to many.

2. Later, many enzymes have been named by adding the suffix '-ase' to the name of the substrate, for example, urease catalyzes the hydrolysis of arginine to ornithine and urea and so on. However, this nomenclature has not been always practicable.

3. Further, many other enzymes have been given chemically uninformative names, e.g., pepsin, trypsin and catalase. With the increasing number of newly discovered enzymes, this kind of nomenclature has not been accepted, this system has its limitations.

4. Finally, a systematic classification of enzymes has been adopted on the recommendation of an International Enzyme Commission as listed in the 1973 edition of Enzyme Nomenclature with a few exceptions. According to this, enzymes have been classified into six major classes and sets of subclasses based on the nature and type of reactions catalyzed. According to this, each enzymes is assigned:

A recommended name: It is usually short and appropriate for everyday use.

A systematic name: It identifies the reaction the enzyme catalyzes, and

A classification number: It is used where accurate and

unambiguous identification of an enzyme is required, as in research journals, abstracts and indexes.

An example is given by the enzymes catalyzing the reaction:

$$\text{ATP + creatine} \rightleftharpoons \text{ADP + phosphocreatine}$$

The recommended name : creatine kinase

The systematic name : ATP : creatine phosphotransferase

Classification number : EC 2.7.3.2, where EC stands for Enzyme Commission, the first digit (2) for the class name (transferases), the second digit (7) for the subclass (phosphotransferases), the third digit (3) for the sub-subclass (phosphotransferases with a nitrogenous group as acceptor) and the fourth digit (2) designates creatine kinase.

Enzyme classification

Enzymes can be classified into six major classes based on the nature and type of reactions catalyzed as given below:

1. Oxidoreductases : These enzymes catalyze oxidation or reduction reactions by transfer of hydrogen or electrons. e.g., succinic dehydrogenase.

$$\underset{\text{Succinic acid}}{\begin{matrix}CH_3 — COOH \\ \\ CH_3 — COOH\end{matrix}} + \text{FAD} \underset{}{\overset{\text{succinic dehydrogenase}}{\rightleftharpoons}} \underset{\text{Fumaric acid}}{\begin{matrix}H \quad COOH \\ C \\ \| \\ C \\ HOOC \quad H\end{matrix}} + \text{FADH}$$

2. Transferases. The enzymes are involved in transferring functional groups between a donor and an acceptor molecule. For example, glutamate: pyruvate amino transferase transfers amino group from glutamate to pyruvate. Glutamate is deaminated to produce 2-oxoglutarate, and pyruvate is aminated to produce alanine;

$$\underset{\text{glutamate}}{\begin{matrix}COOH \\ | \\ H—C—NH_2 \\ | \\ CH_2 \\ | \\ CH_2 \\ | \\ COOH\end{matrix}} + \underset{\text{pyruvate}}{CH_3CO.COOH} \rightleftharpoons \underset{\text{2-oxoglutarate}}{\begin{matrix}COOH \\ | \\ C=O \\ | \\ CH_2 \\ | \\ CH_2 \\ | \\ COOH\end{matrix}} + \underset{\text{alanine}}{CH_3CH(NH_2)COOH}$$

Important groups of enzymes in this class are:

(i) Aminotransferases or transaminases which transfer amino group

(ii) Knases, which transfer phosphoryl group from ATP or another nucleoside triphosphate to alcohol or amino group acceptors.

(iii)Glycosyl transferases which catalyse the transfer of an activated glycosyl residue to a glycogen primer

(iv)Transaldolases and transketolaes which catalyse the transfer of aldol or ketol groups.

3. Hydrolases. These enzymes are special kinds of transferases in which the donor group is transferred to water molecule. In other words, the reaction is that of a hydrolytic cleavage. The bonds which are normally subjected to hydrolytic cleavage are C—O, C—N, O—P and C—S bonds. An example of hydrolase is a lipase which hydrolyses a triacyl glycerol (fat) into fatty acids and glycerol;

$$\begin{array}{ccccccc} CH_2O.COR & & & & CH_2OH & & \\ | & & & & | & & \\ CHO.COR & + & 3\,H_2O & \rightarrow & CHOH & + & 3\,RCOOH \\ | & & & & | & & \text{fatty acid} \\ CH_2O.COR & & & & CH_2OH & & \\ \text{triacly glycerol} & & & & \text{glycerol} & & \end{array}$$

The important types of hydrolases are esterases (e.g. lipase), glycosidases, peptidases, phosphatases, thiolases, phospholipases, amidases, deaminases and ribonucleases.

4. Lyases. This group of enzymes catalyses the addition of groups to double bonds or catalyses cleavage by electronic rearrangement. For example, a decarboxylase catalyses the removal of CO_2 (a cleavage reaction) from keto acid;

$$R.CO.COOH \rightarrow RCHO + CO_2$$

5. Isomerases. These enzymes catalyse isomerization of several types. They include cis-trans, keto-enol and aldose-ketose interconversions. For example, an epimerase which catalyses the conversion of D-xylulose 5-phosphate to D-ribulose phosphate, catalyses the transfer of group at asymmetric carbon atom;

Racemases, epimerases and some mutases are examples of isomerases.

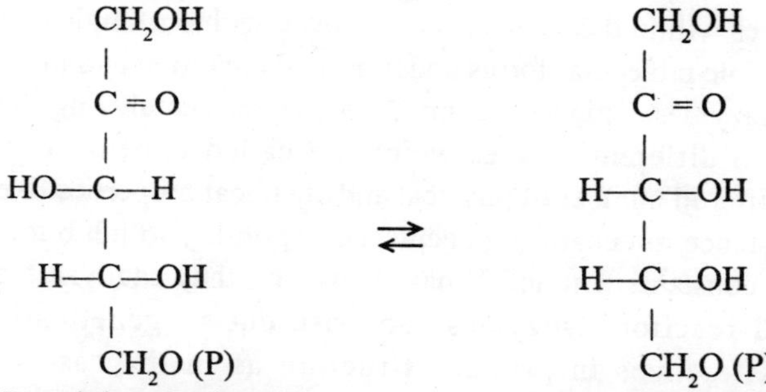

$$\begin{array}{ccc} CH_2OH & & CH_2OH \\ | & & | \\ C=O & & C=O \\ | & & | \\ HO—C—H & \rightleftarrows & H—C—OH \\ | & & | \\ H—C—OH & & H—C—OH \\ | & & | \\ CH_2O\ (P) & & CH_2O\ (P) \\ \text{D-xylulose 5-phosphate} & & \text{D-ribulose 5-phosphate} \end{array}$$

6. Ligases. These enzymes catalyse bond formation and are accompanied by the hydrolysis of ATP or other nucleoside phosphate. These enzymes are also called synthetases. Some carboxylases are also kept in this category. For example, glutamine synthetase catlyses the formation of glutamine from glutamate and NH_3, involving ATP;

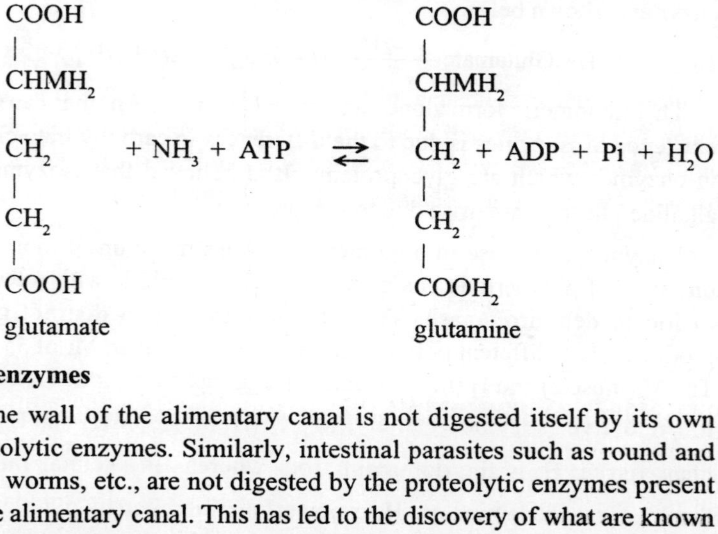

$$\begin{array}{lcl} COOH & & COOH \\ | & & | \\ CHMH_2 & & CHMH_2 \\ | & & | \\ CH_2 \quad + NH_3 + ATP & \rightleftarrows & CH_2 + ADP + Pi + H_2O \\ | & & | \\ CH_2 & & CH_2 \\ | & & | \\ COOH & & COOH_2 \\ \text{glutamate} & & \text{glutamine} \end{array}$$

Antienzymes

The wall of the alimentary canal is not digested itself by its own proteolytic enzymes. Similarly, intestinal parasites such as round and hook worms, etc., are not digested by the proteolytic enzymes present in the alimentary canal. This has led to the discovery of what are known as **antienzymes**. It has been reported that such tissues, or worms posses certain proteins having antagonistic property to the proteolytic enzymes. Reports are also available regarding production of antienzymes in the serum when certain enzyme preparations are given parenterally. The mucous membrane is also believed to contain anti-proteolytic enzymes. *Ascaris* has been shown to contain **anti-trypsin**, **anti-proteolytic enzymes**. The well known antienzymes are **anti-pepsin**, **anti-trypsin**, **anti-rennin** and **anti-urease**.

Isoenzymes (isozymes)

Many enzymes within the same species or even within a single cell can exist in multiple molecular forms under normal condition and these are known as isozymes. This is in contrast to enzymes catalyzing the same reaction in different species which are called alloenzymes. Isozymes can differ in number of physical and chemical properties like net charge, risistance to denaturing agents, susceptibility to inhibitors and the kinetic constants Km and Vmax. However, they catalyze the same chemical reaction. Isozymes can arise due to genetically determined differences in primary structure as in the case of mitochondrial and cytosolic malate dehydrogenases. Enzymes derived from a single precursor, like α and Π-chymotrypsins are also isoenzymes. Monomer, oligomer and polymer forms of an enzyme can also be classified as isozymes. A well known example is liver glutamate dehydrogenase which exists as a hexamer usually, made up of identical polypeptide chains of Mr 60 kDa. The reaction catalyzed by this enzyme is shown below.

$$\text{L} - \text{Glutamate} \xrightarrow{-2\text{H}} \alpha\text{ - Oxoglutarate} + \text{NH}_4^+$$

The monomeric form alone can act on L-alanine. Another cause for existence of isozymes is due to the differences in carbohydrate chains in enzymes which are glycoproteins. It is believed that isozymes of alkaline phosphatase arise due to this property.

Isozymes also arise in oligomeric enzymes made up of more than one type of polypeptide chain. A classical example is with reference to locate dehydrogenase. Vertebrates possess two distinct genes producing two different polypeptide chains each with an Mr of 35 kDa. The M (muscle) and H (heart) froms can aggregate into active tetramers to yield five isozymes namely, M_4, M_3H, M_2H_2, MH_3 and H_4, In heart tissue, H_4 is the dominant from whereas in skeletal muscle, the M_4 form predominates. H_4 is powerfully inhibited by pyruvate the substrate whereas, M_4 form is only weakly inhibited. This is in accordance with the differential metabolism of pyruvate in the two tissues. In aerobic heart tissue, pyruvate takes an alternate route to form acetyl CoA whereas in anaerobic muscle it is converted to lactate. The M polypeptide is basic in nature whereas the H chain is acidic. This property can be used for the separation of lactate dehydrogenase isoenzymes by electrophoresis.

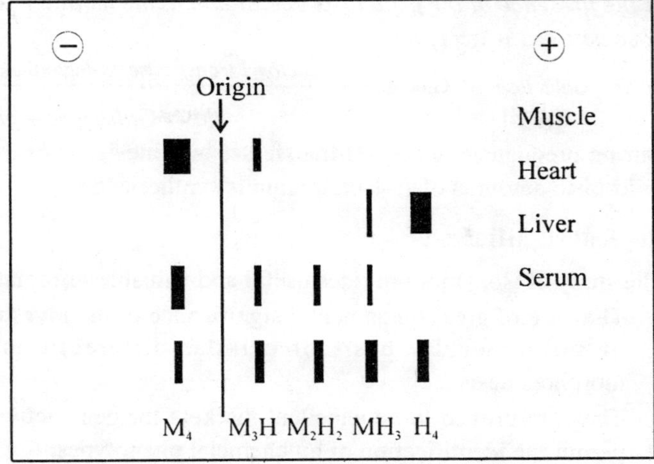

Fig. 6.1 : Lactate dehydrogenase (LDH) isozyme patterns in tissues based on zone electrophoresis at pH 8.0.

Each tissue has a characteristic pattern of isozymes. The relative proportions of M_4 (LDH–5), M_3H (LDH–4), M_2H_2 (LDH–3), MH_3 (LDH–2) and H_4 (LDH–1) in human serum in healthy persons are 5.0%, 8.0%, 27.0%, 35.0% and 25.0% respectively.

Ioszymes of alkaline phosphatase and creatine kinase can also be separated by electrophoresis.

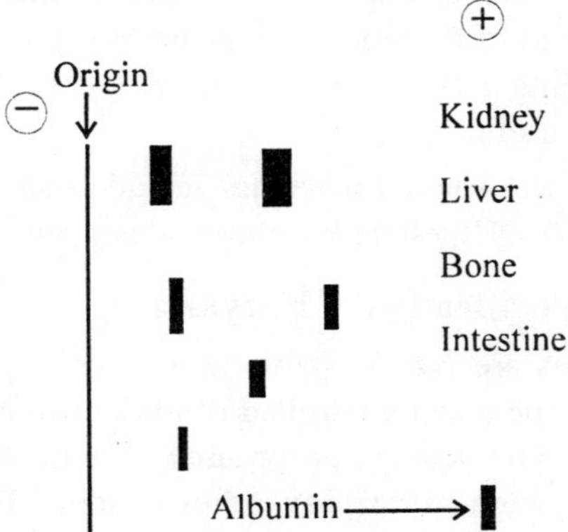

Fig. 6.2 : Alkaline phosphatase isoenzymes in tissues.

UDP – Galactose + N – Acetylglucosamine $\xrightarrow{\text{Galactosyl transferase}}$

(UDP : Uridine diphospho) *Galactose – N–acetylglucosamine*

In the presence of the protein α *-lactalbumin, the modifier protein, glucose acts as the acceptor.*

$$UDP - Galactose + \text{Glucose} \xrightarrow{\text{Calactosyl transferase: } \alpha\text{-lactalbumin}} Galactose - glucose.$$

During pregnancy galactosyl transferase is syntesized. At the time of child-birth amounts of α -lactalbumin is synthesized.

Biological significance

The study of isozymes provides useful and valuable information:

1. They are of great fundamental significance in the investigation of the molecular basis of cellular differentiation and morphogenesis .
2. They are proved to be excellent markers for gene activity and permit the identification of biochemcial phenotypes.
3. They have made an important contribution to the dogma of genetics of one gene-one enzyme to one gene-one polypeptide.
4. They have many practical applications:
   (a) In the diagnosis of heart attack-injured heart muscle releases characteristic patterns of isozymes into the blood stream and these may be readily detected on zymograms of blood. Both lactate dehydrogenase and creatine kinase isozymes are found useful in this diagnosis,
   (b) in the clinical diagnosis of diseases-the measurement of activities of enzymes such as lipases, phosphatases, lactate dehydrogenase, creatine kinase, trypsin, pepsin, etc. are found useful.

Some of the best studied isozymes include that of human lactate dehydrogenases (LDH), esterases, phosphatases, etc.

Isolation and Purification of Enzymes

Since enzymes are very unstable molecules, their isolation from tissues is to be done under controlled conditions of pH, ionic strength and temperature. The standard procedures for extracting and purifying an enzyme are same as for any other protein. Enzyme assay is performed at each step of purification, so as to ensure that enzyme is not lost during purification. Following steps are usually followed in isolation and purification of enzymes:

1. Extraction: The plant or animal tissue from which the enzyme is to be extracted is homogenised in a tissue homogeniser, a blender or with a mortar and pestle. Generally the extraction is done under cold (below 4°C) conditions because most of the enzymes are thermo-labile and they get inactivated at higher temperatures. The extraction is done with a buffer of specific ionic strength and pH. Ethylene diamine tetra acetic acid (EDTA) is generally added in the extraction medium to solubilize the membranes and to chelate heavy metals which could otherwise inhibit enzyme activity. Detergents such as Triton-x are also included sometimes to solubilize the membranes. Many enzymic proteins contain disulfide (S—S) bonds due to the presence of sulfur containing amino acids, which are easily oxidised during enzymes extraction leading to the loss of enzyme structure and activity. To overcome this problem, thiols such as mercaptoethanol or amino acid cysteine is added. Some times proteins such as casein or bovine serum albumin are also added in the extraction medium. These proteins protect the enzyme from the hydrolytic action of endogenously present proteases, which become active during tissue homogenisation.

2. Filtration and Centrifugation. The tissue extract is filtered using cheese cloth or filter paper to remove cell debris and fibers etc. The clear supernatant is then centrifuged in cold using a refrigerated centrifuge. The centrifugation can be used for purification as well as characterisation of proteins. The techniques operates on the principle that molecules of different masses and shape move through a solution at different rates in the presence of the applied gravitational field, obtained by spinning the rotor of the centrifuge. High speed centrifugation (around 20000 × *g*) removes cell debris and also the bigger cell organelles such as chloroplasts and mitochondria. A very high speed centrifugation in ultra-centrifuge at about 100,000 × *g* removes all the cell organelles and only cytosolic proteins remain the supernatant. This supernatant can be used for further purification of cytosolic enzymes.

3. Precipitation. Like any other protein, enzymes are also highly charged molecules and can be precipitated out by proper charge breaking chemicals. Once their charges are neutralized, they form aggregates and settle down as precipitates. Acids, bases, salts and organic solvents are often used to precipitate out proteins. Protein or enzyme precipitation upon addition of concentrated solution of salt is called **salting out**. It is a complex process and involves disruption of various physical forces involved in protein solubilization. The added salt alters the structure of the solvent which can lead to large changes in protein configuration by altering the electrostatic interaction of charged groups on protein surfaces and the solvation of polar uncharged residues exposed to the solvent. The salt may also interfere with the formation of Vander Waal's forces between hydrophobic groups of the amino acids. Further, it may compete with the protein molecules for solvent molecules and thereby lower its solvation. In other words, salt dehydrates protein molecules which then form aggregates and precipitate out. For enzymes, the most commonly used salt precopotant is ammonium sulfate. The salt either as solid powder or as saturated solution is added slowly to the enzyme preparation to achieve desired concentration of the salt. Most proteins are precipitated out between 30 to 70% of ammonium sulfate. The enzymes settle down as precipitate and can be removed by centrifugation.The precipitate is dissolved in small amount of buffer of desired pH and ionic strength and is dialysed to remove excess of ammonium sulfate and other low molecular weight contaminants. About four folds enzyme purification is typically obtained with ammonium sulfate precipitation.

Some organic solvents such as acetone, methanol and ethanol are also used for enzyme precipitation. They are cooled up to – 40°C before their use and precipitation is carried out at O°C. They are added drop by drop to avoid local concentration.

4. Purification of Enzymes. The dialysed enzyme preparation is purified further by a variety of techniques. However, the most commonly used techniques are; chromatography and electrophoresis.

Adsorption or column chromatography is used to separate other proteins from the specific enzyme. Several spleen enzymes of human beings have been separated by this method. The column is packed with hydroxypetite and the enzyme is eluted with phosphate buffer of specific pH and ionic strength. Filtering the enzyme preparation through a column packed with sephadex or any other gel is also used to purify enzymes. This process is called molecular exclusion chromatography or molecular sieve chromatography.

However, many enzymes are purified using affinity chromatography. In this procedure the enzymes are separated according to their biological specificity. The procedure is similar to that of column chromatography. The column is packed with inert material plus with a small molecular weight compound which acts as a ligand to bind with the enzymic protein (Fig;). Generally the solid support and ligand are separated through a spacer arm (chemical). When the enzyme along with other proteins is poured through such a column the enzymic protein binds with the ligand (which is specific for the given enzyme only) and other proteins run off. The bound enzyme then can be recovered by using a suitable eluent, which will break the link between ligand and the protein. The protein (enzyme) will then elute and can be collected from the bottom of the column.

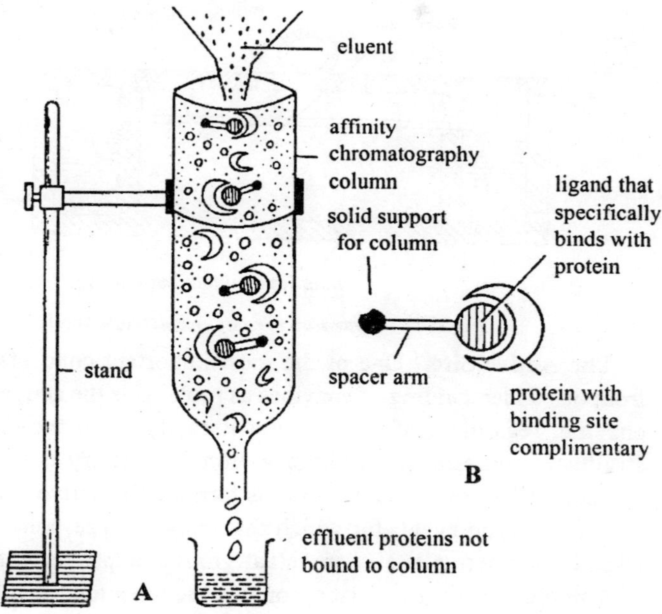

Fig. 6.3 : Affinity chromatography. (The protein to be purified binds specifically to a ligand attached to the solid support in column. Other proteins run off. The protein bound in the column is then eluted with a suitable reagent).

Electrophoretic method is also often used for purifying enzymes and other proteins. It is also used for separating amino acids, nucleosides and nucleotides and also for various isozymes of the same enzyme. There are several types of electrophoretic methods, depending upon the material used for separating proteins. In all however, the principle of separation is based on charge on protein molecules. They are separated according to the charge on them, in a weak electric field. The simplest type of the electrophoresis is the paper electrophoresis in which a paper is used as a support for separating proteins. The protein sample is spotted on a paper strip. The ends of the paper strip are dipped in two electrode tanks filled with a phosphate buffer of specific pH and ionic strength. The electrodes are connected to D.C. power supply of about 6 to 12 volts. When the current is applied to the electrophoresis apparatus through these electrodes, different proteins separate according to their charge (electrophoretic mobility) on the paper. They can be then located and identified using suitable colour reaction of the protein and biological assay.

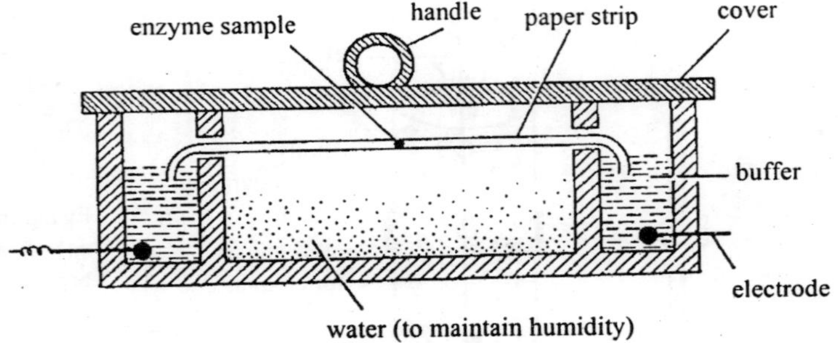

Fig.6.4 : Apparatus for paper electrophoresis

The Active Site. One of the most important concepts to emerge from our understanding of enzymes as proteins is the active site. Every enzyme, regardless of the reaction it catalyzes or the details of its structure, contains somewhere within its tertiary configuration a characteristic cluster of amino acids forming the active site where the actual catallytic event, for which that enzyme is responsible, occurs. Usually, the active site is an actual groove or pocket with chemical and structural properties that accommodate the intended substrate with high specificity. The three-dimensional structure of the active site can be appreciated by the precise fit of the substrate molecule into pockets produced by the characteristic folding of polypeptide chains. (Lysozyme is an enzyme that breaks the glycosidic bond between N-acetylglucosamine [GLcNAc] and N-acetylmuramic acid [murNAc] groups in the peptidoglycan of bacterial cells to lyse [break open] and die. Carboxypeptidase. A an enzyme that breaks down polypeptides by removing one amino acid at a time from the C-terminus of the polypeptide.)

The amino acids that make up the active site of an enzyme are not usually contiguous to one another along the primary sequence of the protein. Instead, they are brought together in just the right conformation by the specific three-dimensional folding of the polypeptide chain. For

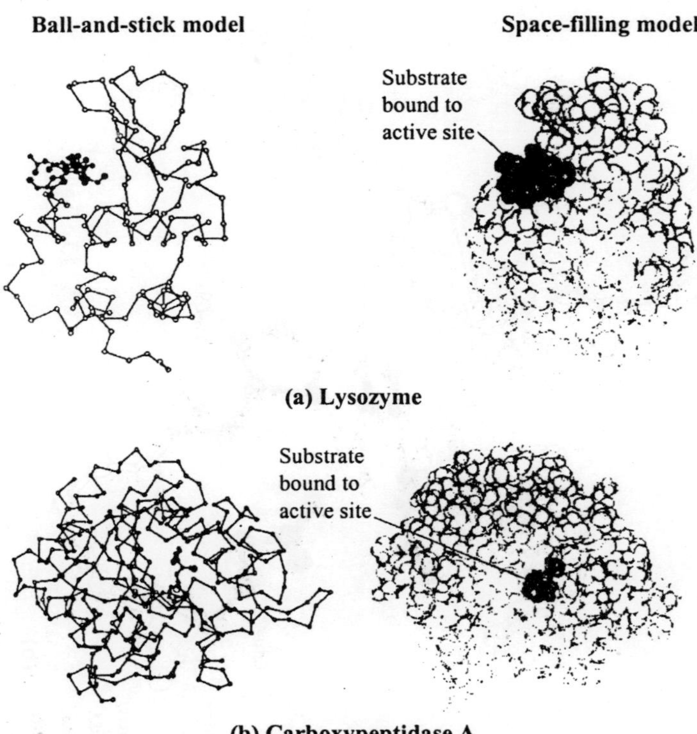

Fig. 6.5 : Molecular Structures of Lysozyme and Carboxypeptidase A.

(a) ***The enzyme lysozyme is represented as a ball-and-stick model (left), showing only the backbone of the protein as deducted by X-ray diffraction studies of protein crystals, and as a space-filling model (right) generated by computer. The molecules shown here is lysozyme from chicken eggs. A substrate molecule is shown in the active site, which appears as a cleft in the side of the enzyme molecule.***

(b) ***The enzyme carboxypeptidase A is represented as a ball-and-stick model (left) and a space-filling model (right), with a substrate molecule bound to the active site.***

the carboxypeptidase molecule, for example, the active site consists of a tightly bound zinc ion bonded to the side chains of the histidine at position 69, the glutamate at position 72, and another histidine at position 196 of the polypeptide chain. When the substrate for the enzyme binds to the active site, the structure of the enzyme changes, bringing the argining at position 145, the glutamate at position 270,

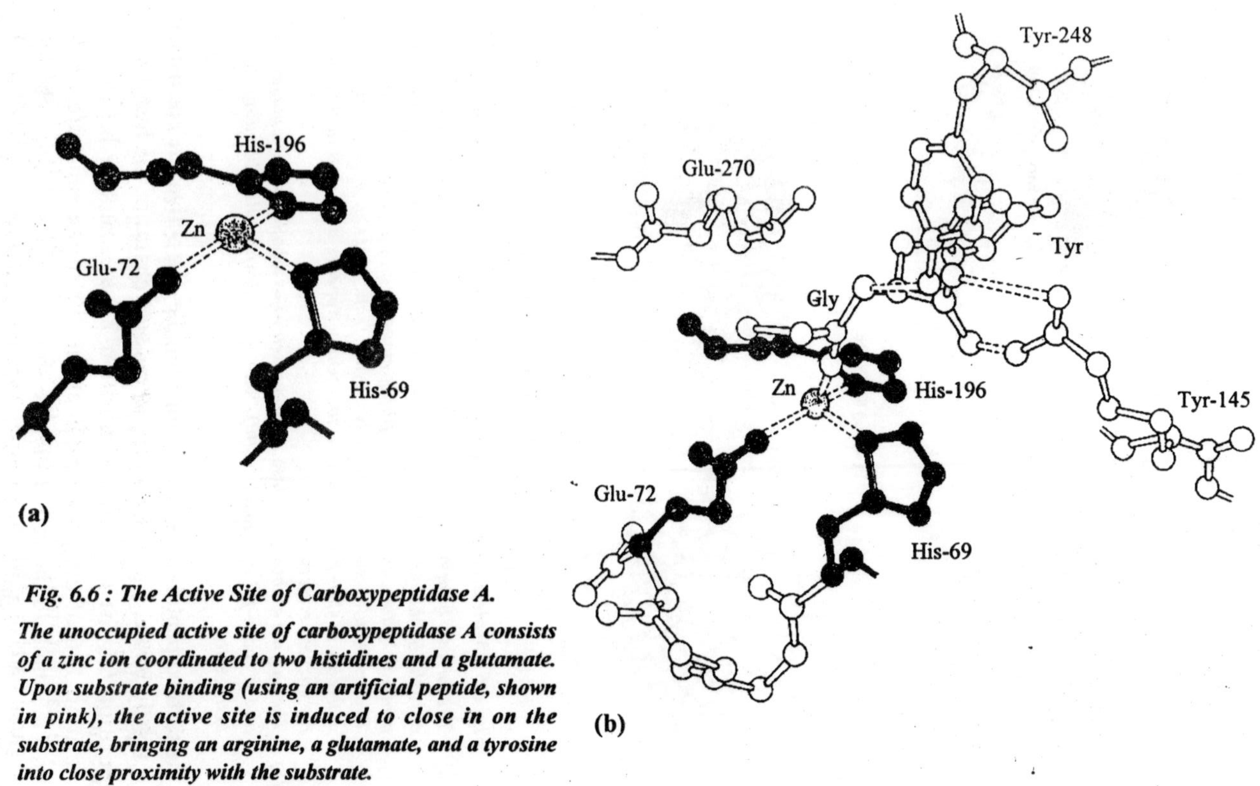

Fig. 6.6 : The Active Site of Carboxypeptidase A.

(a) ***The unoccupied active site of carboxypeptidase A consists of a zinc ion coordinated to two histidines and a glutamate.***

(b) ***Upon substrate binding (using an artificial peptide, shown in pink), the active site is induced to close in on the substrate, bringing an arginine, a glutamate, and a tyrosine into close proximity with the substrate.***

and the tyrosine at position 248 into close proximity with the substrate. Each of these amino acids plays a critical role in the actual catalytic process.

For carboxypeptidase, then, 6 amino acids out of a total chain length of 307 are actually involved at the activesite. This is typical of most enzymes, with the active site usually involving about 5% of the surface area of the enzyme. Also typical is the involvement of amino acids from distant positions along the primary structure of the polypeptide. This involvement underscores the importance of the overall tertiary structure of the protein; only as an enzyme molecule attains its stable three-dimensional conformation are the specific amino acids brought together to constitute the active site.

Of the 20 different amino acids that make up proteins, only a few are actully involved in the active sites of the many proteins that have been studied. In most cases, these are the amino acids cysteine, histidine, serine, aspartate, glutamate, and lysine. All of these can participate in binding or bonding of the substrate to the active site during the catalytic process, and several (histidine,aspartate, and glutamate)also serve as donors or acceptors of protons.

Some enzyms consist not only of one or more polypeptide chains but of specific nonprotein components as well. These components are called **prosthetic groups** and are usually either small organic molecules or metal ions, such as the iron in catalase. Frequently, they function as electron acceptors. Where present, prosthetic groups are located at the active site and are indispensable for the catalytic activity of the enzyme. Carboxypeptidase A is an example of an enzyme with a prosthetic group, containing a zinc atom at its active site.

Enzyme Specificity. A consequence of the structure of the active site is that enzymes display a high degree of substrate specificity, evidenced by an ability to discriminate between very similar molecules. Specificity is probably one of the most characteristic properties of living systems, and enzymes are especially dramatic examples of biological specificity.

We can illustrate their specificity by comparing enzymes with inorganic catalysts. Most inorganic catalysts are quite nonspecific in that they will act on a variety of compounds that share some general chemical feature. Consider, for example the *hydrogenation* of (addition of hydrogen to) an unsaturated C$=$C bond:

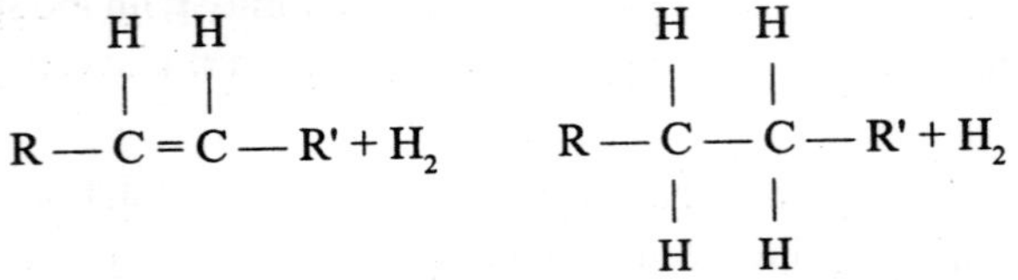

$$R-\overset{H}{\overset{|}{C}}=\overset{H}{\overset{|}{C}}-R' + H_2 \qquad R-\underset{H}{\underset{|}{\overset{H}{\overset{|}{C}}}}-\underset{H}{\underset{|}{\overset{H}{\overset{|}{C}}}}-R' + H_2$$

This reaction can be carried out in the laboratory using a platinum (Pt) or nickel (Ni) catalyst, as indicated. These inorganic catalysts are very nonspecific, however; they can be used to hydrogenate a wide variety of unsaturated compounds. In fact, nickel or platinum is used commercially to hydrogenate polyunsaturated vagetable oils in the manufactuer of solid cooking fats or shortenings. Regardless of the exact structure of the unsaturated compound, it can be effectively hydrogenated in the presence of nickel or platinum.

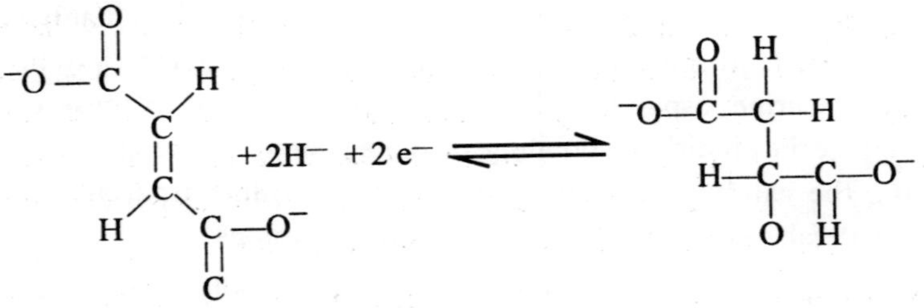

$$^{-}O-\overset{O}{\overset{\|}{C}}-CH=CH-\underset{C}{\underset{\|}{C}}-O^{-} + 2H^{-} + 2\,e^{-} \rightleftharpoons {}^{-}O-\overset{O}{\overset{\|}{C}}-\overset{H}{\overset{|}{C}}H-\underset{O}{\underset{|}{C}}H-\underset{H}{\underset{\|}{C}}-O^{-}$$

Fumarate Succinate

This particular reaction is catalyzed in cells by the enzyme *succinate dehydrogenase* (so named because it normally functions in the opposite direction during energy matabolism). This dehydrogenase, like most enzymes, is highly specific. It will not add or subtract hydrogens from any compounds except those shown in reaction 6-4. In fact, this particular enzyme is so specific that it will not even recognize maleate, a geometric stereoisomer of fumarate.

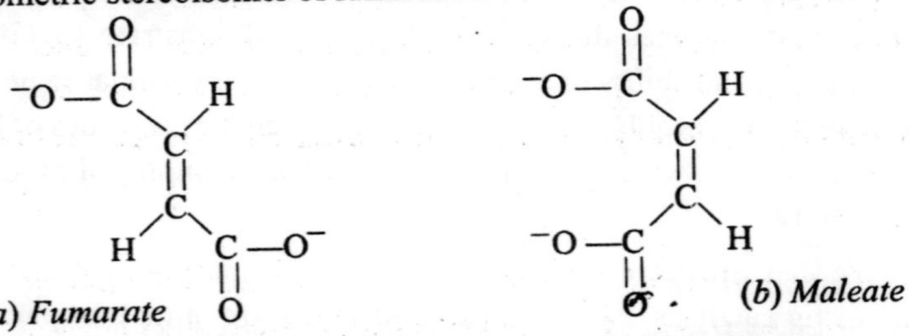

$$^{-}O-\overset{O}{\overset{\|}{C}}-CH=CH-\underset{O}{\underset{\|}{C}}-O^{-} \qquad\qquad ^{-}O-\overset{O}{\overset{\|}{C}}-CH=CH-\underset{O}{\underset{\|}{C}}-O^{-}$$

(*a*) *Fumarate* (*b*) *Maleate*

Fig. 6.7 : The Stereoisomers (a) Fumarate and (b) Maleate

Not all enzymes are quite this specific; some accept a number of closely related substrates , and other accept any of a whole group of substrates as long as they possess some common structural feature. Such **group specificity** is seen most often with enzymes involved in the synthesis or degradation of polymers. The purpose of carboxypeptidase A is to degrade polypeptide chains from the carboxylend ; thus , it makes sense for the enzyme to accept any of a wide variety of polypeptides as substrate, since it would be needlessly extravagant of the cell to require a separate enzyme for every different peptide bond that has to be hydrolyzed in polypeptide degradation.

In general, however, enzymes are highly specific with respect to substrate, such that a cell must possess almost as many different kinds of enzymes as it has reactions to catalyze. For a typical cell, this means that several thousand different kinds of enzymes are necessary to carry out its full metabolic program. At first, that may seem wasteful in terms of proteins to be synthesized, genetic information to be stored and read out, and enzyme molecules to have on hand in the cell. But you should also be able to see the tremendous regulatory possibilities this suggests, a point we will return to later.

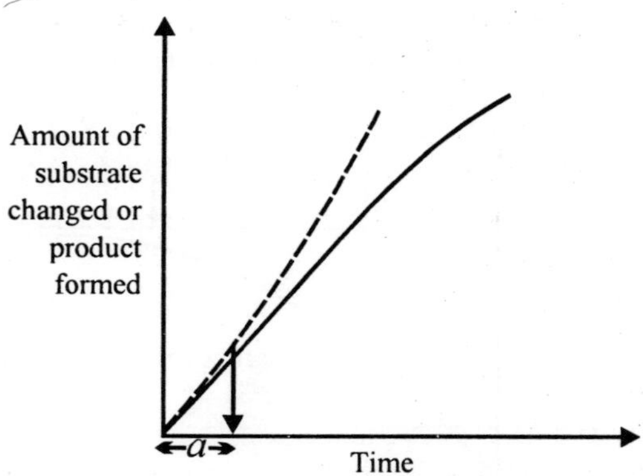

Fig. 6.8 : The rate of an enzyme-controlled reaction

Enzyme concentration

Provided that the substrate concentration is maintained at a high level, and other conditions such as pH and temperature are kept constant, the rate of reaction is proportional to the enzyme concerntration. Normally reactions are catalysed by enzyme concentrations which are much lower than substrate concentrations. Thus as the enzyme concentration is increased, so will be rate of the enzyme reaction.

Substrate concentration

For a given enzyme concentration, the rate of an enzyme reaction increases with increasing substrate concentration. The theoretical maximum rate (V_{max}) is never quite obtained, but there comes a point when any further increase in substrate concentration produces no significant change in reaction rate. This is because at high substrate concentration the active sites of the enzyme molecules at any given moment are virtually saturated with substrate. Thus any extra substrate has to wait until the enzyme/substrate complex has released the products before it may itself enter the active site of the enzyme.

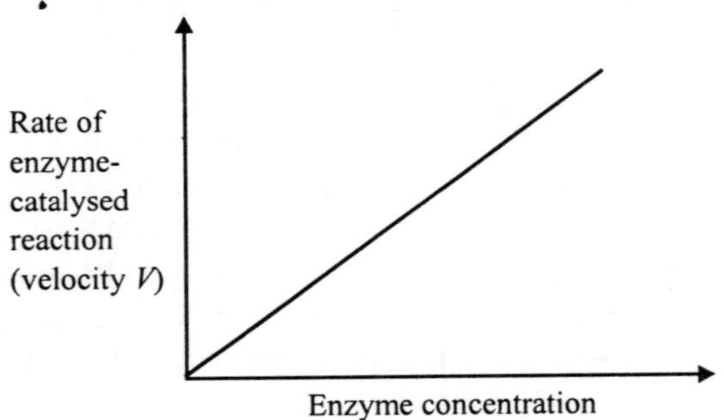

Fig. 6.9. : Relationship between enzyme concentration and the rate of an enzyme-controlled reaction.

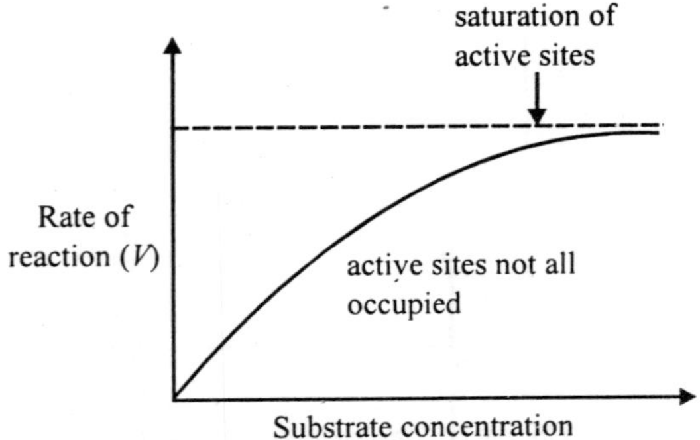

Fig. 6.10 : Effect of substrate concentration on the rate of an enzyme-controlled reaction.

Temperature

Heating incerase molecular motion. Thus the molecules of the substrate and enzyme move quickly and chances of their bumping into each other are increased. As a result there is a greater probability of a reaction occurring. The temperature that promotes maximum activity is referred to as the optimum temperature. If the temperature is increased above this level, then a decrease in the rate of the reaction occurs despite the increasing frequency of collisions. This is because the secondary and tertiary structures of the enzyme have been disrupted, and the enzyme is said to be denatured. In effect, the enzyme unfolds and the precise structure of the active site is gradually lost. The bonds which are most sensitive to temperature change are hydrogen bonds and hydrophobic interactions.

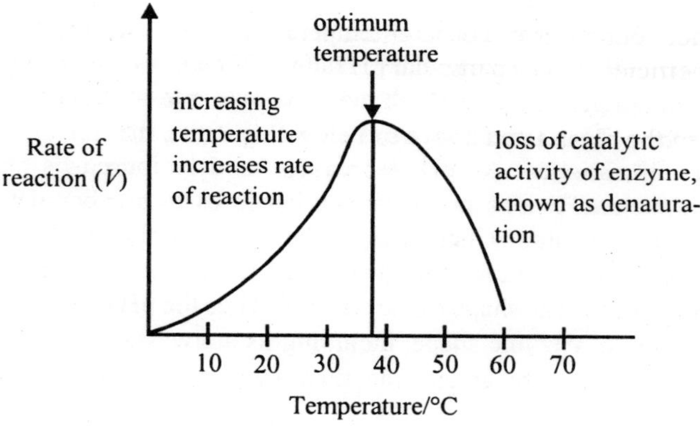

Fig. 6.11 : Effect of temperature on the rate of an enzyme-controlled reaction.

Most mammalian enzymes have a temperature optimum of about 37–40°C but enzymes with higher optima exist. For example, the enzymes of bacteria living in hot springs may have an optimum temperature of 70°C or higher. Such enzymes have been used in biological washing powders for high temperature washes. If temperature is reduced to near or below freezing point, enzymes are **inactivated,** not denatured. They will regain catalytic influence when higher temperatures are restored.

Today techinques of quick –freezing food are in widespread use as a means of preserving food for long periods. This not only prevents growth and multiplication of microorganisms, but also deactivates their digestive enzymes thus making it impossible for them to decompose food. The natural enzymes in the food itself are also inactivated. However, once frozen, it is necessary to keep the food at subzero temperatures until it is to be prepared for consumption.

Temperature cofficirnt, Q_{10}

The effect of temperature on the rate of a reaction can be expressed as the temperature coefficient, Q_{10}.

Q_{10} = rate of reaction at (x + 10) °C/rate of reaction at x°C

Over a range of 0–40°C, Q_{10} for an enzyme–controlled reaction is 2. In other words, the rate of an enzyme–controlled reaction is doubled for every rise of 10 °C.

pH

Under conditions of constant temperature, every enzyme functions most efficiently over a particular pH range. Often this is a narrow range. The optimum pH is that at which the maximum rate of reaction occurs . When the pH is altered above or below this value, the rate of enzyme activity diminishes. As pH decreases, acidity increases and the concentration of H+ ions increases. This increases the number of positive charges in the medium. Change in pH alter the ionic charge of the acidic and basic groups and therefore disrupt the ionic bonding that helps to maintain the specific shape of the enzyme .Thus the pH change leads to an alteration of enzyme shape, including its active site. If extremes of pH are encountered by an enzyme, then it will be denatured.

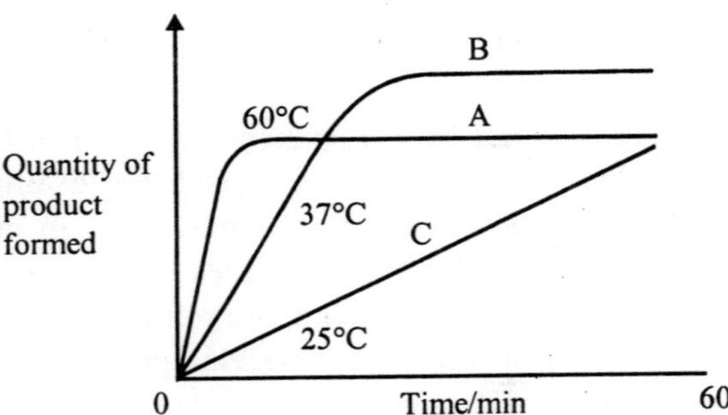

*Fig. 6.12 : **Time course of an enzyme reaction at various temperatures.***

Table 6.1 Optimum pH values for some enzymes.

Enzyme	Optimum pH	Enzyme	Optimum pH
Pepsin	2.00	Catalase	7.60
Sucrase	4.50	Chymotrypsin	7.00–8.00
Enterokinase	5.50	Pancreatic lipase	9.00
Salivary amylase	4.80	Arginase	9.70

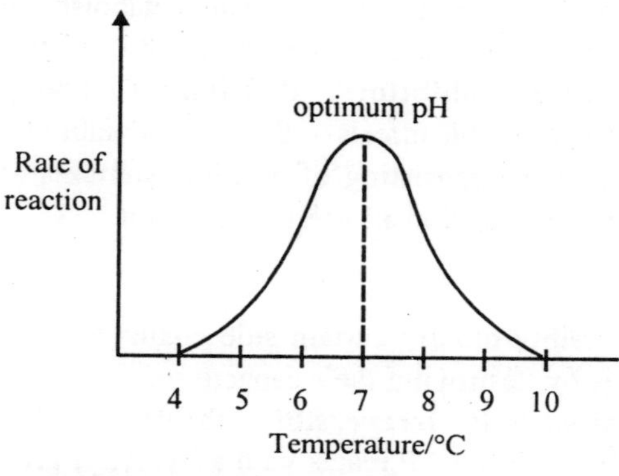

Fig. 6.13 : Effect of pH on the rate of an enzyme-controlled reaction.

Regulation of Enzyme Activity

Various substances, some occurring naturally in cells and others artificial, act upon enzymes to increase or decrease the rates of enzyme–

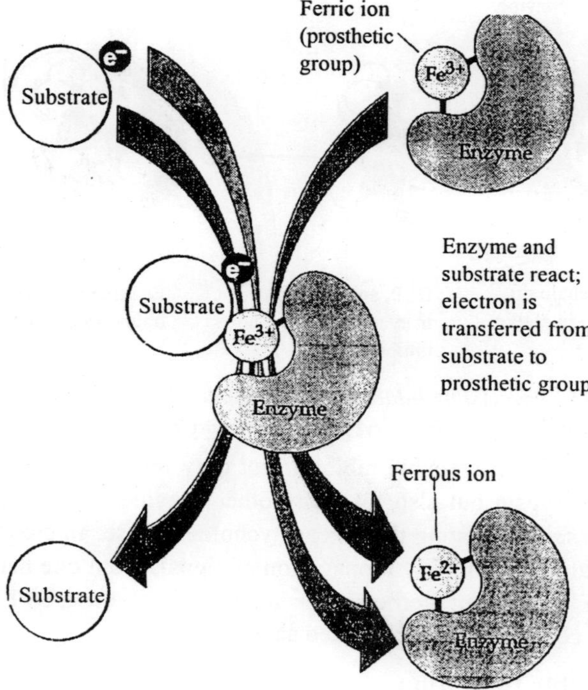

Fig. 6.14 : Metal Ions as Prosthetic Groups : A ferric ion attached to an enzyme as a prosthetic group. In this reaction the ferric ion (Fe^{3+}) withdraws the electron from the substrate and becomes a ferrous ion (FE^{2+}); the substrate moves on, altered by the loss of an electron.

catalyzed reactions. Those that occur naturally regulate metaboism; the artificial ones are used either to treat disease or to study how enzymes work. Some substances,called **inhibitors,** that inhibit enzyme–catalyzed reactions produce reversible effects —that is, these inhibitors can become unbound. Enzymes consisting of multiple subunits are subject to another type of control called allosteric regulation.

Irreversible Inhibition

Some inhibitors irreversibly modify certain side chains at active sites, ruining the enzymes by destroying their capacity to function as catalysts. An example of such an **irreversible inhibitor** is DFP (diisopropylphos-phorofluoridate). DFP reacts with a hydroxyl group belonging to the amino acide serine at the active site of the enzyme trypsin, preventing the use of this side chain in the catalytic machanism.

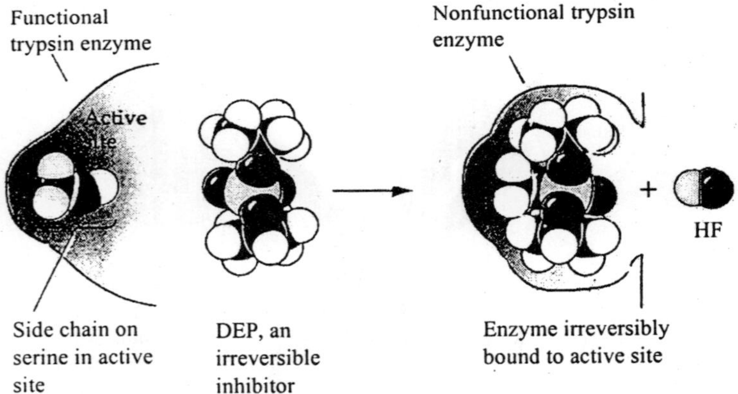

Fig. 6.15 : Irreversible Inhibition DFP disables the digestive enzyme trypsin by reacting with a hydroxyl group belonging to the amino acid serine.

DFP is an irreversible inhibitor not only for the protein–digesting enzyme trypsin but also for many other enzymes whose active sites contain serine. Among these is acetycholinesterase, an enzyme that is essential for the orderly propagation of signals from one nerve cell to another. Because of their effect on acetylcholinesterase, DFP and other similar compounds are classified as nerve gases.

Reversible Inhibition

Not all inhibitory action is irreversible. Some inhibitor molecules are similar enjough to a particular enzyme's natural substrate to bind to the active site, yet different enough that the enzyme catalyzes no chemical reaction. When such a molecule is bound to the enzyme, the natural substrate cannot enter the active site; thus the intruder effectively wastes the enzyme's time, inhibiting its catalytic action. These are called **competitive inhibitors** because they compete with the natural substrate for the active site and block it .The blockage is reversible, however; a competitive inhibitor may become unbound, leaving the active site unchanged. If enough natural substrate molecules are present, they can compete successfully with inhibitor molecules for empty active sites.

Consider the enzyme succinate dehydrogenase, which is subject to competitive inhibition. Recall that this enzyme,found in all mitochondria, removes two hydrogen atoms from succinate to produce fumarate , it then transfers the hydrogens to another molecules. The other molecules shown in the figure are competitive inhibitors of succinate dehydrogenase. They resemble succinate enough that the enzyme is fooled into binding them. Having bound them, however, the enzyme can do nothing more with them, because the inhibitors are the wrong size and shape or have key chemical group in the wrong places. The enzyme molecule cannot bind a succinate molecule until the inhibitor molecule has moved out of the active site.

Dissociation of the inhibitor does occur because binding of a competitive inhabitor is reversible, *as is binding of the substrate.*For example, when the competitive inhibitor malonate is added to a solution containing succinate and succinate dehydrogenase, the reaction of succinate to fumarate is slowed. The effect of malonate can be overcome, however, if enough succinate is added. The relative concentrations of substrate and inhibitor determine which of these is more likely to bind to the active site.

Inhibitors that do not react with the active site are called **noncompetitive inhibitors.** Noncompetitive inhibitors bind to the enzyme at a site away from the active site. Their binding causes a conformational change in the protein that alters the active site. The active site still binds substrate molecules, but the rate of product formation is reduced. Noncompetitive inhibitors can become unbound, so their effects are reversible. Because they do not bind to the active side, their effects do not change as substrate concentration changes.

(a) Competitive inhibition

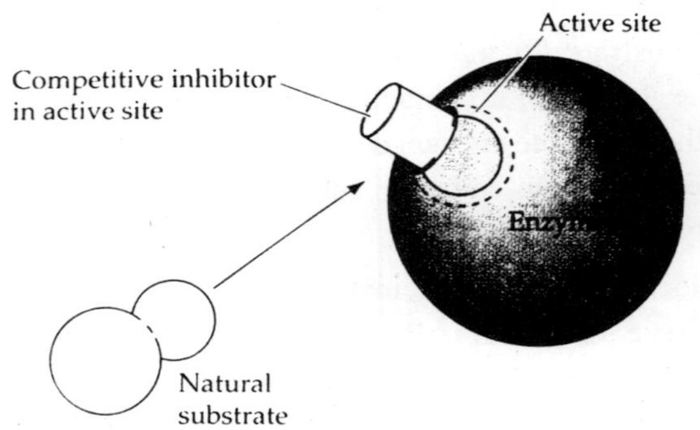

The enzyme molecule's function is disabled as long as the inhibitor remains bound but if the inhibitor becomes unbound, a substrate molecule may bind bind to the active site.

(b) Competitive inhibition of succinate dehydrogenase

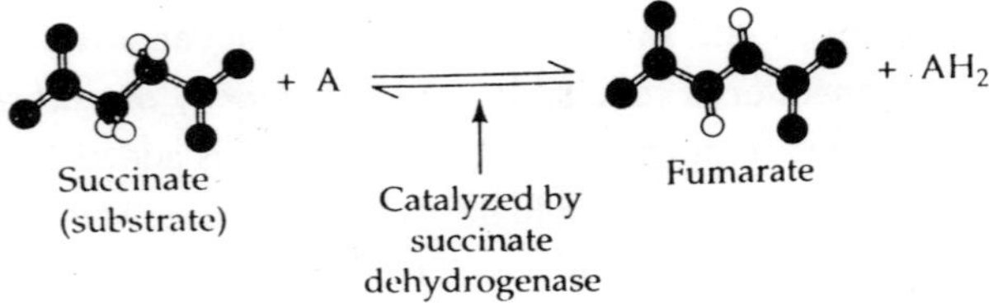

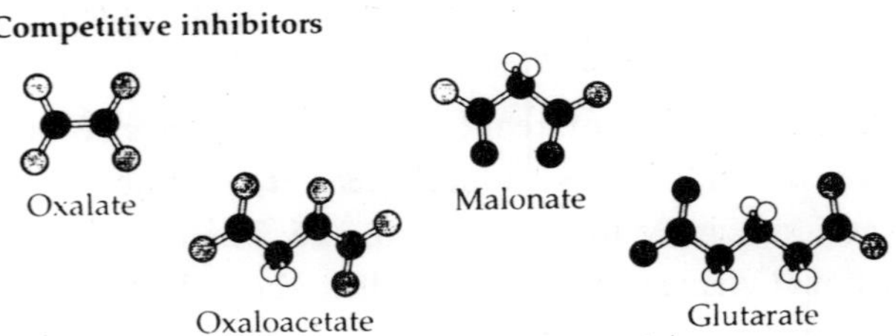

The above series of molecules of increasing length compete with succinate for the enzyme's active site. The similarity that fits them all to the same active site is the presence of two negatively charged carboxyl groups, one at each end of the molecule.

(c) Noncompetitive inhibition

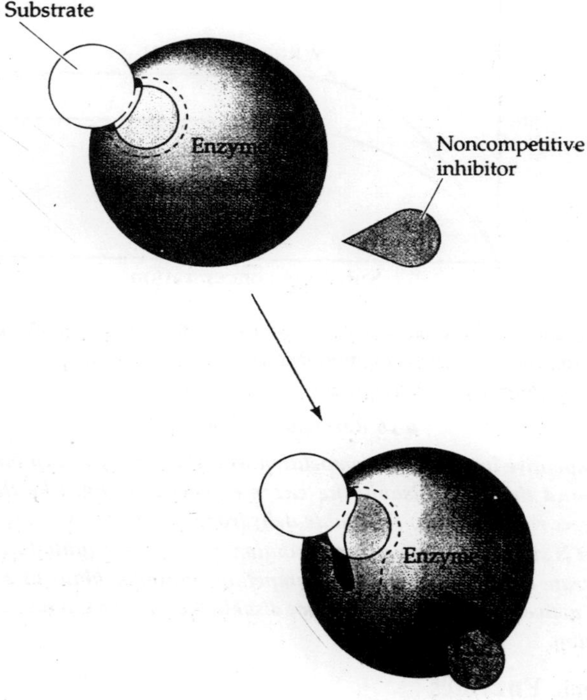

A non competitive inhibitor may not prevent the substrate from binding to the active site, but it modifies the active site.

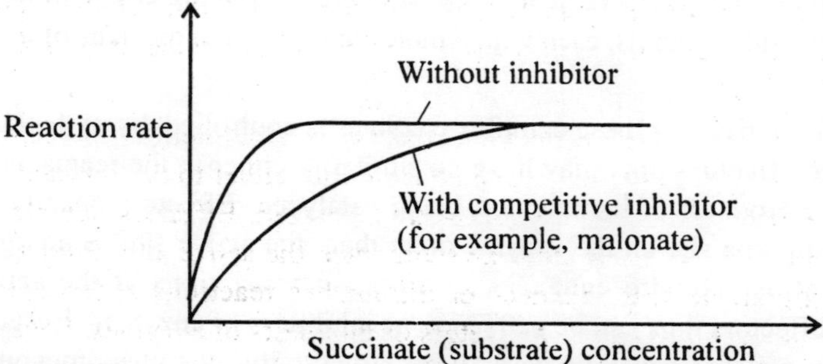

In the absence of a competitive inhibitor, the enzyme increases reaction rate more than it does when one is present. As substrate concentrations increase, however, competitive inhibition become less effective and, eventually, completely ineffective.

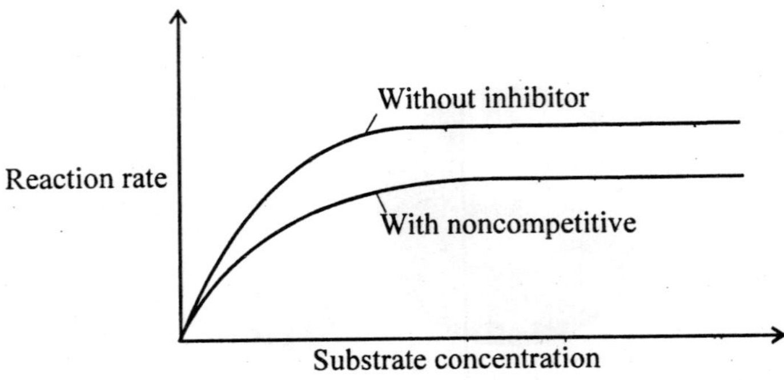

The modification of the active site slows down the rate at which the enzyme catalyzes the reaction. Even high concentrations of substrate do not overcome the inhibitor; compare with graph in (b).

6.16 Reversible Inhibition

(a) A competitive inhibitor, in competition with the substrate, can bind the active site and thus can disable the enzyme. This binding, like that of the substrate, is reversible. (b) Succinate dehydrogenase is an example of an enzyme that is subject to competitive inhibition. Competitive inhibitors resemble the substrate chemically. (c) A noncompetitive inhibitor binds at a site away from the active site. The effect may not disable the enzyme but may slow down the reaction.

Allosteric Enzymes

Many important enzymes are larger and more complex than the ones we have discussed so far, which are individual polypeptides. These complex enzymes have quaternary structures consisting of two or more polypeptide subunits, each with a molecular weight in the tens of thousands.

The activity of these complex enzymes is controlled by molecules, called **effectors** ,that may have no similarity either to the reactants or to the products of the reaction being catalyzed. Effectors operate by binding to a site on the enzyme other than the active site. Binding at this **allosteric site** enhances or diminishes reactions at the active site;effectors thus can be activators or inhibitors of enzymes. Because many effector–substrate pairs differ structurally, this phenomenon is called allostery, meaning "different shape". Enzymes subject to allosteric control are called **allosteric enzymes;**all have two or more sub–units.

Allosteric enzymes and single–subunit enzymes differ greatly in their effects on reaction rates when the substrate concentration is low. Graphs of reaction rates plotted against substrate concentration show this difference. The reaction rate first increases very sharply with increasing substrate concentration, then tapers off to a constant maximum rate as the supply of enzyme becomes saturated with substrate. The plot for an allosteric enzyme is radically different , with a sigmoidal (S–shaped) appearance.The increase in reaction rate with increasing substrate concentration is slight at low substrate concentrations, but there is a range over which the reaction rate is extremely sensitive to reatively small changes in the substrate concentration. Because of this sensitivity, allosteric enzymes are important in fine –tuning the activities of a cell. We can understand this behaviour in terms of the structure of an allosteric enzyme.

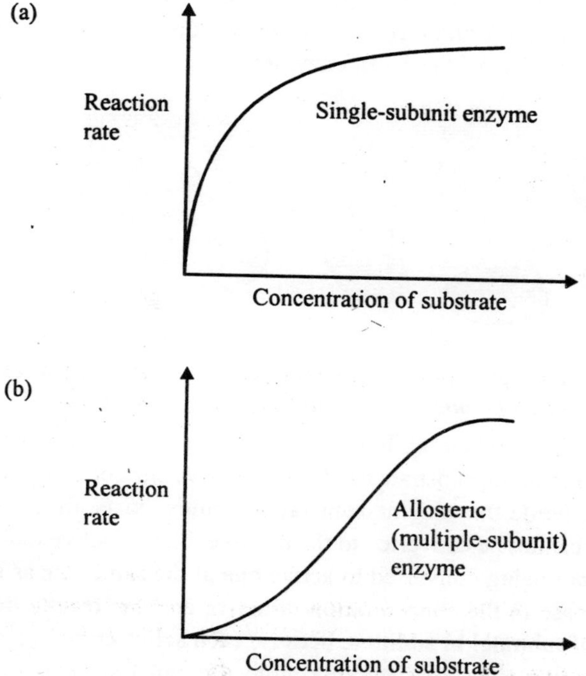

Fig. 6.17 : Allostery and Reaction Rate.

Machanism of Allosteric Effects

An allosteric enzyme has not only more than one subunit, but more than one *type* of subunit: The **catalytic subunit** has an active site that binds the enzyme's substrate; the **regulatory subunit** has one or more allosteric sites that bind specific effector molecules. A molecule of an allosteric enzyme usually consists of two or more catalytic subunits and two or more regulatory subuntis. An allosteric enzyme exists in two or more distinct forms with different catalytic efficiencies, and these forms are in equilibrium with each other. In the simple cases we will examine, the **active form** has full catalytic activity, whereas the **inactive form** is totally without activity. In the active form, the active sites on the catalytic subunits can bind substrate and convert it to product. In the inactive form, the active sites have been distorted in such a way that they cannot bind substrate; however, the allosteric sites are able to bind an effector, which we will consider in this example to be an inhibitor. The regulatory subunits of the active form of the enzyme have deformed allosteric sites and cannot bind effector. When neither substrate nor inhibitor is present, the active and inactive forms convert rapidly back and forth in equilibrium, the equilibrium constant being characteristic of the given enzyme.

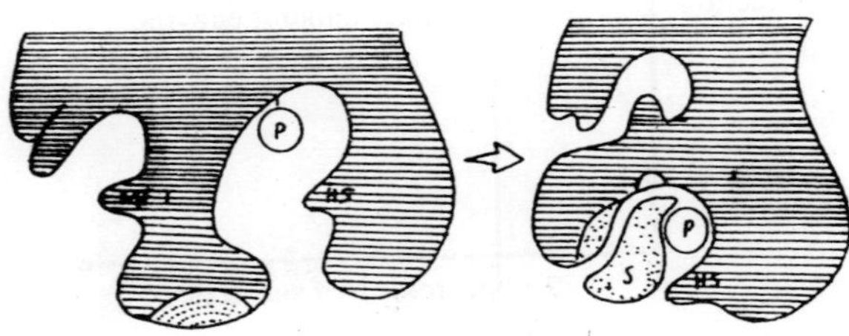

Fig.6.18 : Conformational change in the enzyme as a result of the entry of substrat(s) into the active site; according to Koshland.

What happens when inhibitor or substrate is present? If substrate is present, some of the substrate binds to the active sites of active enzyme molecules; while the enzyme-substrate complex exists, those enzyme molecules cannot be converted to the inactive form. Inactive molecules, however, are being converted to active one at the same rate as before, so an increase in the concentration of active enzyme results from the presence of subtrate! In addition, because each active enzyme molecule has two active sites, one enzyme molecule can bind two substrate molecules and simultaneously catalyze reactions of both of them. This explains the upward curvature at the lower left of a plot of reaction rate versus substrate concentration for an allosteric enzuyme. Increasing the substrate concentration increases the availability of active enzyme

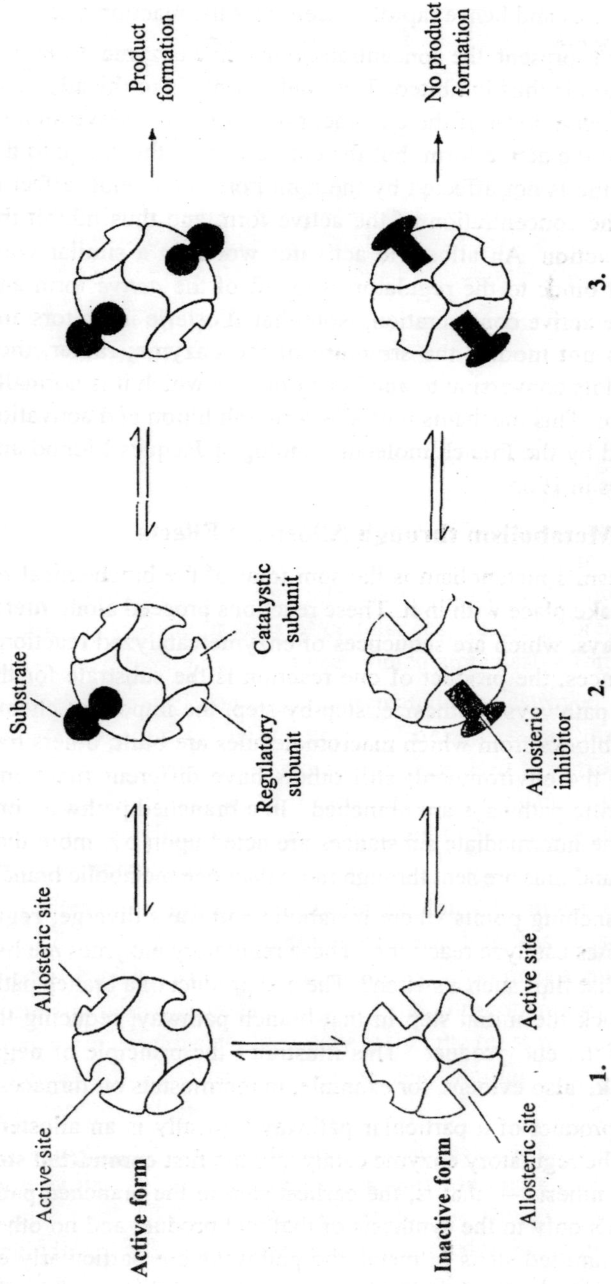

Fig. 6.19 : Allosteric Regulation of Enzymes

The hypothetical enzyme shown here has four subunits, two catalytic (blue) and the other two regulatory. When the enzyme is in its active form the active sites on the catalytic subunits can accept substrate. When the enzyme is in its inactive form, the allosteric sites on the regulatory subunits can accept inhibitor.

and of active sites and hence rapidly accelerates the reaction rate.

If inhibitor is present, the concentration of active enzyme *decreases* and the reaction is thus inhibited. The inhibitor binds to the allosteric site of the *inactive* form of the enzyme, preventing the conversion of the inactive to the active form; but the conversion of the active to the inactive enzyme is not affected by the inhibitor. The overall effect is to decrease the concentration of the active form and thus inhibit the enzymatic reaction. An allosteric activator works in a similar way, except that it binds to the regulatory subunit of the *active* form and holds it in the active configuration. Note that allosteric inhibitors and activators do not modify the structure of the enzyme; rather, they interfere with its conversion to another form with which it is normally in equilibrium. This mechanism of allosteric inhibition and activation was proposed by the French molecular biologist Jacques Monod and his colleagues in 1965.

Control of Metabolism through Allosteric Ellects

An organism's metabolism is the sum total of the biochemical reactions that take place with in it. These reactions proceed along **metabolic pathways,** which are sequences of enzyme-catalyzed reactions. In the sequences, the product of one reaction is the substrate for the next. Some pathways synthesize, step-by-step, the important chemical building blocks from which macromolecules are built; others trap energy from the environment; still others have different functions. Some metabolic pathways are branched. In a branched pathway, one or more of the intermediate substances are acted upon by more than one enzyme and thus are sent through more than one metabolic branch.

At the branching points where metabolic pathways diverge, **regulatory enzymes** catalyze reactions. These regulatory enzymes are like switches. What flips such a switch? The end product of a branch pathway may block the initial step in that branch pathway, reducing the formation of the end product. This illustrates the principle of **negative feedback,** also evident, for example, in thermostats on furnaces.

The end product of a particular pathway typically is an allosteric inhibitor of the regulatory enzyme catalyzing the first **committed step** in its own synthesis — that is, the earliest step in the branched pathway that leads only to the synthesis of that end product and no other. The first committed steps in metabolic pathways are particularly effective points for feedback control. For instance, inhibition of the B -

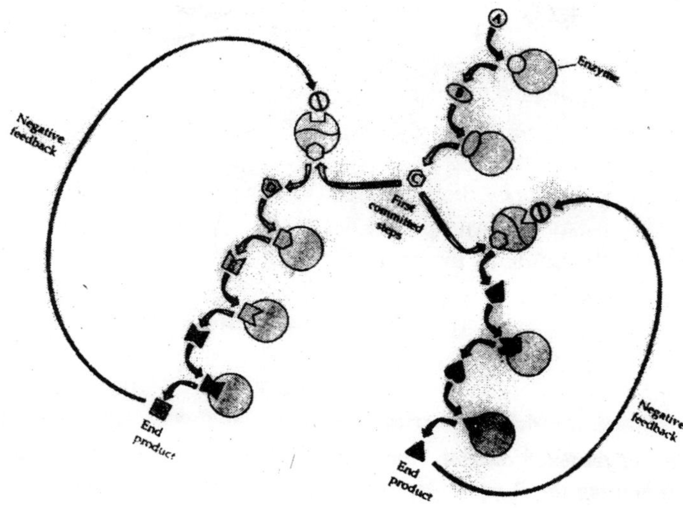

Fig. 6.20 : Feedback in Metabolic Pathways.

The reactions C to D and C to H are the first committed steps in the branch pathways leading to end products G and J, respectively. These end products can block the first committed steps by acting as allosteric inhibitors of the enzymes catalyzing the reactions. Thus, if levels of products G and J build up beyond what the cell needs for other reactions, the extra molecules provide the negative feedback that turns off the synthesis of G and J.

to - B step in Figure shunts all the reactants into the other branch of the pathway, whereas inhibition of the C-to-D reaction, one step later, leads only to a possibly harmful and certainly wasteful buildup of substance C. Living things generally do not accumulate unneeded intermediates.

When two different end products, produced by different branches of a pathway, are both present in excessive concentrations, they may act together to inhibit an earlier branch-point enzyme—one that catalyzes the first committed step for formation of these two products. **Concerted feedback inhibition** like this results in further efficiency: In this example intermediates G and H do not build up as they would if only the steps to I and L were inhibited by their individual end products. Concerted feedback inhibition requires that the enzyme have two allosteric sites, both of which must be bound to inhibitors to stop the enzyme's activity.

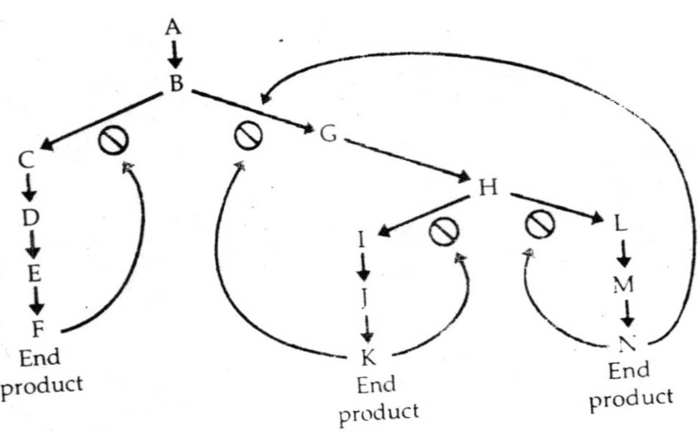

6.21 Concerted Feedback Inhibition

A variety of feedback controls come into play in branching metabolic pathways. Inhibiting the B → C reaction, for example, shunts reactant B into the B → G step. In some cases in which a single compound (H) gives rise to more than one end product (K and N), both end products can join in concerted inhibition of the enzyme catalyzing the first reaction committed to the production of H (B → G); each end product may also inhibit its own branch pathway.

Allosteric regulation is very effective: It allows rapid adjustment to short-term changes in metabolism or in the environment. The activities of enzyme molecules are adjusted by their interactions with small molecules, the end products. If a particular enzyme is not needed, might it not be a goood idea simply to stop making it until it is needed? Wouldn't it be advantageous to be able to regulate *production* as well enzyme *activity?* This is indeed the case, and the regulation of enzyme synthesis plays an important role in controlling development and metabolism.

Chemical Kinetics

Before we proceed to examine catalysis of reactions by enzymes, some relationships and terms employed in measuring and expressing the rates of chemical reactions must be outlined. Chemical reactions may be classified on the basis of the number of molecules that must ultimately react to form the reaction products. Thus, we may have *monomolecular, bimolecular,* and *termol-ecular* reactions.

Chemical reactions may also be classified on a kinetic basis, by *reaction order*. Thus, we may have zero-order, first-order, second-order, and third-order reactions, depending on how the reaction rate is influenced by the concentration of the reactants under a given set of conditions.

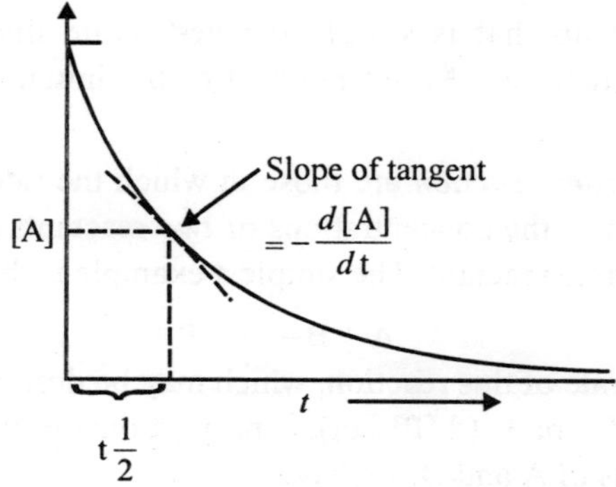

Fig. 6.22 : Plot of the course of a first reaction. The half-time ($t\frac{1}{2}$) is the time required for one-half of the initial reactant to be consumed.

First-order reactions are those which proceed at a rate exactly proportional to the concentration of *one* reactant. The simplest example is when the rate of the reaction

$$A \longrightarrow P$$

is exactly proportional to the rate of disappearance of A (or the rate of appearance of P). For such a case, the raction rate at any time t is given by

$$\frac{-d[A]}{dt} = k[A]$$

in which [A] is the molar concentration of A and $-d[A]/dt$ is the rate at which the concentration of A decreases. The proportionality constant k is called the *rate constant* or *specific reaction rate.* First-order rate constants have the dimensions of reciprocal time, i.e.,sec $^{-1}$ or min^{-1}.

The integrated form of this equation, which is more useful for carrying out kinetic calculation, is

$$\log_{t0}\frac{[A]}{[A]} = \frac{kt}{2.303}$$

in which $[A_o]$ is the concentration of A at zero time and [A] is the concentration at time t.

In first-order reactions, the half-time ($t\frac{1}{2}$) of the reaction is given by

$$t\frac{1}{2}=\frac{0.693}{k}$$

a relationship that is simply derived. Note that in first–order reactions the half–time is independent of the initial concentration of substrate.

Second –order reaction are those in which the rate is proportional to the product of the concentrations of *two* reactants or to the second power of a single reactant. The simplest example is the reaction

$$A + B \longrightarrow P$$

When the rate of this reaction, which may be designated as – (d[A]/dt), –(d [B] /dt), or + (d [P] / dt), is proportional to the product of the concentrations of A and B, we have

$$\frac{-d[A]}{dt} = k[A][B]$$

the second–order rate equation, in which k is the second –order rate constant. If the reaction has the from

$$2A \longrightarrow P$$

and its rate is proportional to the product of the concentration of the two reacting molecules, the second-order rate equation is

$$\frac{-d[A]}{dt} = k[A][A] = k[A]^2$$

The rate constants of second–order reactions have the dimensions 1/concentration × time ,or M^{-1} sec^{-1}. The integrated form of the second-order expression is

$$t = \frac{2.303}{k([A_0]-[B_0])}\log_{t0}\frac{[B_0][A]}{[A_0][B]}$$

where[A_o] and [B_o] are initial concentrations and [A] and [B] the concentrations at time *t*.

It is important to note that the reaction

$$A + B \longrightarrow P$$

which we have taken as an example, is not mecessarily a second–order reaction under all circumstances. Under some conditions this bimolecular reaction can be a first–order reaction. For example, if the concentration of A were very high and that of B very low, this reaction might be first–order with respect to B , since its rate is then proportional to the concentration of only one reactant (in this case, B).

This order of a reaction is thus determined by the conditions under which it is taking place and is not automatically a reflection of whether the reaction is monomolecular, bimolecular, or termolecular.

Third–order reactions, which are relatively rare, are those whose velocity is proportional to the product of three concentration terms. Some chemical reactions are independent of the concentration of any reactant ; these are called *zero–order reactions.* Many catalyzed reactions are zero–order with respect to the reactants. When this is the case, the rate of reaction depends on the concentration of catalyst or on some factor other than the concentration of the molecular species undergoing reaction. Reaction rates need not necessarily be pure first–order ; often mixed–order reactions are observed under certain conditions.

Catalysis

A chemical reaction such as A $\longrightarrow$ P takes place because a certain fraction of the population of A molecules at any given instant possesses much more energy than the rest of the popu-lation—enough energy to attain an "activated state," in which a chemical bond may be made or broken to form the product(s) P. The term *activation energy* refers to

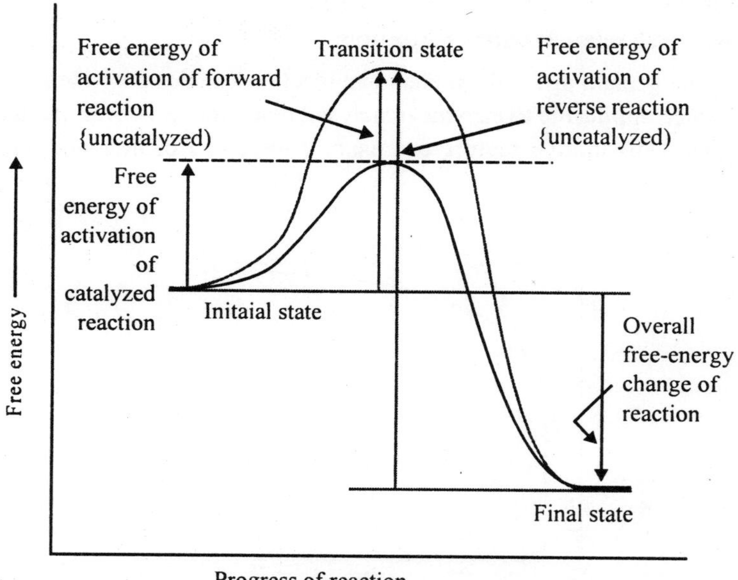

Fig. 6.23 : Energy diagram for a chemical reaction, uncatalyzed and catalyzed.

the amount of energy, in calories, required to bring all the molecules of 1 mole of a substance at a given temperature to this activated state.

In every chemical reaction there is a *transition state,* which is defined as the energy–rich state of the interacting molecules at the top of the activation barrier

$$\underset{\text{Reactant}}{A} \rightleftharpoons [\text{transition state}] \rightleftharpoons \underset{\text{Product (s)}}{P}$$

The rate of a reaction is proportional to the concentration of the transition–state species. A rise in temperature, because it increases thermal motion and energy, increases the number of molecules capable of entering the transition state; in many reactions, the reaction rate is approximately doubled by a 10°C rise in temperature.

Catalysts accelerate chemical reaction by lowering the free energy of activation. They combine with the reactants to produced a transition state having less free energy than the transition state of the uncatalyzed reaction. When the reaction products are formed, the free catalyst is regenerated.

Kinetics of Enzyme–Catalyzed Reactions: the Michaelis–Menten Equation

The general principle of chemical reaction kinetics described above are also applicable to enzyme–catalyzed reactions. However, the latter show a distinctive feature not usually observed in non–enzymatic

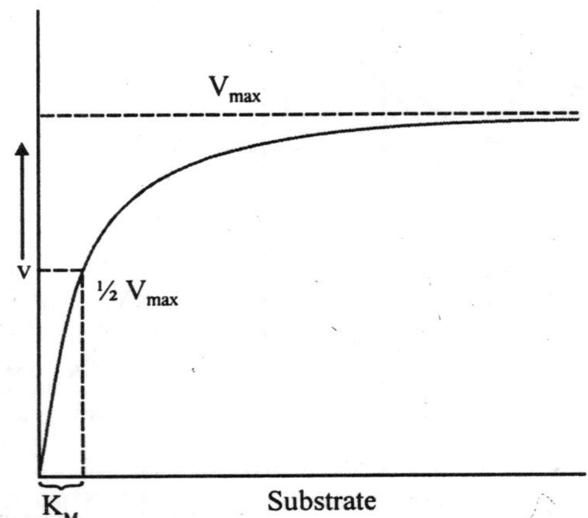

Fig. 6.24 : Effect of substrate concentration on the rate of an enzyme-catalyzed reaction.

reactions, namely, the phenomenon of *saturation* with substrate. In Figure, we see the effect of substrate concentration on the rate of the enzymatic reaction A $\rightarrow$ P. At low substrate concentration, the reaction velocity v is proportional to the substrate concentration and the reaction is thus first–order with respect to the substrate. However, as the substrate concentration is increased , the reaction rate falls off and is no longer proportional to the substrate concentration; in this zone, the reaction is mixed–order. On further increase in substrate concentration, the rate becomes constant and independent of substrate concentration. In this range of substrate concentration, the reaction is zero–order with respect to the substrate; the enzyme is then saturated with its substrate. All enzymes show this saturation effect, but they vary widely with repect to the substrate concentration required to produce saturation.

The saturation effect led L. Michaelis and M.L.Menten in 1913 to a general theory of enzyme action and kinetics, which was later extended by G.E. Briggs and J.B.S.Haldane. According to this theory, which is basic to the quantitative analysis of all aspects of enzyme kinetics and inhibition, the enzyme E first reacts with the substrate S to form the enzyme–substrate complex ES, which then breaks down in a second step to form free enzyme and the product(s) P:

$$E + S \underset{k_2}{\overset{k_1}{\rightleftharpoons}} ES \qquad (1)$$

$$ES \underset{k_2}{\overset{k_1}{\rightleftharpoons}} E + P \qquad (2)$$

Both reactions are considered to be reversible; k_1, k_2, k_3 and k_4 are specific rate constant for the reactions designated.

In the following derivation of the Michaelis–Menten equation, which is that of Briggs and Haldane,[E] represents the total enzyme concentration (the sum of the free and combined enzyme),[ES] is the concentration of the enzyme–substrate complex, and [E]–[ES] represents the concentration of free, or uncombined, enzyme.[S] represents the substrate concentration, which is ordinarily far greater than[E], so that the amount of S bound by E at any given time is negligible compared with the total concentration of S. The derivation begins by considering the rates of formation and breakdown of ES. The rate of formation of ES from E + S is given by

$$\frac{d[\mathrm{ES}]}{dt} = k_1([\mathrm{E}]-[\mathrm{ES}])[\mathrm{S}] \tag{3}$$

The rate of formation of ES from E + P is very small and may be neglected.

Similarly, the rate of breakdown of ES is given by

$$\frac{-d[\mathrm{ES}]}{dt} = k_2[\mathrm{ES}]+k_3[\mathrm{ES}] \tag{4}$$

When the rate of formation of ES is equal to its rate of breakdown, i.e., when the reaction system is in a steady state, with the ES concentration remaining constant,

$$k_1([\mathrm{E}]-[\mathrm{ES}])[\mathrm{S}] = k_2[\mathrm{ES}]+k_3[\mathrm{ES}] \tag{5}$$

Rearranging this expression , we have

$$\frac{[\mathrm{S}][[\mathrm{E}]-[\mathrm{ES}]]}{[\mathrm{ES}]} = \frac{k_2+k_3}{k_1} = K_M \tag{6}$$

The "lumped" constant K_M which replaces the term $(k_2-k_3)/k_1$, is called the *Michaelis–Menten constant.* From this equation, the steady–state concentration of the ES complex may be obtained by solving for [ES]:

$$[\mathrm{ES}] = \frac{[\mathrm{E}][\mathrm{S}]}{K_M+[\mathrm{S}]} \tag{7}$$

Since the intial rate v of an enzymatic reaction is proportional to the concentration of the ES complex, we can write

$$v = k_3[\mathrm{ES}] \tag{8}$$

When the substrate concentration is so high that essentially all the enzyme in the system is present as the ES complex, we reach the *maximum velocity* V_{max} for which we can write

$$V_{max} = k_3[\mathrm{ES}] \tag{9}$$

in which [E] is the total enzyme concentration.

We can now substitute for the term [ES] in equation(8) its value from equation (7):

$$v = k_3\frac{[\mathrm{E}][\mathrm{S}]}{K_M+[\mathrm{S}]} \tag{10}$$

If we now divide this euqation by equation (9), we obtain

$$\frac{v}{V_{\max}} = \frac{k_3 \frac{[E][S]}{K_M + [S]}}{k_3[E]} \quad (11)$$

Solving for *v*,

$$v = \frac{V_{\max}[S]}{K_M + [S]} \quad (12)$$

This is the *Michaelis-Menten equation* ; it defines the quantit-ative relationship between the enzyme reaction rate and the substrate concentration [S] if both $V_{\max}$ and K_M are known.

An important numerical relationship emerges from the Michaelis- Menten equation in the special case when $v = \frac{1}{2} V_{\max}$. We then have

$$\frac{V_{\max}}{2} = \frac{V_{\max}[S]}{K_M + [S]}$$

If we divide by $V_{\max}$, we obtain

$$\frac{1}{2} = \frac{[S]}{K_M + [S]}$$

Rearranging,

$$K_M + [S] = 2[S]$$

$$K_M = [S]$$

We may therefore conclude that K_M *is equal to the substrate concentration at which the velocity is half maximal.* K_M has the dimensions moles liter^{-1}.

Figure shows that K_M can be extrapolated graphically from data on the effect of substrate concentration on the reaction

Table : K_M for some eczymes

Enzyme and substrate	K_M *(mM)*
Catalase	25
H_2O_2	
Hexokinase	
Glucose	0.15
Fructose	1.5
Chymotrypsin	
N–Benzoyltyrosinamide	2.5
N–Formyltyrosinamide	12.0
N–Acetyltyrosinamide	32
Glycyltyrosinamide	122
Carbonic anhydrase	
HCO_3^-	9.0
Glutamate dehydrogenase	
Glutamate	0.12
*–Ketoglutarate	2.0
NH_4^-	57
NAD_{ox}	0.025
NAD_{red}	0.018

Table: *Effect of substrate structure on V_{max} for D–amino acid oxidase (Data are relative to D-alanine =100)*

Substrate	Relative V_{max}
D-Tyrosine	297
D-Proline	231
D-Methionine	125
D-Alanine	100
D-Valine	55
D-Histidine	9.7
Glycine	0.0

velocity; its magnitude is independent of enzyme concentration . Note that K_M is not a fixed value. It may vary with the structure of the substrate, with pH, and with temperature. In those enzymes having more than one substrate, each substrate has a characteristic K_M. Under intracellular conditions, enzymes are not necessarily saturated with their substrates.

The maximum velocity V_{max} varies widely from one enzyme to another. It also varies with the structure of the substrate, with pH, and with temperature.

The Michaelis constant, as we noted above, is an experimentally determined quantity, which in the idealized case is represented by

$$K_M = \frac{k_2 + k_3}{k_1} \tag{6}$$

However, in many enzymatic reactions, k_2 and k_1 may be very large compared with k_3. In such reactions, the rate-limiting step in the overall reaction is the slow step, $ES \xrightarrow{k_3} P$. k_3 is then negligibly small, and equation (6) simplifies to the expression

$$K_M = \frac{k_2}{k_1}$$

Under these conditions, K_M is evidently the dissociation constant of the ES complex and is replaced by the expression K_s:

$$K_S = \frac{[E][S]}{[ES]}$$

Unfortunately, K_M and K_S are frequently used interchangeably. K_M should not be regarded as the dissociation constant of the ES complex unless specific information is available that k_3 is very low compared with k_2 and k_1.

The Michaelis-Menten equation is fundamental to all quantitative treatment of enzyme action. Its derivation from first principles, as shown above, leads to many other useful relationships. But it must be stressed that most enzymes show kinetic behaviour that is much more complex than the idealized case we have just treated . For one thing, our formulation assumed that there is but one enzyme-substrate complex. However, it now appears likely that most enzyme-catalyzed reactions involve two or three enzyme-substrate complexes, acting in the following sequence:

$$E + S \rightleftharpoons ES \rightleftharpoons EZ \rightleftharpoons EP \rightleftharpoons E - P$$

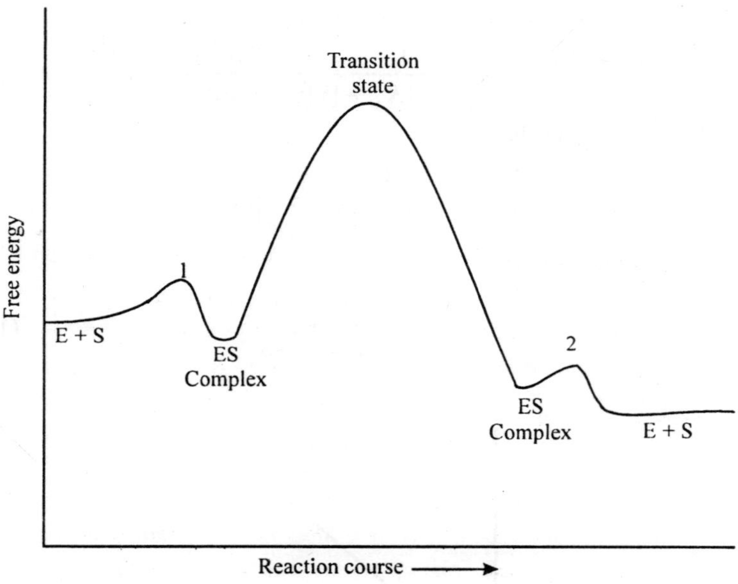

Fig. 6.25 : Energy diagram for an enzyme-catalyzed reaction. Small energy barriers exist at points 1 and 2:

in which EZ is the true transition-state complex and EP an enzyme-product complex. Furthermore, it must be pointed out that in most enzymatic reactions there is more than one substrate molecule and there may be two or more products. In a reaction with two substrates S_1 and S_2, there may be three enzyme-substrate complexes, namely, ES_1, ES_2, and ES_1S_2. If the reaction has two products P_1 and P_2 , there may be at least three additional complexes EP_1, EP_2 and EP_1P_2. Many intermediate steps occur in such reactions, each having its own rate constant. Kinetic analysis of enzymatic reactions involving two or more reactants can sometime be exceedingly complex and may require computer solutions. Nevertheless, the starting point for analysis of the kinetics of all enzymatic reactions is the Michaelis-Menten relationship as derived above.

Transformations of the Michaelis-Menten Equation

The Michaelis-Menten equation can be transformed algebraically into other forms that are more useful in plotting experimental data. One of the most widely used transformations is derived simply by taking the reciprocal of both sides of the Michaelis-Menten [equation(12)]:

$$\frac{1}{v}=\frac{1}{V_{\max}[\mathrm{S}]/\left(K_M+[\mathrm{S}]\right)}=\frac{K_M+[\mathrm{S}]}{V_{\max}[\mathrm{S}]}$$

Rearranging, we have

$$\frac{1}{v}=\frac{K_M}{V_{\max}[\mathrm{S}]}+\frac{[\mathrm{S}]}{V_{\max}[\mathrm{S}]}$$

which reduces to

$$\frac{1}{v}=\frac{K_M}{V_{\max}}\cdot\frac{1}{[\mathrm{S}]}+\frac{1}{V_{\max}} \tag{13}$$

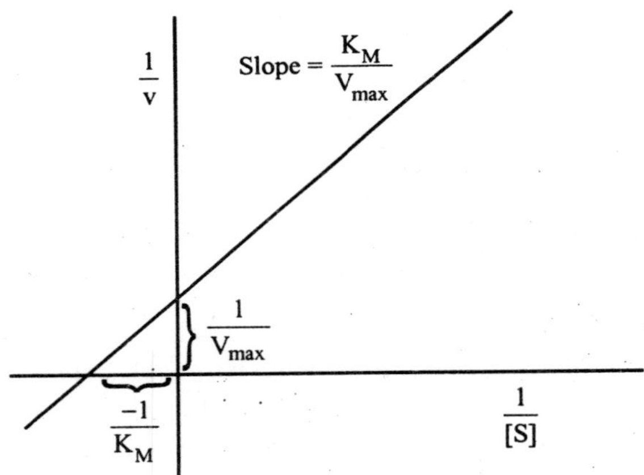

Fig. 6.26: Lineweaver-Burk plot.

Equation (13) is the *Lineweaver-Burk equation,* which represents a straight line with a slope of $K_M/V_{\max}$ and an intercept of 1/ V_{max} on the 1/ v axis. This line is obtained by plotting 1/ v vs. 1/[S]. Such a "double-reciprocal" plot has the advantage that V_{max} can be more accurately arrived at than from the simple plot *v vs.* [S]; in the latter, $V_{\max}$ is approached asymptotically and its value therefore is uncertain. The intercept on the abscissa of the Lineweaver-Burk plot is $-1/K_M$. The Lineweaver-Burk plot can also give valuable information on enzyme inhibition, as we shall see below.

An other useful transformation of the Michaelis-Menten equation is obtained by multiplying both sides of equation (13) by $V_{\max}$ (v) and rearranging to yield the equation

$$v = -K_M\left(\frac{v}{[S]}\right) + V_{max}$$

When v is plotted against $v/[S]$, the plot shown in Figure 6.27 results. This plot (the *Eadie- Hofstee plot*) not only yields V_{max} and K_M in a very simple way but also magnifies departures from linearity which might not be seen in a Lineweaver-Burk plot.

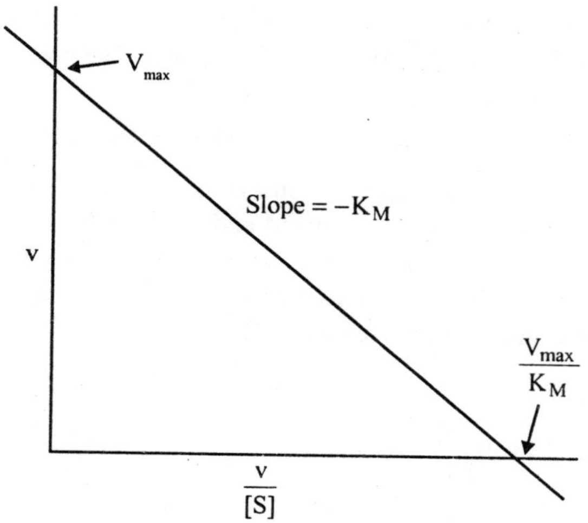

Fig. 6.27 : Eadie-Hofstee plot.

Dependence of Enzyme Reaction Rates on Cafactores

The Michaelis-Menten equation not only defines the quantitative relationship between substrate concentration and enzyme reaction rate but can also be applied to the relationship between cofactor concentration and reaction rate, in the case of those enzymes requiring a cofactor for activity. The following type of equilibtium is assumed:

$$\underset{\text{(Inactive)}}{\text{E + coenzyme}} \rightleftharpoons \underset{\text{(Active)}}{\text{E – coenzyme}}$$

The enzyme-coenzyme complex may then bind the substrate to form the enzyme-coenzyme-substrate complex

$$\text{E – coenzyme + S} \rightleftharpoons \text{S – E – coenzyme}$$

whose concentration determines the overall reaction rate. It is evident from these equilibria that such an enzyme will show a saturation phenomenon not only with its substrate but also with its cofactor.

Just as we may have a Michaelis constant K_M for substrate affinity, defined as the concentration of substrate at which half-maximal velocity is observed, we may also have a Michaelis constant to express affinity of the enzyme for its cofactor, similarly defined as the concentration of cofactor at which half-maximal activity of the enzyme is observed. To estimate the K_M for a cofactor, the substrate concentration is held constant at a saturating level and the effect of cofactor concentration on velocity is measured.

Table lists K_M values for the enzyme glutamate dehydrogenas, which requires the cofactor NAD to accept H atoms. The reaction catalyzed by this enzyme is

$$\text{Glutamate} + \text{NAD}_{\text{ox}} \rightleftharpoons \alpha\text{-ketoglutarate} + \text{NAD}_{\text{red}} + \text{NH}_3$$

The two substrates glutamate and α - ketoglutarate have characteristic K_M values, as do the oxidized and reduced forms of NAD , the cofactor. In this and many similar enzymatic reactions involving coenzymes, the coenzyme participates in the reaction as though it were a substrate, and it can be treated by the same kinetic formalism.

CHAPTER 7
BIOENERGETICS

All living things require energy because life processes involve work.It may seem obvious that cells need energy to grow and reproduce, but even non-growing cells need energy simply to maintain themselves.

The sun is the ultimate source of almost all the energy that powers life. Plants and other photosynthetic organisms capture a tiny portion of the sun's energy and, in the process of photosynthesis, convert it to chemical energy in organic molecules. When plants, animals, or other organisms need the energy stored in these organic molecules, they commonly use the process of cellular respiration to break them apart and convert their energy to more immediately usable forms.

Because energy can be neither created nor destroyed, cells have no way of producing new energy. Energy is captured from the environment, temporarily stored and then used to perform biological work. However, not all of the captured energy can be used; at every step some inevitably becomes converted to heat and is dispersed back into the environment.

Cells obtain energy in many forms, but seldom can that energy be used directly to power cellular process. For this reason metabolic mechanisms have evolved that enable cells to convert energy from one form to another. Because most of the components of these energy conversion systems evolved very early , the most fundamental aspects of energy metabolism tend to be very similar in a wide range of different organisms.

This chapter focuses on some of the basic principles that govern how cells capture, transfer, store, and use energy. We discuss the function of ATP and other molecules used in energy conversions, including those that transfer electrons in redox reactions. We also pay particular attention to the essential role of enzymes in cellular energy dynamics.

Biological Work Requires Energy

Energy may seem to be an abstract concept, but it helps to remember that energy can be understood in the context of matter (anything that has mass and takes up space.) This is because energy can be defined as the capacity to do work, which is any change in the state or motion of matter. Biologists generally express energy in units of work **(Kilojoules, KJ)** or units of heat energy **(Kilocalories, Kcal)**. One kilocalorie equals 4.184 kilojoules. Because heat energy cannot do cellular work (see *Making the Connection: Energy, Work, and Heat*), the kilojoule is the unit preferred by most biologists today. However, we will use both because references to the kilocalorie are common in the scientific literature.

Much of the work an organism does is mechanical work. At this very moment you are expending considerable energy to carry out such activities as breathing and circulating your blood. In these examples , it is evident that the state or motion of matter is being changed in some way. However, all these rather; obvious forms of mechanical work are the consequence of work being performed on a microscopic scale by your individual cells. For example, the cells of the heart muscle use a great deal of energy to contract, thereby pumping the blood through your body. As we shall see, however, not all of the work of cells is mechanical. A great deal of it is chemical work. For example , heart muscle cells expend energy to synthesize the proteins required for contraction. Energy can be converted to may different forms, including not only mechanical and chemical energy but also heat energy and radiant energy (the energy of electromagnetic waves, such as radio waves , visible light, x-rays, and gamma rays).

Occurrence and Properties of ATP and ADP

ATP was first isolated from acid extracts of muscles in 1929 by Fiske and Subbarow. It structure was deduced some years later by degradation experiments and ultimately confirmed by total chemical synthesis by Todd and his colleagues in 1948. From its first discovery ATP was suspected to play a role in cellular energy transfer, but it was not until 1939-1941 that Lipmann proposed it serves as a principal means of transfer of chemical energy in the cell. ATP, ADP, and AMP are not trace substances; the sum of their concentrations in the aqueous phase of various types of intact cells is between 2 and 15 mM. The concentration of ATP usually greatly exceeds the sum of concentration of the other two;

the AMP concentration is usually much the lowest of the three. These nucleotides are present not only in the soluble cytoplasm but also within organelles such as mitochondria and nuclei. The intracellular compartmentation of the ATP system is an important feature in cellular regulation of metabolism.

At pH 7.0, both ATP and ADP are highly charged anions. ATP has four ionizable protons in its triphosphoric acid group. Three have low pK′ values of about 2 to 3 and are thus completely dissoiated at pH 7.0; the fourth has a pK′ of 6.50 and therefore is about 75 percent dissociated at pH 7.0 ADP has three ionizable protons; two are completely dissociated at pH 7.0 and the third, which has pK' of 7.2, is about 39 percent diss-ociated at pH 7.0 The high concentration of negative charges around the tripho-sphate group of ATP is an important factor in its high -energy nature, as will be seen. In the intact cell, very little ATP and ADP exist as free anions; they are largely present as the 1:1 Mg ATP^{2-} and $MgADP^{-}$ complexes, because of the high affinity of the pyrophosphate groups for binding divalent cations and the high concentration of Mg^{2+} in intracellular fluid. The affinity of ATP for Mg^{2+} is about 10 times as great as that of ADP. In most enzymatic reactions in which ATP participates as phosphate donor, its active from

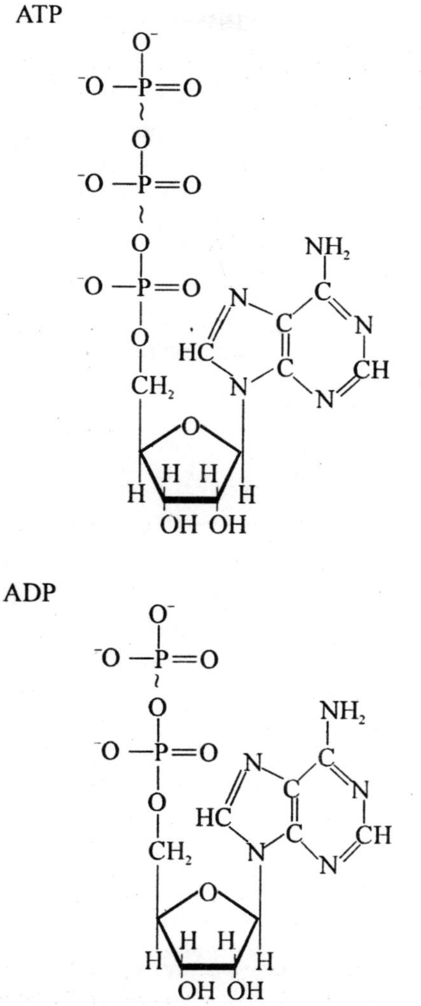

ATP

ADP

Fig. 7.1 : Structures of ATP and ADP

is the $MgATP^{2-}$complex. In other reactions, however, the Mn ATP complex appears to be the more active substrate.

$MgATP^{2-}$

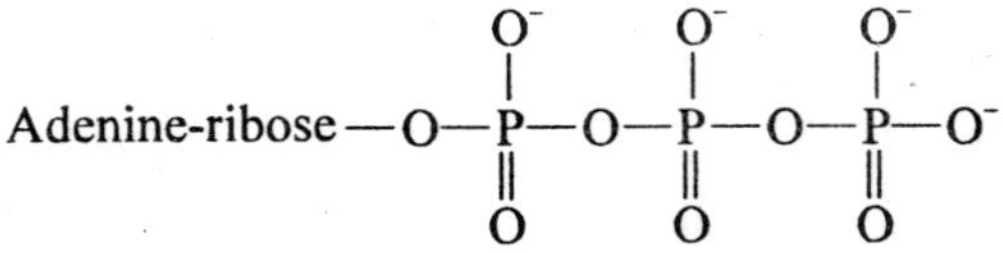

$$\text{Adenine-ribose}-O-\overset{O^-}{\underset{O}{P}}-O-\overset{O^-}{\underset{O}{P}}-O-\overset{O^-}{\underset{O}{P}}-O^-$$

$MgADP^-$

$$\text{Adenine-ribose}-O-\overset{O^-}{\underset{O}{P}}-O-\overset{O^-}{\underset{O}{P}}-O^-$$

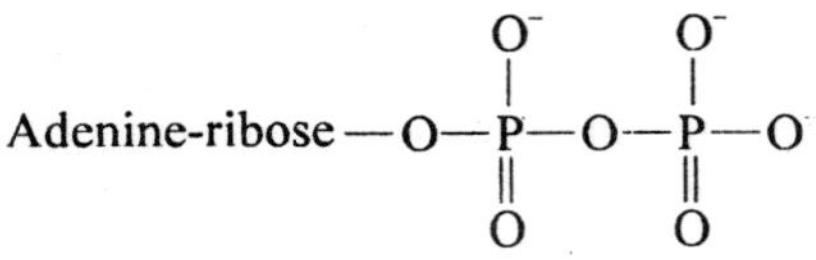

$MnADP^-$

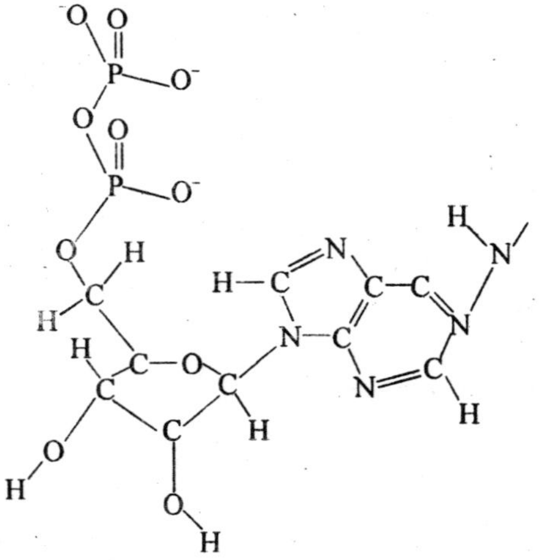

Fig. 7.2 : Metal complexes of ATP and ADP. In the Mg^{2-} complexes the two terminal phosphate groups are the ligands. In the Mn^{2-} complexes the 7 nitrogen atom of the odenine ring is also a ligand.

ADP and ATP are easily separated and measured by paper electrophoresis or thin-layer chromatography. Both show the characteristic ultraviolet absorption peak at 260 nm given by the adenine moiety. The last two phosphate groups of ATP and the terminal phosphate group of ADP can be hydrolyzed by boiling in 1 N HCI for

7 min; the ester linkage between the remaining phosphate groups and ribose is stable to this treatment. Specific enzymes hydrolyzing or transferring the therminal phosphate group of ATP are discussed below.

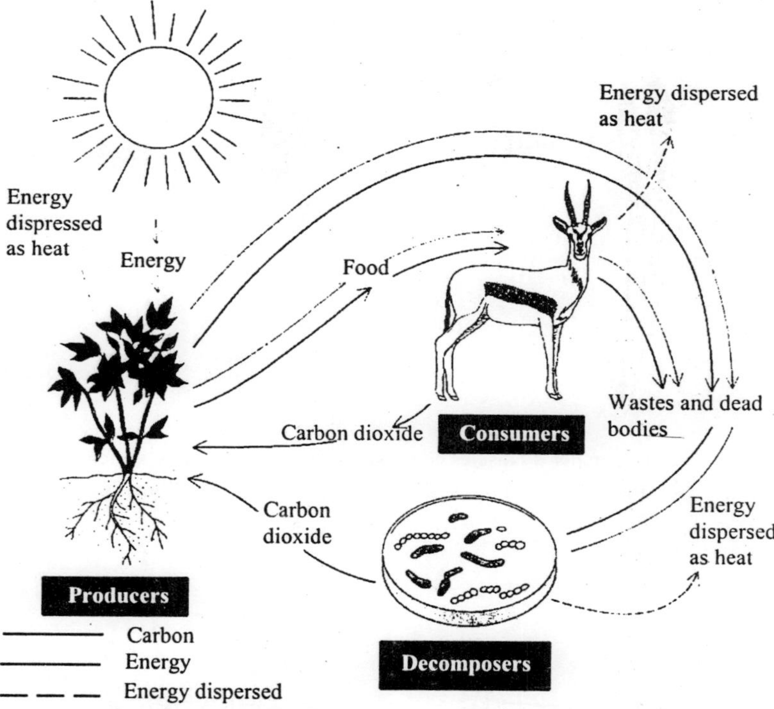

Fig. 7.3 : Flow of energy through the ecoshphere.

Measuring Energy

To study energy transformations, scientists must be able to measure energy. How is this done? Heat is a convenient form in which energy can be measured because all other forms of energy can be converted into heat. In fact, the study of energy has been named **thermodynamics**, that is, heat dynamics. Although several units may be used in measuring energy, the most widely used unit in biological systems is the **kilocalorie (kcal)**. A kilocalorie is equal to 1000 calories. A **calorie** is the heat required to raise the temperature of 1 gram of water from 14.5° to 15.5° C. Nutritionists use the kcal in measuring the potential energy of foods and usually refer to this unit as a Calorie (with a capita C).

The Laws of thermodynamics

All activities of our universe—from the life and death of cells to the life and death of stars—are governed by two laws of energy.

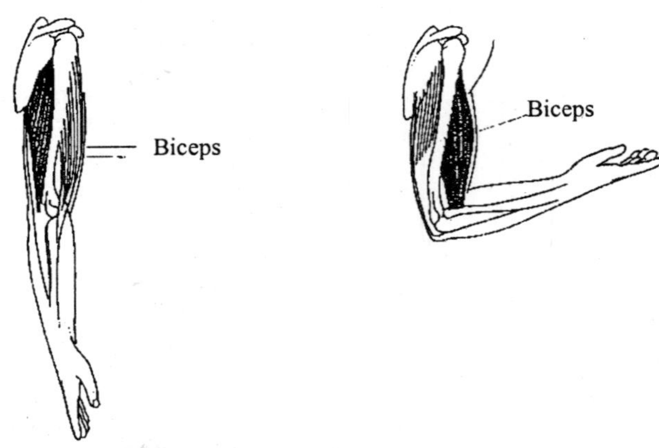

a. The archer's muscles contract, pulling on the bones, which then move and transmit the muscular force to the bow.

b. The archer has drawn the bow. The energy stored in the drawn bow is potential mechanical energy. When the archer releases her fingers the string moves forward and the arrow is propelled toward a target.

Fig.: 7.4 : See opposite page for legend.

The First Law

The **First law of thermodynamics**, known also as the law of conservation of energy, states that during ordinary chemical or physical processes energy can be transferred and changed in form. But can be neither created nor destroyed. The universe is a closed system when it comes to energy. As far as we know, the energy present when the universe formed an estimated 20 billion years ago is all that can ever exist.

Although an organism can neither make nor destroy energy, it can capture some from the environment and use it for its own needs. Organisms can also transform energy to electrical energy and then to chemical energy stored in chemical bonds. Some of that chemical energy may later be transformed by an animal that eats the plant or alga to the mechanical energy of muscle contraction or to another needed form. As these many transformations occur, some of the energy is converted to heat energy and dissipated to the environment. Although this energy can never again be used by the organism, it is not really "lost" because it can be accounted for in the surrounding physical environment.

The Second Law

Given all possible forms of arrangement, order is an extremely unlikely state. From your own experiences you might have observed that creating order requires an input of energy—or work. For example, if you spill the many pieces of jigsaw puzzle on to the floor, it is highly unlikely that the pieces will spontaneously reassemble themselves into the original picture. Similarly, if you drop a crystal vase, the fragments of glass will not suddenly jump back into place to reconstruct the vase. In fact, it would be very difficult for you to gather every small shard of glass and fit these fragments together to mend the vase. Out of the multitude of possible ways in which the pieces can be assembled, only one represents the highly ordered form that is the vase or the finished puzzle.

The **second law of thermodynamics** states that disorder in the universe is continuously increasing. This law explains that in any process requiring or releasing energy, some energy is dissipated as heat and is no longer available to do work. Although the total amount of energy in the universe is not decreasing with time, the energy is continuously degraded to heat, a less useful form of energy. In fact, heat is actually the energy of random molecular motion. Heat can be made to do work when there is a temperature gradient, that is, a difference in potential.

1. An automobile travels down the highway. Its energy is highly directional.

2. As the car travels, the surface of the road heats up, along with the surrounding air and the car's tires.

3. The engine of the car is turned off and the car begins to slow

4. All the energy of the car's forward motion has been converted into heat and the car is now at a halt.

Fig. 7.5 : The concentrated energy of motion is converted into the dispersed energy of heat—also basically motion but that of millions of randomly moving molecules. Though all the car's kinetic energy is lost to use, none of it is lost in the sense of being destroyed—it is merely dispersed and diluted past the point of recovery.

The term **entropy** refers to the energy that has become so randomized and uniform throughout a system that it is no longer available to do work. It is a measure of the disorder of a system. According to the second law of thermodynamics, entropy in the universe is continuously increasing. Eventually all energy will be random and uniform in distribution. With only this useless form of energy, no work could be performed. The universe will have run down. We are in no immediate danger, however, because this depressing state is not scheduled to occur for several billion years.

Because of the second law of thermodynamics, no process requiring energy is ever 100% efficient. Much of the energy is dispersed as heat and is transferred to the environment, so that there is an increase in entropy. Cellular energy utilization is about 55% efficient, with the other 45% of the energy being lost as heat. Such biological processes are actually quite efficient compared with most machines made by human beings. For example, a gasoline engine is only about 17% efficient.

Because they are highly organized, living organisms are very unstable. In fact, life is a constant struggle against the second law of thermodynamics. Survival individual organisms, as well as ecosystems, depend upon continuous energy input. Thus, producers must carry on photosynthesis, and consumers and decomposers must eat.

Metabolic Reactions involve Energy Transformations

The myriad chemical reactions of an organism that enable it to carry on its activities-to grow, move, maintain and repair itself, reproduce, and respond to stimuli together make up its metabolism. *Metabolism* is the sum total of all the chemical and physical changes that take place in an organism, interacting to make up what has been called the **metabolic web**. An organism's metabolic web consists of many intersecting series of reactions, or pathways, which are of two main types. **Anabolism** refers to the various metabolic pathways in which complex molecules are synthesized from simpler substances, such as the linking of amino acids to germ proteins. **Catabolism** includes the pathways in which larger molecules are broken down into smaller ones, such as the degradation of starch to form monosaccharides. As we shall see, these changes not only involve alterations in the arrangement of atoms, but also various energy transformations. Catabolism and anabolism are complementary processes; catabolic pathways involve an overall release of energy, some of which is used to power the anabolic pathways, which have an overall energy requirement. In the following sections we will discuss how to predict whether a particular chemical reaction requires energy or releases it.

Enthalpy (H) is the total potential energy of a system

In the course of any chemical reaction, including the metabolic reactions of a cell, chemical bonds break and new and different bonds may form. Every specific type of chemical bond has a certain amount

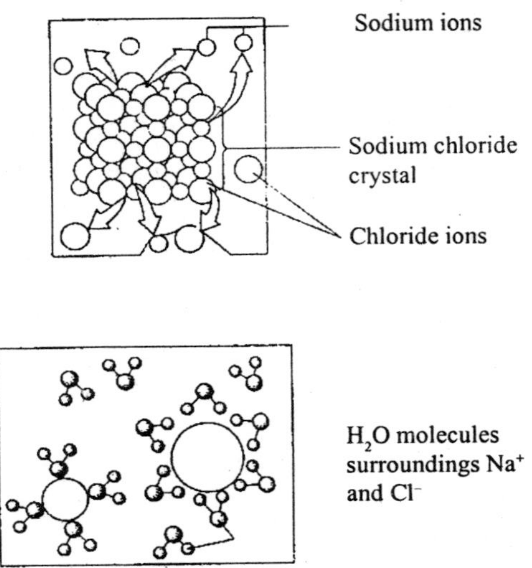

Figure 7.6 : **Entropy is a measure of disorder**
As particles leave a crystal to go into a solution, they become more disordered. The entropy (randomness) of this system increases during the process.

of bond energy defined by chemists as the energy required to break that bond. The total bond energy is essentially equivalent to the total potential energy of the system, a quantity known as enthalpy, H. Because energy can be conveniently measured as heat, enthalpy is often referred to as the heat content of the system.

Free energy is available to do cellular work

Entropy and enthalpy are related by a third dimension of energy, termed free energy, which can be expressed in kilojoules or kilocalories per mole. Free energy is available energy; it is the component of the total energy of a system that is available to do work under defined conditions of constant temperature and pressure. Changes in pressure are not generally important in biochemistry, but changes in temperature are. For our purposes, then, we can consider the free energy of a system to be the *maximum amount of energy that is available to do work, without a change in temperature*. The requirement that there be no change in temperature is important because heat energy cannot to cellular work. Free energy, the only kind of energy that can do cellular work, is the aspect of thermodynamics of greatest interest to a biologist.

Entropy, represented by the letter *S*, and free energy , represented by *G*, are related inversely; as entropy increases, the amount of free energy decreases. The two are related by the following equation:

$$G = H - TS$$

in which H is the enthalpy of the system, T is the absolute temperature (expressed in degrees Kelvin; K = °C + 273), and S is entropy. If we assume that the entropy is Zero, the free energy is simply equal to the total potential energy (enthalpy); entropy reduces the free energy. What, then, is the significance of the temperature (T) ? Remember that as the temperature increases, there is an increase in random molecular motion that contributes to disorder and multiplies the effect of the entropy term.

Chemical reactions involve changes in free energy

Biologists need ways of analyzing the role of energy in the many

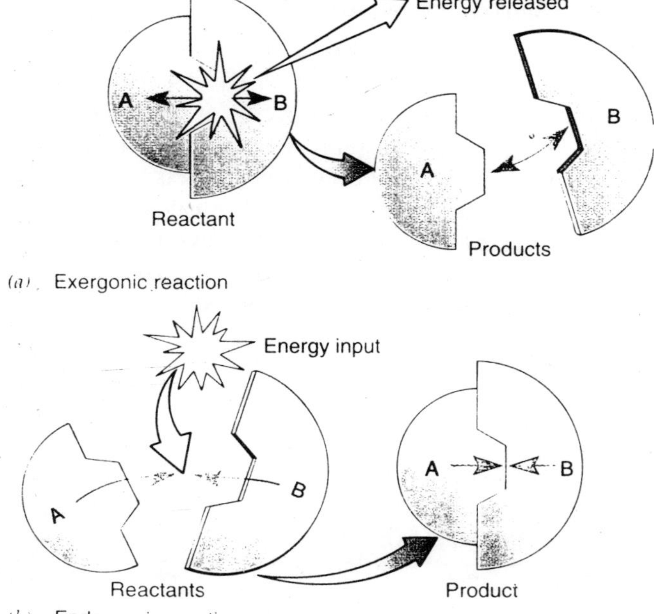

Fig. 7.7 : An exergonic reaction involves a loss of free energy; free energy is gained in an endergonic reaction.

reactions that make up cellular metabolism. Although the total free energy of a system (G) cannot be effectively measured, the equation G = H – TS is nevertheless useful because it can be extended to predict whether any particular chemical reaction will release energy or require an input of energy. This is because changes in free energy can be measured.

We use the Greek letter delta (Δ) to denote any change that occurs in the system between its *initial state* and its *final state* . To express what happens with respect to energy in a chemical reaction, the equation becomes:

$$\Delta G = \Delta H - T\Delta S$$

Notice that the temperature does not change; it is held constant during the reaction. Thus, the change in free energy (Δ *G*) during the reaction is equal to the change in enthalpy (Δ H) minus the product of the absolute temperature (T) multiplied by the change in entropy (Δ *S*). Δ *G* and Δ *H* are expressed in kilojoules or kilocalories per mole ; Δ *S* is expressed in kilojoules per degree or in kilocalories per degree.

Exergonic reactions do not require outside energy

In accordance with the second law of thermodynamics, no chemical reaction is 100% efficient. No reaction can take place without a decrease in enthalpy, an increase in entropy, or both (see *Making the Connection: Energy and Diffusion*). For this reason, the total free energy of the system in its final state is always less than the total free energy of the system in its initial state. When calculated in this way, Δ *G* whether expressed in kilojoules per mole or kilocalories per mole, is a negative number. Such a reaction, with a negative value of Δ G (– Δ *G*), is referred to as an **exergonic reaction**.

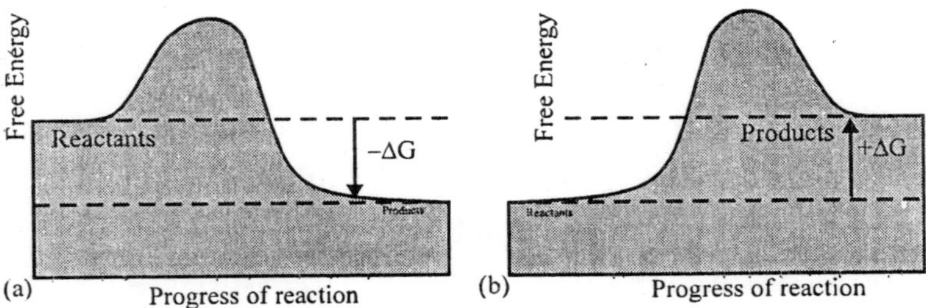

Fig. 7.8 : Exergonic and endergonic reactions can be distinguished by comparing the free energy change that takes place during the reaction.

An exergonic reaction releases energy and is said to be a **spontaneous** or a " downhill " reaction. The term *spontaneous* may give the false impression that such reactions are always instantaneous. In fact, spontaneous reactions do not necessarily occur readily; some are extremely slow. This is because energy, known as activation energy, is required to initiate every reaction, even a spontaneous one.

All reactions have a required energy of activation

The energy necessary to begin a reaction is the required energy of activation or activation energy. This is the amount of energy needed to break the existing chemical bonds and initiate the reaction. The activation energy may be supplied as heat, which raises the temperature.

For bonds to break, the reacting molecules must come together and be in the right orientation. The molecules have a certain average kinetic energy that depends on the temperature. As the temperature increases, random motions of the molecules increase. Furthermore, the internal vibrations in each molecule increase, causing the bonds to become less stable and more reactive. The reactants move faster and collide with one another more frequently and more forcefully. Some of these collisions are in the proper orientation and occur with a force sufficient to break the weakened chemical bonds.

For example, molecular hydrogen and molecular oxygen can react explosively to form water:

$$2H_2 + O_2 \longrightarrow 2H_2O$$

This reaction is spontaneous (exergonic), yet hydrogen and oxygen can be safely mixed as long as all sparks are kept away. This is because the required energy of activation for this particular reaction is relatively high. A tiny spark provides the activation energy that allows a few molecules to react. Their reaction liberates so much heat that the rest react, producing an explosion . Later in this chapter we will see how cells use enzymes to lower the activation energy barrier for specific reactions and thereby regulate their metabolism.

An endergonic reaction requires an energy source

It is possible to write an equation for an endergonic reaction - that is , a reaction in which there is a gain of free energy. Because the free energy of the products is greater than the free energy of the reactants, ΔG has a positive value. Such a reaction, cannot actually take place in isolation. Instead, it must occur in such a way that energy can be supplied from the surroundings. Of course, many energy-requiring reactions take place in cells, and, as we shall see, metabolic mechanisms have evolved that supply the energy needed to "drive" these nonspontaneous cellular reactions in a particular direction.

Actual free energy changes depend on the concentrations of reactants and products

In order to compare different reactions, chemists measure free energy changes under a defined set of standard conditions that would never exist in an organism. This standard free energy change, which can be thought of as a measure of the intrinsic free energy difference between the reactants and the products, depends mainly on the difference in their bond energies (enthalpy, H). The actual free energy change in a cellular reaction depends only partly on the standard free energy change; it is also affected by the relative concentrations of reactants and products.

In most biochemical reactions there is little standard free energy difference between reactants and products. Such reactions are reversible, a fact that is indicated by drawing double arrows ($\rightleftharpoons$) between the reactants and the products.

At the beginning of a reaction, only the reactant molecules may be present. These molecules move about and collide with one another with sufficient energy and in the right orientation to react. Thus the concentration of the reactant molecules decreases while the concentration of the product molecules increases. The product molecules collide more frequently as their concentration increases, and some have sufficient energy to initiate the reverse reaction. The reaction thus proceeds in both directions simultaneously; if undisturbed it could eventually reach a state of *dynamic equilibrium*, in which the rate of the reverse reaction is equal to the rate of the forward reaction. At equilibrium there is no net change in the system; every forward reaction is balanced by a reverse reaction.

Keep in mind that the knowledge that a system is at equilibrium tells us nothing about the relative concentrations of reactants and products. If the reactants have much greater intrinsic free energy than the products, the reaction goes almost to completion; that is, it reaches equilibrium at a point at which most of the reactants have been converted to products. This fact is indicated by drawing the arrows unequally, with arrow pointing to the products (A $\rightleftharpoons$ B). Reactions in which the reactants have much less intrinsic free energy than the products reach equilibrium at a point where very few of the reactant molecules have been converted to products (A $\rightleftharpoons$ B). If the arrows are equal (A $\rightleftharpoons$ B), we assume that there is very little standard free energy difference between the reactants and the products and that their concentrations are therefore about equal at equilibrium.

When a reaction is equilibrium, even if the standard free energy difference between the reactants and products is great, the actual free energy difference is zero. Therefore, a system at equilibrium can do no work. The system will stay at equilibrium unless an external stress is applied to it , at which time the system reacts to partially alleviate that stress. According to this rule, know as **Le Chatelier's principle**, a change that affects the reacting system may cause the equilibrium to shift temporarily until a new equilibrium is established.

When little intrinsic free energy difference exists between the reactants and the products (A $\rightleftharpoons$ B), the overall direction of the reaction is determined mainly by their relative initial concentrations. This concept, which is actually part of Le Chatelier's principle, is sometimes referred to as the **Law of Mass Action**.

If we increase the initial concentration of A, then the equilibrium will "shift to the right" and more A will be converted to B. A similar effect can be obtained if B is removed from the reaction mixture. An opposite effect (a "shift to the left") occurs if the concentration of B is increased, or if A is removed. The actual free energy change that occurs during a reaction is defined mathematically to include these mass action effects, which are a consequence of the relative initial concentrations of reactions and products.

A cell maintains its reactions far from equilibrium

Mass action effects are of profound importance in biology. The cell uses its ordered components and energy derived from the surroundings to manipulate the relative concentrations of reactants and products of almost every reaction. Cellular reactions are virtually never at equilibrium. *By displacing its reactions far from equilibrium, a cell is able to supply energy to endergonic reactions and direct its metabolism in accordance with its needs.*

Cells drive endergonic reactions by coupling them to exergonic reactions

Many metabolic reactions in a cell—protein synthesis, for

example—are anabolic and endergonic. Because an endergonic reaction cannot take place without an input of energy, endergonic cellular reactions are coupled to exergonic cellular reactions. In energy coupling, the thermodynamically favourable exergonic reaction provides the energy required to drive the thermodynamically unfavourable endergoinc reaction. The endergonic reaction can proceed only if it absorbs less free energy than is released by the exergonic reaction to which it is coupled.

Mass action effects play important roles in energy coupling

Consider the standard free energy changes, ΔG, in the following reactions:

(1) $A \longrightarrow B \quad \Delta G = +16.7$ kJ/mole (+4 Kcal/mole)

(2) $B \longrightarrow C \quad \Delta G = -41.8$ kJ/mole (– 10 Kcal/mole)

Reaction 1, with a positive value of ΔG is endergonic. Its product, B, is the reactant in Reaction 2. Reaction 2 has a negative value of ΔG and is exergonic.

In analyzing free energy changes we consider the initial and final states of reactions 1 and 2 we obtain:

(1) $A \longrightarrow B \quad \Delta G = +16.7$ kJ/mole (+ 4 Kcal/mole)

(2) $B \longrightarrow C \quad \Delta G = -41.8$ kJ/mole (– 10 Kcal/mole)

$A \longrightarrow C \quad \Delta G = -25.1$ kJ/mole (– 6 Kcal/mole)

The pathway for the conversion of Reactant A to the final product, C,is exergonic overall. Keep in mind that the relative initial concentrations of reactants and products can have a significant effect on the actual free energy change. Reaction 2, which is highly exergonic, can, under the right conditions, dramatically lower the concentration of Product B of Reaction 1, thus shifting that reaction to the right. This mass action effect can be sufficient to make the actual free energy change of Reaction 1 negative ($-\Delta G$), such that it becomes thermodynamically favourable.

Reactions can be coupled through an "energized intermediate"

Consider the standard free energy change $-\Delta G$, in the following reaction:

(3) $A \longrightarrow B \quad \Delta G = +20.9$ kJ/mole (+ 5 Kcal/mole)

Because ΔG has a positive value, we know that the product of this reaction has more free energy than the reactant. This is an endergonic reaction. It is not spontaneous and does not take place without an energy source. By contrast, consider the following reaction:

(4) $C \longrightarrow D \quad \Delta G = -33.5$ kJ/mole (– 8 Kcal/mole)

The negative value of ΔG tells us that the standard free energy of the reactant is greater than the free energy of the product. This exergonic reaction can proceed spontaneously.

We can sum up Reactions 3 and 4 as follows:

(3) $A \longrightarrow B \quad \Delta G = +20.9$ kJ/mole (+ 5 Kcal/mole)

(4) $C \longrightarrow D \quad \Delta G = -33.5$ kJ/mole (– 8 Kcal/mole)

Overall $\quad \Delta G = -12.6$ kJ/mole (–3 Kcal/mole)

Because thermodynamics considers the overall changes in these two reactions, which show a net negative value of ΔG, two reactions taken together are exergonic.

The fact that we can write reactions this way is a useful bookkeeping device, but it does not mean that an exergonic reaction can mysteriously transfer energy to an endergonic "bystander" reaction. However, these reactions can be coupled if their pathways are altered such that they are linked by a common intermediate. It is important to remember that the free energy change in a reaction depends only on the difference in free energy between the reactant(s) and the products(s). For this reason, changing the reaction pathways does not alter the overall change in free energy. Reactions 3 and 4 might be coupled in the following way:

(5) $A - C \longrightarrow I \quad \Delta G = -8.4$ kJ/mole (– 2 Kcal/mole)

(6) $I \longrightarrow B + D \quad \Delta G = -4.2$ kJ/mole (– 1 Kcal/mole)

Overall $\quad \Delta G = -12.6$ kJ/mole (– 3 Kcal/mole)

"I" refers to a transitory common intermediate. It can be called an "energized" intermediate because it is unstable and can enter into reactions that would be impossible for Reactant A. Note that Reactions 5 and 6 are sequential. Thus the reaction pathways have changed, but overall the reactants and products are the same and the overall free energy change is the same.

Generally, for each endergonic reaction occurring in a living cell, there is a coupled exergonic reaction to derive it. Often, the exergonic chemical reaction involves the breakdown of adenosine triphosphate (ATP) . We will examine specific example of the role of ATP in energy coupling in the following section.

An Overview of Oxidative Respiration

Cells are able to make ATP in two quite different ways. In the first, called **substrate-level phosphorylation**, ATP is formed by directly

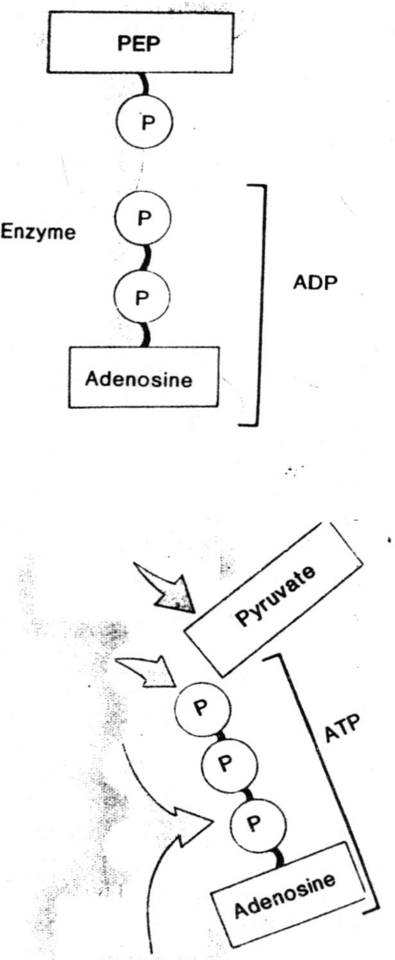

Fig. 7.9 : Substrate-level phosphorylation.

transferring a phosphate group to ADP from a phosphate-bearing intermediate. During glycolysis, the chemical bonds of glucose are shifted around in reactions that provide the energy required to form ATP. While this process is inefficient, it is the only way in which many cells derive their ATP. In the second, electrons are harvested and transferred along the election transport chain. Eukaryotes produce the majority of their ATP from glucose in this way. In most organisms, these two processes are combined in oxidative respiration, which is carried out as a complex series of enzyme-catalyzed reactions that can be broken down into four stages.

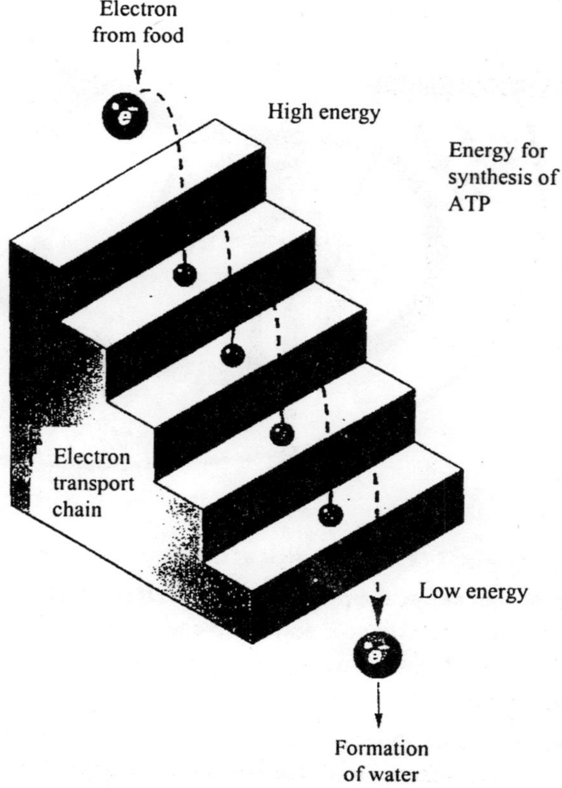

Fig.7.10 : **How electron transport works.** *This schematic diagram shows how ATP is generated with the subsequent transfer of electrons from one energy level to another. Rather than a single explosive burst of energy, electrons "fall" to lower and lower energy levels in steps, releasing stored energy as they tumble to the lowest (most electronegative) electron acceptor.*

The *first* stage of extracting energy from glucose is a 10-reaction biochemical pathway called glycolysis . The enzymes that catalyze the glycolytic reactions are present in the cytoplasm of the cell, not bound to any membrane or organelle . In glycolysis, two ATP molecules are used up in preparing the glucose, and four ATP molecules are formed by substrate-level phosphorylation, for a net yield of two ATP molecules for each molecules of glucose catabolized. In addition, four electrons are harvested as NADH that can be used to form ATP by oxidative respiration. Still, the total yield of ATP is small. When the glycolytic process is completed, two molecules of pyruvate that are formed still contain most of the energy that was present in the original glucose molecule.

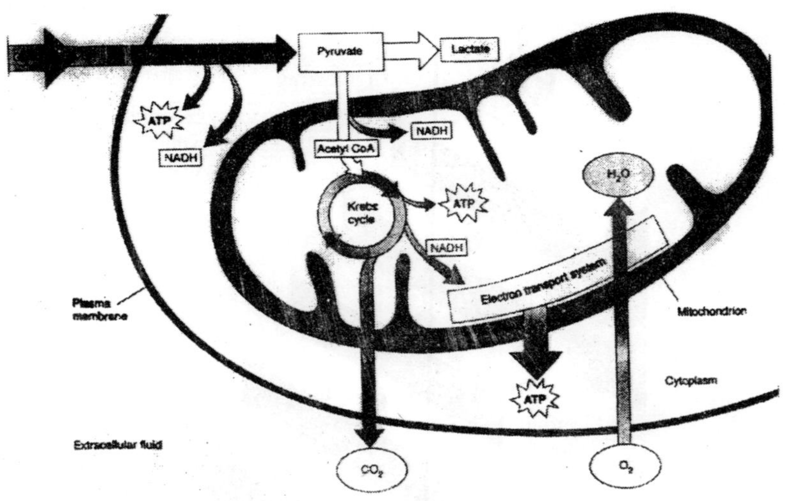

Fig. 7.11 : An overview of cellular respiration.

In the *second* stage, the pyruvate that is left over from glycolysis is converted into carbon dioxide and a two-carbon molecule, acetyl-CoA. For each molecule of pyruvate that is converted, one molecule of NADH is synthesized.

The *third* stage introduces acetyl-CoA into a cycle of nine reactions, the **Krebs cycle** , named after the British biochemist, Sir Hans Krebs, who discovered it. In the Krebs cycle, two more ATP molecules are extracted by substrate-level phosphorylation, and a large number of electrons are removed as NADH.

In the *fourth* stage, the energetic electrons carried by NADH are employed to drive the synthesis of ATP by the electron transport chain. It is this process that leads to the greatest formation of ATP.

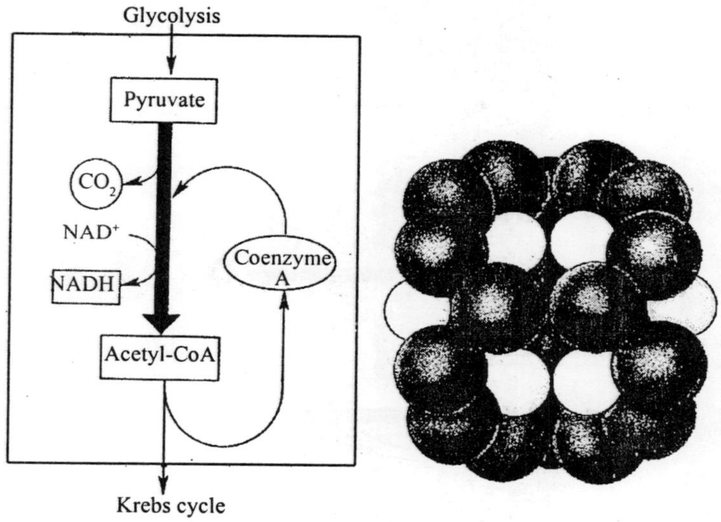

Fig. 7.12 : Pyruvate dehydrogenase and the oxidation of pyruvate.

Glycolysis occurs in nearly all organisms, and the glycolytic reactions that form ATP by substrate-level phosphorylation can occur whether or not oxygen is present. In animals, however, the harvesting of energetic electrons for the electron transport chain cannot take place indefinitely in the absence of oxygen because oxygen serves as the final acceptor of the electrons harvested from glucose. Without oxygen, animal cells are restricted to substrate-level phosphorylation to obtain ATP. Some organisms respire using different electron acceptors. For example, many bacteria use sulfur, nitrate, or other inorganic compounds as the electron acceptor in place of oxygen.

Pyruvate oxidation, the reactions of the Krebs cycle, and ATP production by electron transport chain occur inside the mitochondria of all eukaryotes, as well as within many forms of bacteria. Because plants and algae can produce ATP by photosynthesis, it is sometimes easy to lose sight of the fact that they also produce ATP by oxidative respiration, just as animals and other nonphotosynthetic eukaryotes do.

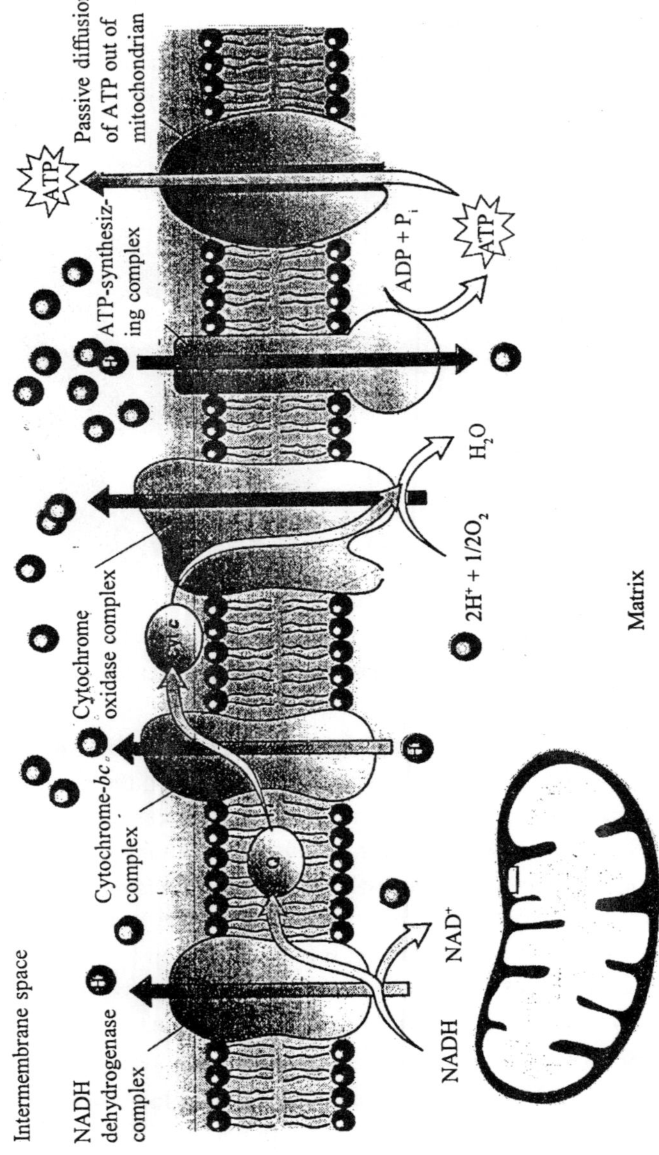

Fig.7.13 : The electron transport chain: how mitochondria use electrons to make ATP

Stage one : Glycolysis

The metabolism of primitive organisms focused on glucose. Glucose molecules can be dismantled in many ways, but primitive organisms evolved the ability to do it by a process that releases enough free energy to drive the synthesis of ATP in coupled reactions. This process, called glycolysis, involves a sequence of ten reactions that convert glucose into two three-carbon molecules of pyruvate. For each molecule of glucose that passes through this transformation, the cell nets two ATP molecules by substrate-level phosphorylation.

An Overview of Glycolysis

The first half of glycolysis is composed of five sequential reactions in which one molecule of glucose is converted into two molecules of the three-carbon compound, glyceraldehyde-3-phosphate (G3P). These reactions involve the expenditure of ATP, so they are an energy-requiring process. In the second half of glycolysis, five more reactions convert G3P into pyruvate in an energy-yielding process that generates ATP. Overall, then, glycoysis is a series of ten enzyme-catalyzed reactions in which some ATP is invested in order to produce more. The ten reactions of glycolysis proceed in four steps.

Step A, glucose priming . Three reactions change glucose into a compound that can be cleaved readily into three carbon phosphorylated molecules. Two of these reactions requires the cleavage of ATP, So this step require the cell to use two ATP molecules.

Step B, cleavage and rearrangement. The six-carbon product of Step A is split into two three-carbon molecules. One is G3P, and the other is converted to G3P by another reaction.

Step C, oxidation. Two electrons and one proton are transferred from G3P to NAD^+, forming NADH. Note that NAD^+ is an ion, and that *both* electrons in the new covalent bond come from G3P.

Step D, ATP generation. Four reactions convert G3P into another three-carbon molecule, pyruvate, and in the process generate two ATP molecules.

Because each glucose molecule is split into *two* G3P molecules, the overall reaction sequence yields two molecules of ATP, as well as two molecules of NADH and two of pyruvate:

4 ATP (2 ATP for each of the 2 G3P molecules in step D)
– 2 ATP (two reactions in step A require ATP)

---

2 ATP

Since under the nonstandard conditions that occur within a cell, each ATP molecule that is produced in a cell represents the capture of about 12 Kcal of energy per mole of glucose, rather than the 7.3 traditionally quoted for standard conditions, glycolysis harvests about 24 Kcal/mole. This is not a great deal of energy. The total energy content of the chemical bonds of glucose is 686 Kcal per mole, so glycolysis harvests only 3.5 % of the chemical energy of glucose.

Although far from ideal in terms of the amount of energy it releases, glycolysis does generate ATP, and for more than a billion years during the anaerobic first stages of the history of life on earth, it was the primary way for heterotrophic organisms to generate ATP from organic molecules. Like many biochemical pathways, glycolysis is believed to have evolved backward, with the last steps in the process being the most ancient. Thus, the second half of glycolysis, the ATP -yielding breakdown of G3P , may have been the original process used by early heterotrophs to generate ATP. The synthesis of G3P from glucose would have appeared later, perhaps when alternative sources of G3P were depleted.

All Cells Use Glycolysis

The glycolytic reaction sequence is thought to have been among the earliest of all biochemical processes to evolve. It uses no molecular oxygen and occurs readily in an anaerobic environment. All of its reactions occur free in the cytoplasm; none is associated with any organelle or membrane structure. Except for a few bacteria, every living creature is capable of carrying out glycolysis. Most present-day organisms, however, are able to extract considerably more energy from glucose through oxidative respiration.

Why is glycolysis maintained even now, considering that its energy yield in the absence of oxygen is comparatively so paltry? The answer to this question is that evolution is an incremental process. Change occurs during evolution by improving on past successes. In catabolic metabolism, glycolysis satisfied the one essential evolutionary criterion: it was an improvement. Cells that could not carry out glycolysis were at a competitive disadvantage, and only those cells that were capable of glycolysis survived the early competition of life. Later improvements in catabolic metabolism built on this success. Glycolysis was not discarded during the course of evolution; rather, it was used as the starting point for the further extraction of chemical energy. Metabolism evolved as one layer of reactions was added to another, just as

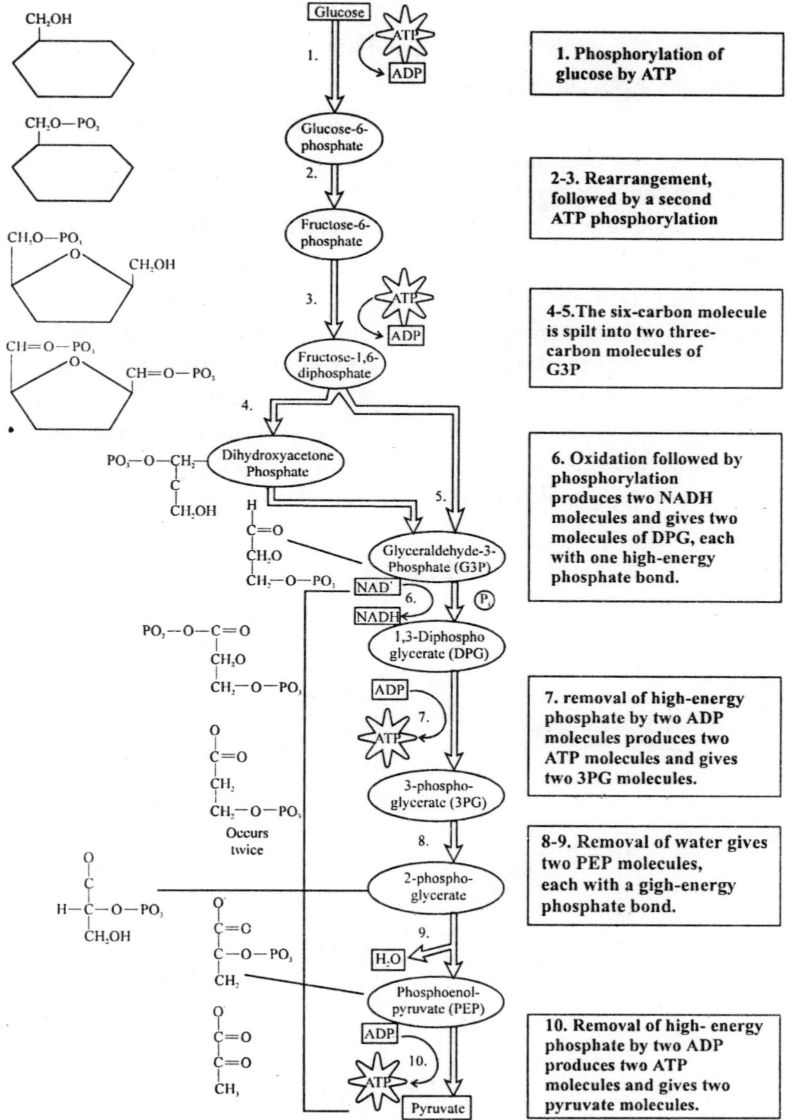

Fig. 7.14 : The glycolytic Pathway.

successive layers of paint can be found on the walls of an old building. Nearly every present-day organism carries out glycolysis as a metabolic memory of its evolutionary past.

Closing Metabolic Circle: The Regeneration of NAD^+

Inspect for a moment the net reaction of the glycolytic sequence:

$$\text{Glucose} + 2\ \text{ADP} + 2\text{P}_i + 2\ \text{NAD}^+ \longrightarrow 2\ \text{Pyruvate} + 2\ \text{ATP} + 2\ \text{NADH} + 2\ \text{H}^+ + 2\ \text{H}_2\text{O}$$

You can see that three changes occur in glycolysis: (1) glucose is converted into two molecules of pyruvate; (2) two molecules of ADP are converted into ATP; and (3) two molecules of NAD^+ are converted into NADH.

As long as foodstuffs are available that can be converted into glucose, a cell can continually churn out ATP to drive its activities. In doing so, however, it accumulates NADH and depletes the pool of NAD^- molecules. A cell does not contain a large amount of NAD^-, and for glycolysis to continue, NADH must be recycled into NAD^+. Some other molecule must accept the hydrogen atom taken from G3P and be reduced. In general, there are two ways this can happen.

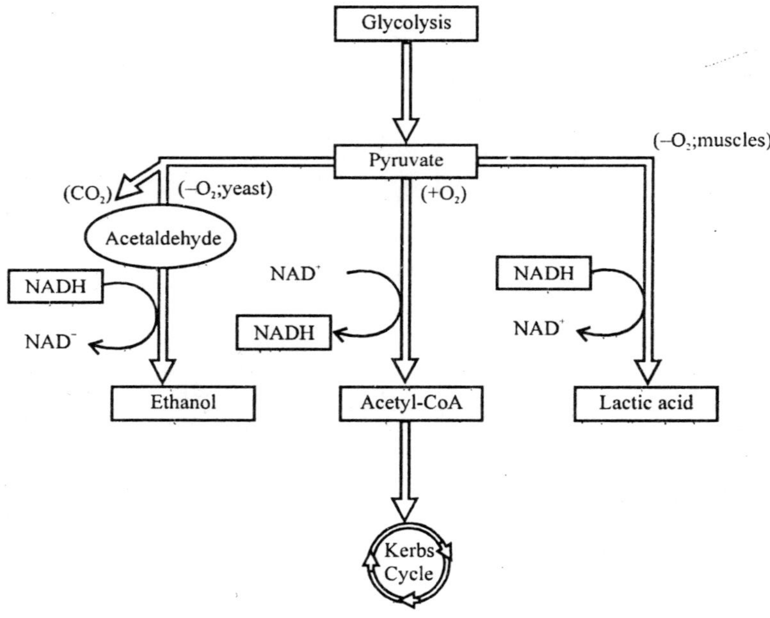

Fig. 7.15: What happens to pyruvate, the product of glycolysis?

1. *Oxidative Respiration*. Oxygen is an excellent electron acceptor, and through a series of electron transfers, the hydrogen atom taken from

G3P can be donated to oxygen, forming water. This is what happens in the cells of eukaryotes in the presence of oxygen. Because air is rich in oxygen, this process is also referred to as aerobic metabolism.

2. *Fermentation*. When oxygen is not available, another organic molecule can accept the hydrogen atom instead. Fermentation, which is also called anaerobic metabolism, plays an important role in the metabolism of most organisms, even those capable of oxidative respiration.

Stage Two: The Oxidation of Pyruvate

In all aerobic organisms, the oxidation of glucose that begins in glycolysis is continued where glycolysis leaves off - with pyruvate. As

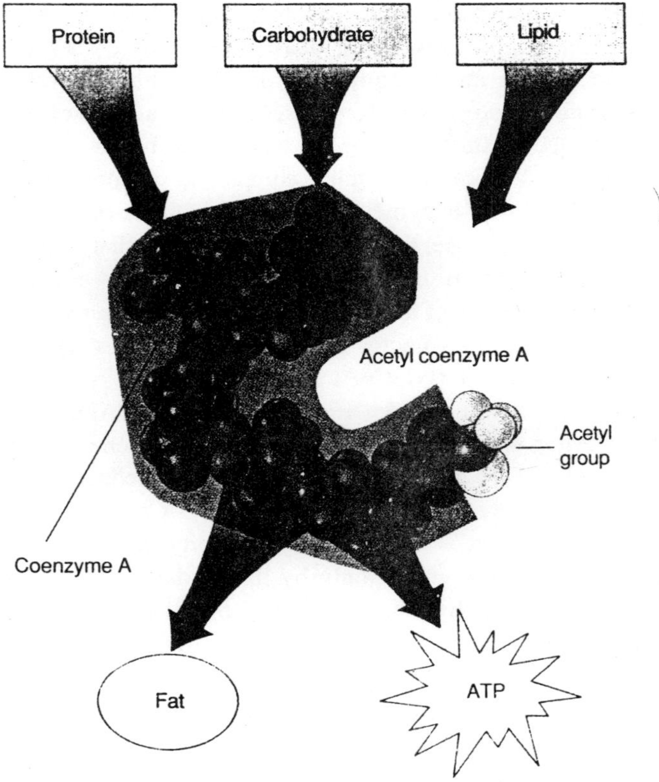

Fig. 7.16 : Acetyl–CoA is the central molecule of energy metabolism.

is almost always the case in evolution, the development of this new biochemical process was conservative; it was simply added to the old one. In eukaryotic organisms, the further extraction of energy from pyruvate takes place exclusively inside mitochondria, which probably originated as symbiotic bacteria with ability to catabolize pyruvate. The continued harvesting of pyruvate's considerable energy takes place in two steps: the oxidation of pyruvate to form acetyl–CoA, and the subsequent oxidation of acetyl–CoA in the Krebs cycle. Let us begin with the first step.

The oxidation of pyruvate is accomplished in a single "decarboxylation" reaction, in which one of the three carbons of pyruvate is cleaved off, departing as CO_2. This reaction produces a two-carbon fragment called an actyl group, as well as a pair of electrons and their associated hydrogen, which reduce NAD^+ to NADH. The reaction is complex, involving three intermediate stages, and is catalyzed within mitochondria by an assembly of enzymes, a **multienzyme complex**. Such complexes organize a series of enzymatic steps so that the chemical intermediates do not diffuse away or undergo other reactions. Within the complex, component polypeptides pass the substrates from one enzyme to the next, without releasing them. Pyruvate dehydrogenase, the complex of enzymes that removes CO_2 from pyruvate, is one of the largest enzymes known — it contains 72 subunits! In the course of the reaction, the acetyl group removed from pyruvate is added to a cofactor called coenzyme A (CoA), forming a compound known as acetyl–CoA:

$$\text{Pyruvate} + NAD^+ + \text{CoA} \longrightarrow \text{Acetyl–CoA} + \text{NADH} + CO_2$$

This reaction produces a molecule of NADH, which is later used to produce ATP. Of far greater significance than the reduction of NAD^+ to NADH, however, is the production of acetyl–CoA. Acetyl–CoA is important because it is generated by so many different metabolic processes. It is produced not only by the oxidation of pyruvate, an intermediate in carbohydrate catabolism, but also by the metabolic breakdown of proteins, fats, and other lipids. Thus, it provides a focus for the many catabolic processes of the eukaryotic cell, whose resources are channeled into this single metabolite.

Although acetyl–CoA is formed by many processes in the cell, there are only a limited number of processes that use acetyl–CoA. Most of it is either directed towards energy storage (lipid synthesis, for example) or oxidized to produce ATP. Which of these two options occurs depends on the level of ATP in the cell. When ATP levels are high, the oxidative pathway is inhibited and acetyl–CoA is channeled into fatty acid biosynthesis. This explains why many animals (humans included)develop fat reserves when they consume more food than they require for energy. Alternatively, when ATP levels are low, the oxidative pathway is stimulated and acetyl-CoA flows into energy-producing oxidative metabolism.

Stage Three: The Kerbs Cycle

The third stage of extracting energy from glucose is the oxidation of the acetyl–CoA produced from pyruvate in series of nine reactions called the Krebs cycle. In this cycle, the acetyl group of acetyl–CoA is combined with a four-carbon molecule called oxaloacetate. The resulting six-carbon molecule then goes through a sequence of

The Reactions of the Krebs Cycle

The Krebs cycle consists of nine sequential reactions that cells use to extract energetic electrons and drive the synthesis of ATP. A two carbon molecule, acetyl–CoA, enters the cycle at the beginning, and two CO_2 molecules and several electrons are given off during the cycle.

Reaction 1, Condensation. Acetyl–CoA is joined to a four-carbon molecule, oxaloacetate, to form a six-carbon molecule, citrate. This condensation reaction is irreversible, committing acetyl–CoA to the Krebs cycle. The reaction is inhibited when the cell's ATP concentration is high an stimulated when it is low. Hence, when the cell possesses ample amounts of ATP, the Krebs cycle is shut off and acetyl-CoA is channeled instead into fat synthesis.

Reactions 2 and 3, Isomerization. Before the oxidation reaction can begin, the hydroxyl (–OH) group of citrate must be repositioned. This is done in two steps: first, a water molecule is removed from one carbon; then, water is added to a different carbon. The result is that an –H group and an –OH group change positions. The product is an isomer of citrate called isocitrate.

Reaction 4, The First Oxidation. In the first energy-yielding step of the cycle, isocitrate undergoes and oxidative decarboxylation reaction, like the one involved in the conversion of pyruvate to acetyl–CoA. First, isocitrate is oxidized, yielding a pair of electrons that reduce a molecule of NAD^+ to NADH. Then the oxidize intermediate is decarboxylated; the central carbon atom is split off to form CO_2 yielding a five-carbon molecule called α -ketoglutarate.

Reaction 5, The Second Oxidation. α -ketoglutarate is then decarboxylated by a multienzyme complex similar to pyruvate dehydrogenase. The succinyl group left after the removal of CO_2 is joined to a molecule of coenzyme A, forming succinyl–CoA. In the process, two electrons are extracted, which reduce a molecule of NAD^+ to NADH.

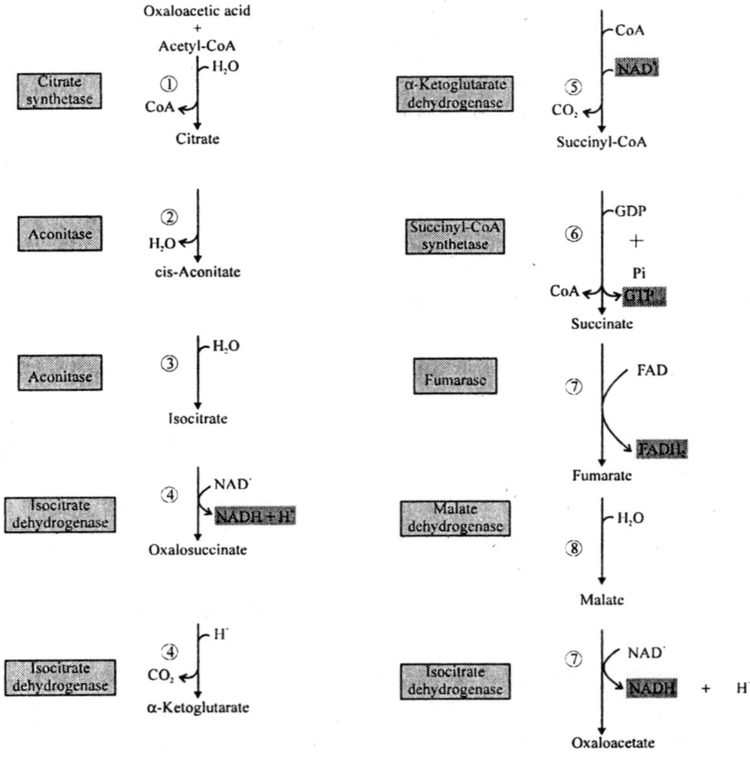

Fig. 7.17 : The reactions of the Kerbs cycle.

Reaction 6, Substrate-Level Phosphorylation. The linkage between the four carbon succinyl group an CoA is a high energy bond, like the phosphate linkages in ATP. In a coupled reaction similar to those that take place in glycolysis, this bond is cleaved, and the energy that is released drives the phosphorylation of guanosine diphosphate, forming guanosine triphosphate (GTP). The GTP that results is readily converted into ATP. The four-carbon fragment that remains at the completion of this reaction is called succinate.

Reaction 7, The Third Oxidation. Succinate is oxidized to fumarate. The free energy change in this reaction is not large enough to reduce NAD^+. Instead, **flavin adenine dinucleotid (FAD^+)**, is used as the electron acceptor. Unlike NAD^+, FAD^+ is not free to diffuse within the mitochondrion; it is an integral part of the inner mitochondrial membrane. Its reduced form, $FADH_2$, contributes electrons directly to the electron transport chain in the membrane.

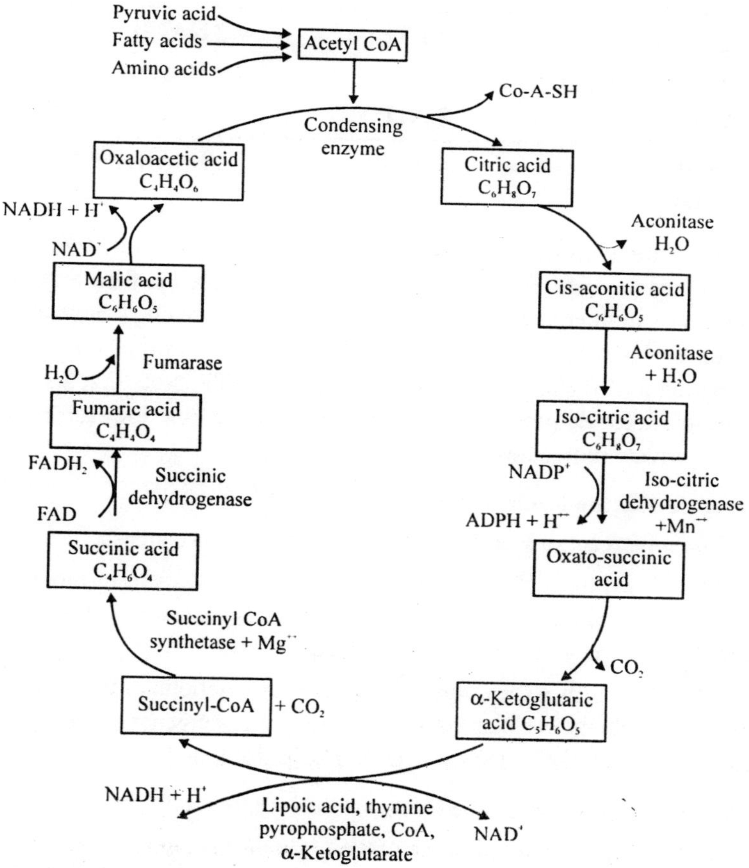

Fig. 7.18 : Kerbs cycle or tricarboxylic acid cycle.

Reaction 8 and 9, Regeneration of Oxaloacetate. In the final two reactions of the cycle, a water molecule is added to fumarate, and the resulting malate is oxidized. The oxidation yields a four-carbon molecule of oxaloacetate and two electrons that reduce a molecule of NAD^+ to NADH. Oxaloacetate, the molecule with which the cycle began, is now free to combine with another molecule of acetyl-CoA and reinitiate the cycle.

Table 7.1 : The Output of Aerobic Metabolism

	Substrate-Level phosphorylation	Oxidation
Glycolysis	2 ATP	2 NADH
Oxidation of pyruvate	—	2 NADH
Krebs Cycle •	2 ATP	6 NADH + 2 $FADH_2$
Total	**4 ATP**	**10 NADH + 2 $FADH_2$**

Electron-yielding oxidation reactions, during which two CO_2 molecules are split off, restoring oxaloacetate, which is then recycled to bind to another acetyl group. In each turn of the cycle, a new acetyl group replaces the two CO_2 molecules lost, and more electrons are extracted to drive proton pumps that generate ATP. Note that a single glucose molecule produces two turns of the cycle, one for each of the two pyruvate molecules generated by glycolysis.

Overview of the Krebs Cycle

The nine reactions of the Krebs cycle occur in two steps:

Step A, priming. Three reactions prepare the six-carbon molecule for energy extraction. In the first reaction, acetyl–CoA joins the cycle, and then chemical groups are rearranged.

Step B , energy extraction. Four of the six reactions in this step are oxidations in which electrons are removed, and one generates and ATP equivalent directly by substrate level phosphorylation.

The Products of the Krebs Cycle

In the process of aerobic respiration, glucose is entirely consumed. As we have already described, the six-carbon glucose molecule is first cleaved into a pair of three-carbon pyruvate molecules during glycolysis. One of the carbons of each pyruvate is then lost as CO_2 in the conversion of pyruvate to acetyl–CoA, and the other two carbons are lost as CO_2 during the oxidations of the Krebs cycle. All that is left to mark the passing of the glucose molecule into six CO_2 molecules is its energy, some of which is preserved in four ATP molecules and in the reduced state of 12 electron carriers. Ten of these carriers are NADH molecules; the other two are molecules of $FADH_2$.

Stage Four: The Electron Transport Chain

The NADH and $FADH_2$ molecules formed during the first three stages of oxidative respiration each contain a pair of electrons that were

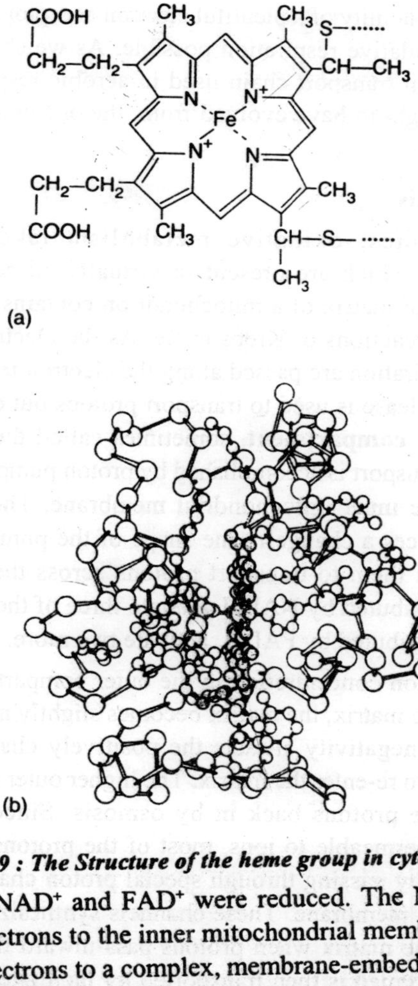

COOH CH₃ CH₃ S—......

CH_2—CH_2 —CH—CH_3

N N⁺ Fe N⁺ N

CH_2—CH_2 —CH_3

COOH CH_3 CH—S— ......

CH_3

(a)

(b)

Fig.7.19 : The Structure of the heme group in cytochrome c.

gained when NAD^+ and FAD^+ were reduced. The NADH molecules carry their electrons to the inner mitochondrial membrane, where they transfer the electrons to a complex, membrane-embedded protein called **NADH dehydrogenase**. The electrons are then passed sequentially along a series of **cytochromes** and other carrier molecules, losing much of their energy in the process by driving several transmembrane proton pumps. This series of membrane-associated electron carriers is collectively called the electron transport chain. $FADH_2$, which is always attached to the inner mitochondrial membrane, also feeds its electrons into the electron transport chain. At the terminal step of the chain, the electrons are passed to the **cytochrome c oxidase complex**, which uses four of the electrons to reduce a molecule of oxygen and form water.

$$O_2 + 4H^+ + 4e^- \longrightarrow 2\ H2O$$

It is the availability of a plentiful electron acceptor (namely, oxygen) that makes oxidative respiration possible. As we shall see in chapter 10, the electron transport chain used in aerobic respiration is similar to, and is thought to have evolved from, the one employed in aerobic photosynthesis.

Chemiosmosis

In eukaryotes, oxidative metabolism takes place within mitochondria, which are present in virtually all cells. The internal compartment or matrix of a mitochondrion contains the enzymes that carry out the reactions of Krebs cycle. As the electrons harvested by oxidative respiration are passed along the electron transport chain, the energy they release is used to transport protons out of the **matrix** and into the **outer compartment**, sometimes called the inter-membrane space. This transport is accomplished by proton pumps, transmembrane proteins in the inner mitochondrial membrane. The flow of excited electrons induces a change in the shape of the pump proteins, which in turn causes them to transport protons across the membrane. The electrons contributed by NADH activate three of these proton pumps, and those contributed by $FADH_2$ activate two more.

As the proton concentration in the outer compartment rises higher than that in the matrix, the matrix becomes slightly negative in charge. This internal negativity attracts the positively charged protons and induces them to re-enter the matrix. The higher outer concentration also tends to drive protons back in by osmosis. Since membranes are relatively impermeable to ions, most of the protons that re-enter the matrix do so by passing through special proton channels in the inner mitochondrial membrane. These channels synthesize ATP from ADP + P, within the matrix when protons pass inward through them. The ATP that is formed is then transported by facilitated diffusion out of the mitochondrion and into the cell's cytoplasm. Because the chemical formation of ATP is driven by diffusion force similar to osmosis, this process is referred to as **chemiosmosis**.

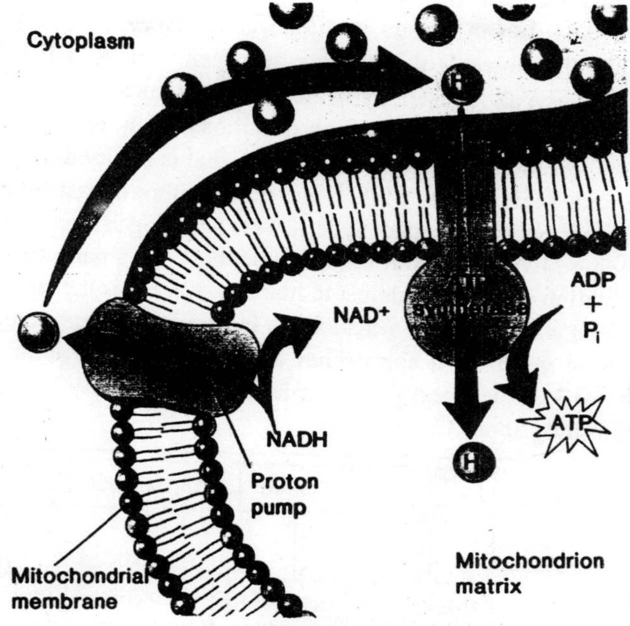

Fig. 7.20 : Chemiosmosis

Thus, the electrons harvested in oxidative respiration are used to pump a large number of protons across the inner mitochondrial membrane. Their re-entry into the mitochondrial matrix drives the synthesis of ATP.

Living without Oxygen: Fermentation

In the absence of oxygen, oxidative respiration cannot occur, and cells, must rely exclusively on glycolysis to produce ATP. Under these conditions, the hydrogen atoms generated by glycolysis are donated to organic molecules in a process called **fermentation**.

Bacteria carry out more than a dozen kinds of fermentation, all of which use some form of organic molecule to accept the hydrogen atom from NADH and thus reform NAD^+:

$$\begin{array}{ccc} \text{Organic molecule} & \longrightarrow & \text{Reduced organic molecule} \\ + \text{NADH} & & + \text{NAD}^+ \end{array}$$

Often the resulting reduced compound is an organic acid, such as acetic acid, butyric acid, propionic acid, or lactic acid.

Eukaryotic cells are capable of only a few types of fermentations. In one type, which occurs in single-celled fungi called yeast, for example, the molecule that accepts hydrogen from NADH is the end product of glycolysis itself, pyruvate. Yeast enzymes remove a terminal CO_2 group from pyruvate through decarboxylation, producing a two-carbon molecule, acetaldehyde. The CO_2 that is released causes bread made with yeast to rise, while bread made without yeast (unleavened bread) does not. The acetaldehyde accepts a hydrogen atom from NAD, producing NAD^- and **ethanol** (ethyl alcohol). This particular type of fermentation is of great interest to humans, since it is the source of the ethanol in wine and beer. Ethanol is a by-product of fermentation that is toxic to yeast; as it approaches a concentration of about 12%, it begins to kill the yeast. That explains why naturally brewed wine contains only about 12% ethanol.

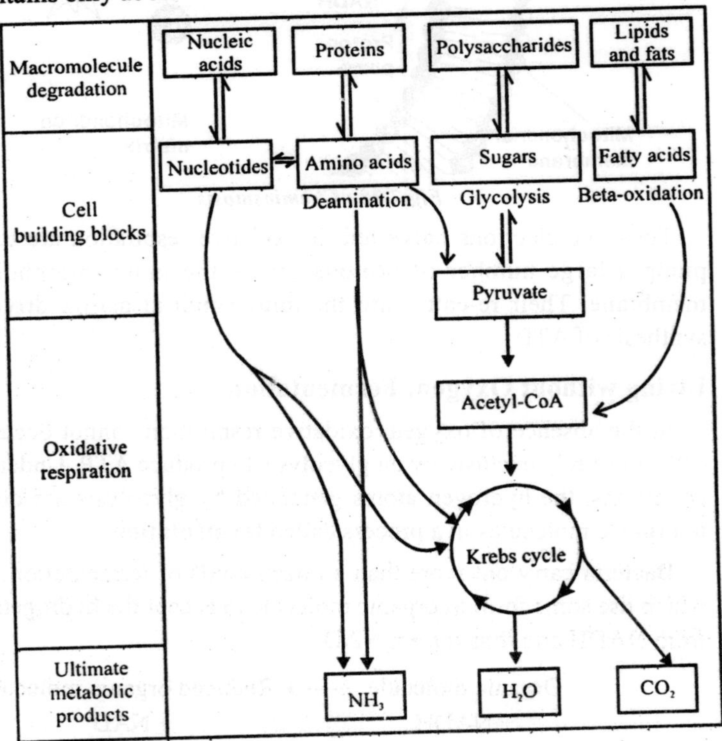

Fig.: How cells extract chemical energy.

Most animal cells regenerate NAD$^-$ without decarboxylation. Muscle cells, for example, use an enzyme called lactate dehydrogenase to transfer a hydrogen atom from NADH back to the pyruvate that is produced by glycolysis.

This reaction converts pyruvate into lactate and regenerates NAD$^+$ from NADH. It therefore closes the metabolic circle, allowing glycolysis to continue as long as glucose is available. Circulating blood removes excess lactate from muscles, but when lactate removal cannot keep pace with its production, the accumulating lactate interferes with muscle function and contributes to muscle fatigue.

CHAPTER 8
CARBOHYDRATES

MONOSACCHARIDES

Structure of Glucose

Since glucose occupies an exceptional position in carbohydrate metabolism , it is described in some detail. Much of this discussion applies to other sugars as well.

A knowledge of the molecular formula, $C_6H_{12}O_6$, for glucose and of many of its chemical properties permits us to assign, as a first approximation, the following structural formula, in which the C atoms are numbered in the traditional fashion:

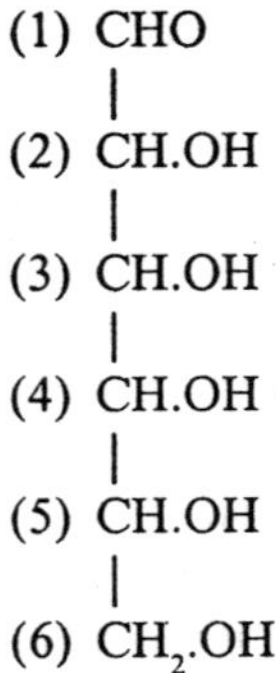

Isomers of Glucose

An examination of its formula reveals that glucose has four asymmetric carbon atoms (at position 2,3,4,and 5), each carbon being attached to four different groups. For example, the carbon at position 2 may be shown thus:

$$\begin{array}{c} R \\ | \\ H - C - OH \\ | \\ R' \end{array}$$

where R stands for the CHO group and R′ for everything below carbon 2. According to Van't Hoff, the number of possible isomers is given by the formula $I = 2^n$, where n represents the number of asymmetric carbon atomes. Since glucose has four such asymmetric carbon atoms, the number of isomers should equal 16 (2^4).

The Spatial Arrangement of the Group in Glucose. Proof for the absolute stereochemical configuration of each ismer of glucose is beyond the scope of this book. The isomers are traced back to some simple compounds, the constitution of which is known. We may regard

Configuration of the D-Aldoses

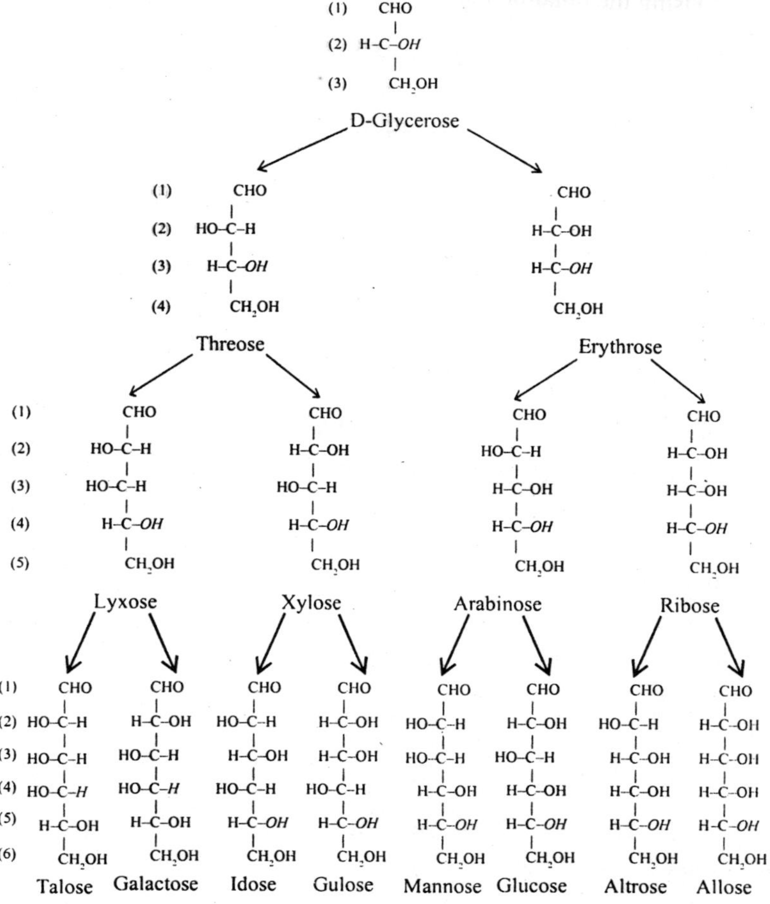

these isomers as being derived from glyceraldehyde:

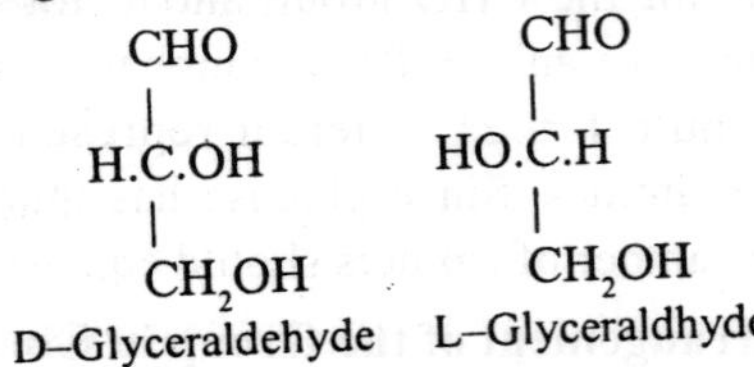

D–Glyceraldehyde L–Glyceraldhyde

From each isomer there may be derived eight isomers of glucose. For example, the accompanying chart shows the derivation and configuration of eight aldhexoses, all members of the D–series, since the OH group at the 5–carbon is derived from the 2–carbon of D–glyceraldehyde:

Using the notation that

$$\text{H}\cdot\overset{|}{\underset{|}{\text{C}}}\cdot\text{OH} = * \text{ and } \text{HO}\cdot\overset{|}{\underset{|}{\text{C}}}\cdot\text{H} = *$$

then D–glucose,

$$\begin{array}{c} \text{CHO} \\ | \\ \text{H}\cdot\text{C}\cdot\text{OH} \\ | \\ \text{HO}\cdot\text{C}\cdot\text{H} \\ | \\ \text{HCOH} \\ | \\ \text{HCOH} \\ | \\ \text{CH}_2\text{OH} \end{array}$$

can be represented in shorthand form as

and L–glicose,

$$\begin{array}{c} \text{CHO} \\ | \\ \text{HO}\cdot\text{C}\cdot\text{H} \\ | \\ \text{H}\cdot\text{C}\cdot\text{OH} \\ | \\ \text{HO}\cdot\text{C}\cdot\text{H} \\ | \\ \text{HO}\cdot\text{C}\cdot\text{H} \\ | \\ \text{CH}_2\text{OH} \end{array}$$

can be represented as

The two end carbons are ignored in this notation. We can represent the sixteen posible isomers as follows:

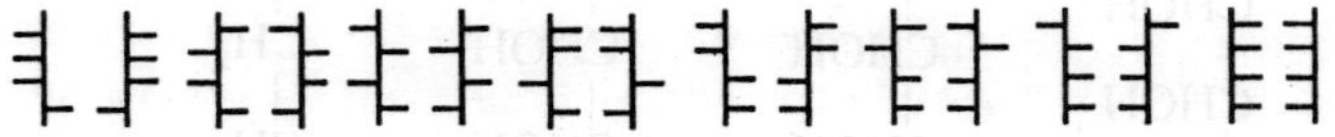

The first of each pair represents the D–series (related to D–glucose and D–glyceraldehyde), and the second, the L–series.

In the D–series, the hydroxy1 group next to the primary alcohol group is written to the right, whereas the reverse is true in the L–series.

The Cyclic Structure for Glucose

The straight chain structure for glucose still does not explain all the facts. To begin with, glucose, which is pictured as aldehyde, falls somewhat short of certain common aldehydic properties. For example, glucose fails to give a Schiff's test (the formation of reddish–violet colour with magenta solution that has been decolourized with SO_2), nor does it form a stable addition compound with sodium bisulfite. Hydroxy acides of the γ – or δ – variety, similar in general structure to glucose, form lactones very readily, and these are cyclic in structure—for example, γ – hydroxybutyric acid ($CH_2OH\ CH_2CH_2COOH$) ischanged to γ –butyrolactone,

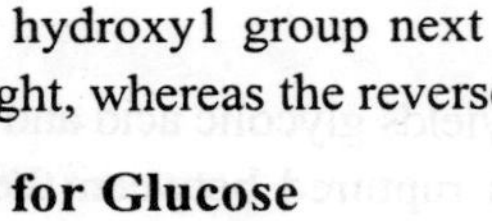

O

$CH_2 \cdot CH_2 \cdot CH_2 \cdot CO$

The Structure of Fructose

Fructose, a naturally occuring sugar, is levorotatory. The sugar is known as D–fructose because it is related structurally to D–glucose. It has the same molecular formula as glucose($C_6H_{12}O_6$), forms a pentaacety1 derivative, and has the properties of a carbony1 compound. When fructose is treated with HCN, it forms an addition compound, which upon hydrolysis and subsequent reduction with HI gives methylbutylactic acid and this means that fructose is a ketone and not an aldehyde. (Under similar conditions, glucose yields a seven–carbon straight–chain, and not a branched–chain, compound.)

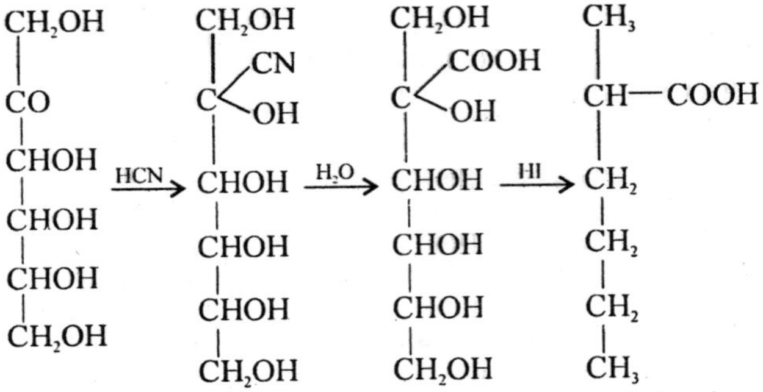

$$\begin{array}{lclclcl}
CH_2OH & & CH_2OH & & CH_2OH & & CH_3 \\
| & & |\quad CN & & |\quad COOH & & | \\
CO & & C\langle\ OH & & C\langle\ OH & & CH-COOH \\
| & & | & & | & & | \\
CHOH & \xrightarrow{HCN} & CHOH & \xrightarrow{H_2O} & CHOH & \xrightarrow{HI} & CH_2 \\
| & & | & & | & & | \\
CHOH & & CHOH & & CHOH & & CH_2 \\
| & & | & & | & & | \\
CHOH & & CHOH & & CHOH & & CH_2 \\
| & & | & & | & & | \\
CH_2OH & & CH_2OH & & CH_2OH & & CH_3 \\
\text{Fructose} & & & & & & \text{Methylbutyl-acetic acid}
\end{array}$$

Oxidation of fructose yields glycolic acid and trihydroxybutyric acid, proving that the bond is ruptured between the 2–carbon and the 3–carbon.

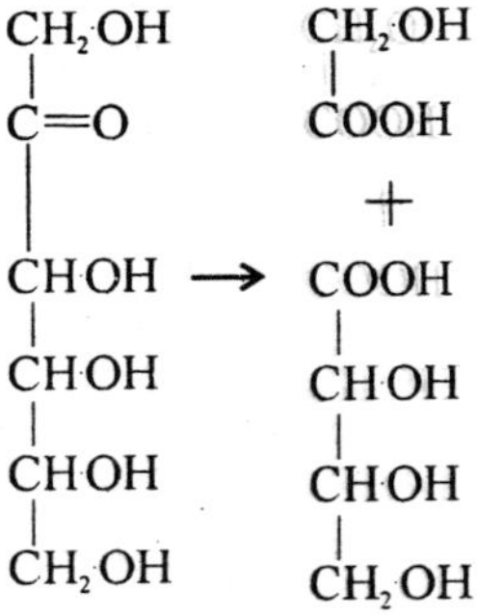

$$\begin{array}{lcll}
CH_2OH & & CH_2OH & \text{Glycollic acid} \\
| & & | & \\
C=O & & COOH & \\
| & & + & \\
CHOH & \longrightarrow & COOH & \\
| & & | & \\
CHOH & & CHOH & \text{Trihydroxybutyric acid} \\
| & & | & \\
CHOH & & CHOH & \\
| & & | & \\
CH_2OH & & CH_2OH &
\end{array}$$

Fructose has three asymmetric C atoms (3,4,and5), which can be proved to have the same configuration as the corresponding atoms in glucose since both compounds yeil˙ identical osazones on heating with solutions of phenylhydrazine:

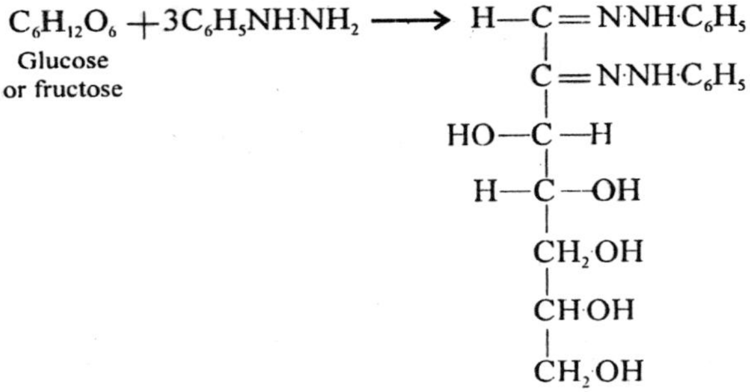

$$\begin{array}{lcc}
C_6H_{12}O_6 + 3C_6H_5NH{\cdot}NH_2 & \longrightarrow & H-C=N{\cdot}NH{\cdot}C_6H_5 \\
\text{Glucose or fructose} & & | \\
 & & C=N{\cdot}NH{\cdot}C_6H_5 \\
 & & | \\
 & & HO-C-H \\
 & & | \\
 & & H-C-OH \\
 & & | \\
 & & CH_2OH \\
 & & | \\
 & & CHOH \\
 & & | \\
 & & CH_2OH \\
 & & \text{Glucosazone or fructosazone}
\end{array}$$

Since the two osazones are identical, only the group attached to 1–carbon and the 2–carbon differ:

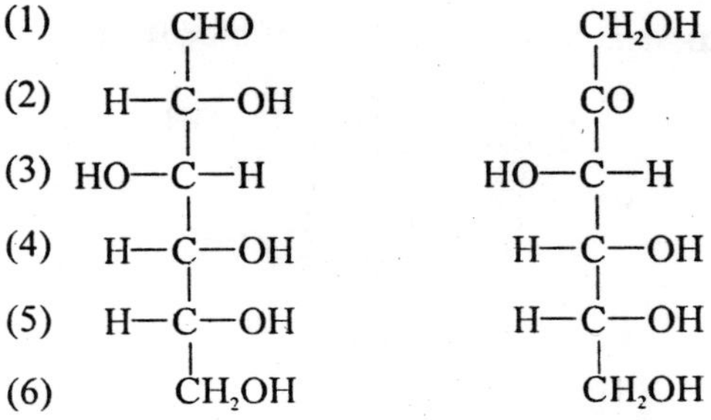

	D-Glucose	D-Fructose
(1)	CHO	CH_2OH
(2)	H—C—OH	CO
(3)	HO—C—H	HO—C—H
(4)	H—C—OH	H—C—OH
(5)	H—C—OH	H—C—OH
(6)	CH_2OH	CH_2OH

Like glucose, fructose exhibits the property of mutarotation, but it normally forms a 2:6 oxygen bridge linkage, giving a fructopyranose. These structures may be summarized as follows:

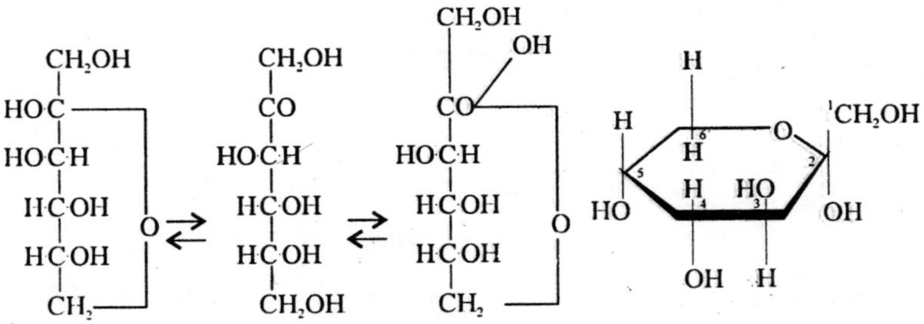

CH_2OH, HOC, HOCH, HCOH, HCOH, CH_2 (O bridge) ⇄ CH_2OH, CO, HOCH, HCOH, HCOH, CH_2OH ⇄ CH_2OH, CO (OH), HOCH, HCOH, HCOH, CH_2 (O bridge)

H, H, O, [1]CH_2OH, H, 6, 2, 5, H, HO, 4, 3, HO, OH, OH, H

β-Fructose Ketose form of fructose α-Fructose α-D-Fructopyranose

Galactose

Mutarotation ($\beta-$ and α –forms) with pyranose formation also occurs in galactose.

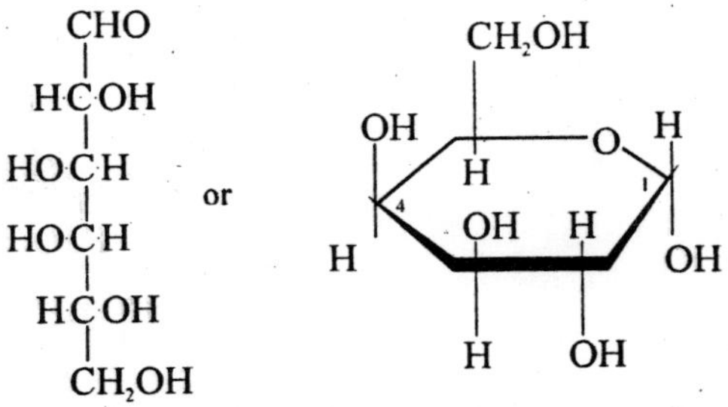

CHO, HCOH, HOCH, HOCH, HCOH, CH_2OH

Galactose

or

CH_2OH, OH, O, H, H, 1, 4, OH, H, H, OH, H, OH

α-D-Fructopyranose

Note that in galactose the position of the OH group at the 4–carbon differs from that in glucose.

Pentose Sugars

Several aldo– and ketopentoses are of biochemical importance:

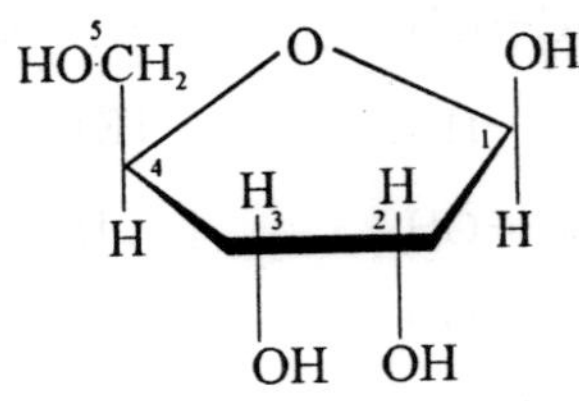

D–Ribose
(β–D–ribofurance)

$HOCH_2$ O OH

H H H H

OH H

2–Deoxy–D–ribose
(β–2–deoxyribofuranose)

Examples of a ketopentose (D–ribulose) and of a ketoheptose (sedoheptulose) follow:

CH_2OH
$C{=}O$
$HCOH$
$HCOH$
CH_2OH

D-Fructose

CH_2OH
$C{=}O$
$HOCH$
$HCOH$
$HCOH$
$HCOH$
CH_2OH

D-Sedoheptulose

Amino Sugars

A number of sugars in which a hydroxyl group of the aldose has been replaced by an amino or acetylamino group occur in nature. The structures of several are shown in the following diagrams:

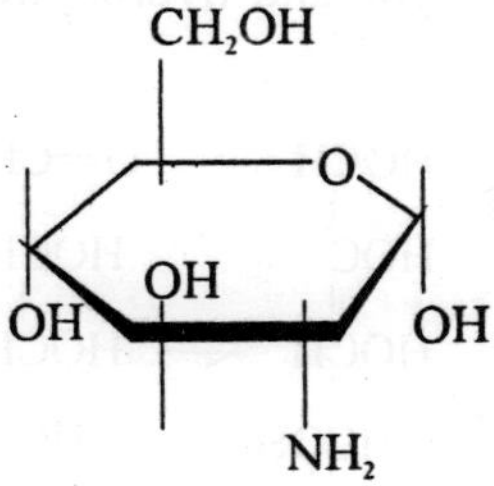

D–Glucosamine

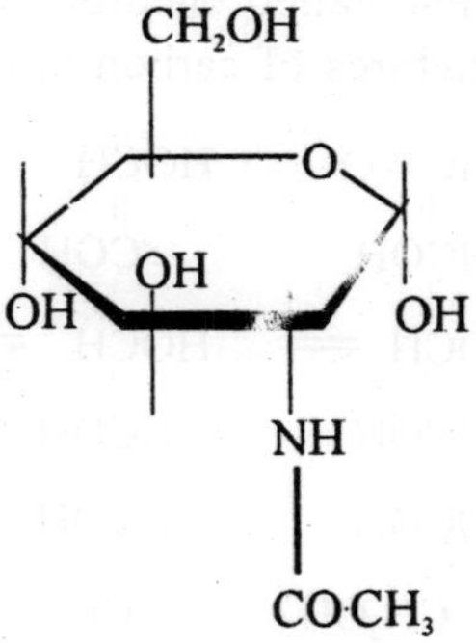

D–Acetylglucosamine

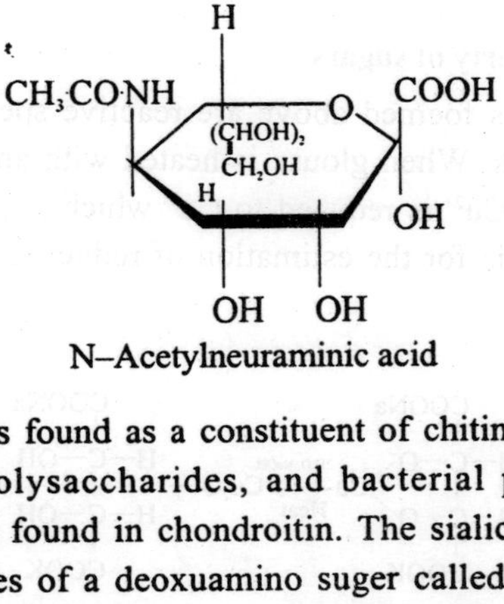

N–Acetylneuraminic acid

Glucosamine is found as a constituent of chitin, hyaluronic acid, heparin, mucopolysaccharides, and bacterial polysaccharides. Galactosamine is found in chondroitin. The sialic acids are *N*–and *O*–acyl derivatives of a deoxuamino suger called neuraminic acid. They are found as constituents of lipids, polysaccharides, and mucoproteins in avariety of tissues.

Reactions of sugars

The carbohydrates, because of the various reactive groups present in the molecule, undergo a large number of chemical reactions.

Reactions of the carbonyl group

With dilute alkali

Dilute aqueous bases at room temperature cause rearrangements about the anomeric carbon atom and its adjacent carbon atom without affecting substituents at other carbon atoms. Treatment of D–glucose, with dilute alkali yields an equilibrium mixture of D–glucose, D–fructose and D–mannose. This reaction involves intermediate enol forms, called enediols of the hydroxyaldehyde and hydroxyketone structures of carbon atoms 1and 2.

D-Glucose		trans -Endiol		D-Fructose		cis -Enediol		D-Mannose
HC=O		HOCH		$HOCH_2$		HOCH		O=CH
HCOH		‖ COH		C=O		HOC		HOCH
HOCH	⇌	HOCH	⇌	HOCH	⇌	HOCH	⇌	HOCH
HCOH		HCOH		HCOH		HCOH		HCOH
HCOH		HCOH		HCOH		HCOH		HCOH
CH_2OH		CH_2OH		CH_2OH		CH_2OH		CH_2OH

Reducing property of sugars

The enediols foemed above are reactive species. They are good reducing agents. When glouse is heated with an alkaline solution of Cu^{2+} ions, the Cu^{2+} is reduced to Cu^{+} which is precipitated as Cu_2O. This is the basis for the estimation of reducing sugars.

D-Glucose		Sodium Potassium tartrate-copper complex		Cuprous oxide (red)		Sodium potassium tartrate		Sodium D-Gluconate
CHO		COONa				COONa		COONa
H—C—OH		H—C—O⟍				H—C—OH		H—C—OH
HO—C—H		Cu	$\xrightarrow[\text{Heat}]{2H_2O,\ NaOH}$	Cu_2O	+		+	HO—C—H
H—C—OH	+	H—C—O⟋				H—C—OH		H—C—OH
H—C—OH		COOK				COOK		H—C—OH
H—C—OH								H—C—OH
H								H

$\xrightarrow{\text{Further Oxidation}}$

Reduction to alcohols

The carbonyl group of monosaccharides can be reduced by hydrogen gas in the presence of a metal catalysts or by sodium amalgam in water to form the corresponding sugar alcohols. For example, D–glucose on reduction yields D–glucitol (also called as L–sorbitol) while D–mannose yields D–mannitol.

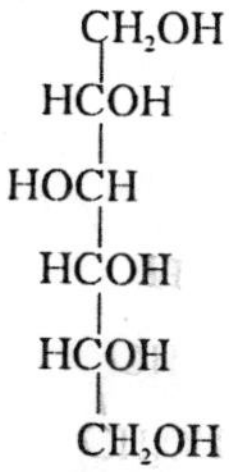

CH_2OH
HCOH
HOCH
HCOH
HCOH
CH_2OH
D-Glucitol
(L-sorhbitol)

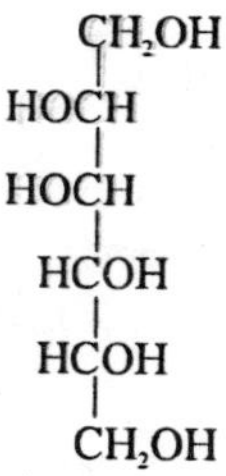

CH_2OH
HOCH
HOCH
HCOH
HCOH
CH_2OH
D-Mannitol

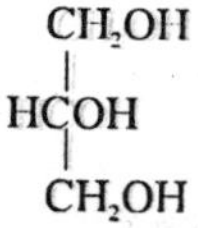

CH_2OH
HCOH
CH_2OH
Glycerol

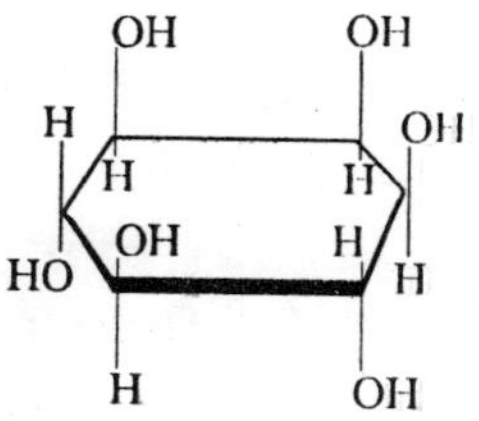

OH OH H OH H H OH H HO H H OH
myo-Inositol

Sugar alcohols occur in nature, particularly in plants. One such alcohol, glycerol is an essential component of lipids, myo–inositol, a stereoisomer of inositol is found as a component of phosphoglycerides and also in phytic acid, the hexaphosphoric ester of inositol.

Reactions with carbonyl reagents

The carbonyl group of monosaccharide reacts with carbonyl reagents like hydrazine, phenylhydrazine, hydroxylamine or semicarbazide to yield crystalline hydrazone, phenylhydrazone, oxime or semicarbazide, respectively. Further, reaction of hydrazone or phenylhydrazone with excess of the reagent gives rise to osazones whose characteristic crystalline structures are used to identify the sugars.

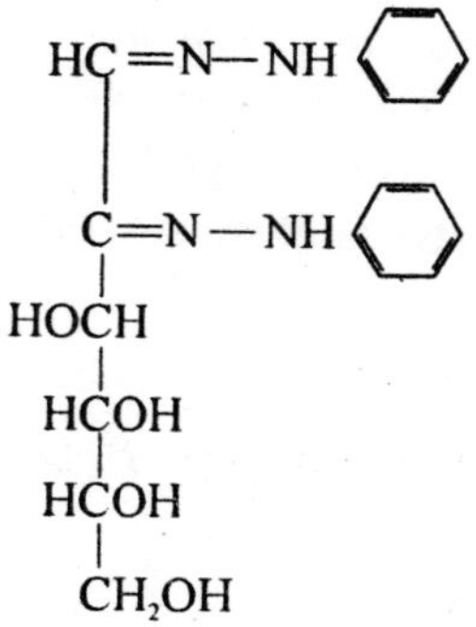

HC=N—NH—C_6H_5
C=N—NH—C_6H_5
HOCH
HCOH
HCOH
CH_2OH

Reactions with concentrated acids

Monosaccharides are generally stable to hot dilute mineral acids even on heating. When aldohexoses are heated with strong mineral acids, however, they are dehydrated, and 5–hydroxymethyl furfural is formed:

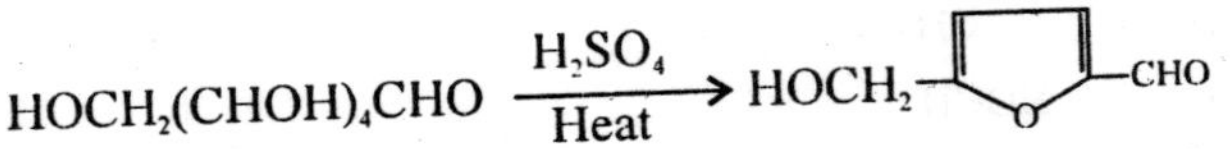

$$HOCH_2(CHOH)_4CHO \xrightarrow[Heat]{H_2SO_4} HOCH_2\text{-furan-}CHO$$

Hydroxymethyl furfural

Under the same condition,pentoses yield furfural:

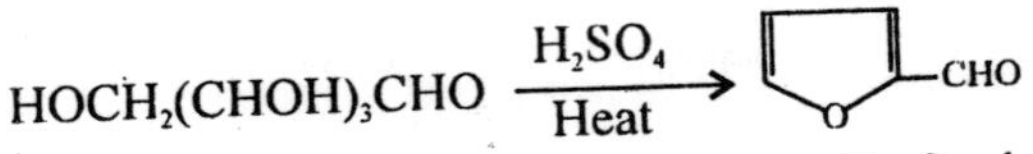

$$HOCH_2(CHOH)_3CHO \xrightarrow[Heat]{H_2SO_4} \text{furan-}CHO$$

Furfural

This dehydration reaction is the basis of certain qualitative tests for sugars, since the furfurals can be reacted with α –naphthol and other aromatic compounds to form characteristic coloured products.

Glycoside formation

Sugar hemiacetals and hemiketals can react with alcohol in the presence of a mineral acid to form anomeric α – and β – glycosides. Glycosides are asymmetric mixed acetals formed by the reaction of the anomeric carbon atom of the intramolecular hemiacetal or pyranose form of the aldohexose with a hydroxyl group of an alcohol. This is called a glycosidic bond. The anomeric carbon in such glycosides is asymmetric.

The glycoside linkage is also formed by the reaction of the anomeric carbon of a monosaccharide with a hydroxyl group of another monosaccharide to yield a disaccharide. Oligo–and polysaccharides are chains of monosaccharides joined by glycosidic linkages.

D–glucose with methanol in the presence of HCL yields methyl–α –D–glucopyranoside or methyl–β–D–glucopyranoside.

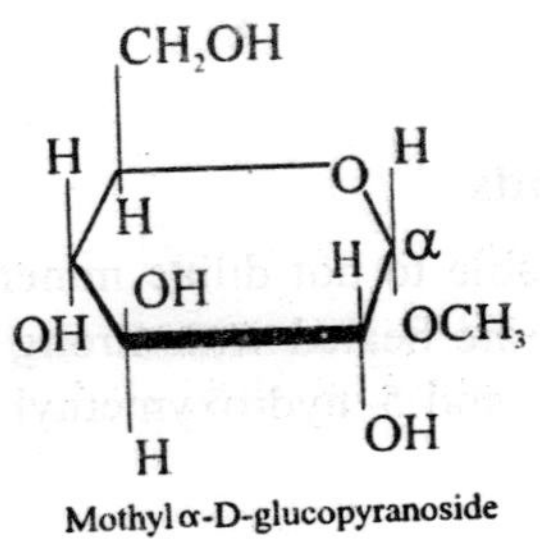

CH_2OH, H, O, H, H, α, H, OH, OH, OCH_3, H, OH

Mothyl α-D-glucopyranoside

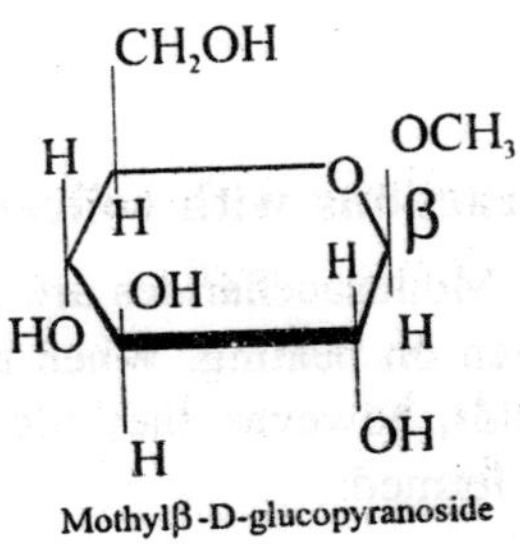

CH_2OH, OCH_3, H, O, H, β, H, OH, HO, H, H, OH

Mothylβ-D-glucopyranoside

Reactions of the alcoholic group

Acylation

The free hydroxyl group of monosaccharides and polysaccharides can be acylated by reaction with an acyl chloride or acetic anhydride to yield O–acyl derives which are useful in determining the structure. Treatment of α –D–glucose with excess of acetic anhydride yields penta–O–acetyl α –D–glucose, The resulting esters can be hydrolyzed by alkali.

CH_2OCOCH_3
H O H
H
$OCOCH_3$ H
CH_3COO $OCOCH_3$
H $OCOCH_3$

Methylation

The methylation of the hydroxyl group on the anomeric carbon atom occurs readily with methanol in the presence of acide to yeild methyl glycosides, which are acetals. The remaining hydroxyl groups of monosaccharides require much more drastic conditions for methylation, e.g.treatment with dimethyl sulfate or methyl iodide and silver oxide which yield methyl ethers not methyl acetals. Methylation of all the free hydroxyl groups of a carbohydrate is called exhaustive methylation. These ether derivatives are resistant to hydrolysis unlike the glucosides. Methyulation studies are important in structural analysis of carbohydrates.

Methylation of methyl α –D–glucopyranoside for exammple, yields methyl–2,3,4,6–tetra–O–methyl–D–glucoopyranoside.

$HCOCH_3$
HCOH
HOCH O
HOCH
HC
CH_2OH
Methyl -α-D-
glucopyranoside

$\xrightarrow{CH_3I}$

$HCOCH_3$
$HCOCH_3$
CH_3OCH O
$HCOCH_3$
HC
CH_2OCH_3
Methyl 1-2,3,4,5-tetra-O-methyl
D-glucopyranoside

Periodate oxidation

Periodic acid (HIO_4) will oxidize and cleave the C–C bonds which contain adjacent free hydroxyl groups or an hydroxyl group and a carbonyl grouping. For example, in the case of glycerol,C–2 is oxidized two times so that it is converted into formic acid while C–1 and C–3 are converted only to formaldehyde. In case of free sugars, like glucose, all the bonds are broken by periodate oxidation.

$$\underset{\text{Glycerol}}{\mathrm{CH_2OH{-}CH(OH){-}CH_2OH}} \xrightarrow{2HIO_4} \underset{\text{Formaldehyde}}{2\,\mathrm{H{-}CH{=}O}} + \underset{\text{Formic acid}}{\mathrm{H{-}C(=O){-}OH}}$$

Periodate oxidation studies like the exhaustive methylation technique have been useful in the elucidation of the struture of carbohydrates.

$$\underset{\alpha\text{-D- glucopyranoside}}{\mathrm{HCO{-}HCOH{-}HOCH{-}HCOH{-}HC(O){-}CH_2OH}} \xrightarrow{5HIO_4} \underset{\text{Formaldehyde}}{\mathrm{H{-}CH{=}O}} + \underset{\text{Formic acid}}{5\mathrm{H{-}C(=O){-}OH}}$$

Sugar derivatiives

Sugar acids

There are three important types of sugar acids : aldonic, aldaric and uronic acids

1. Aldonic acids : These are obtained by the mild oxidation of aldoses when only the aldehyde group is oxidized. Halogens at pH 5.0 oxidize aldoses to the corresponding lactones which then hydrolyze slowly to give monocarboxylic acids. For example, D–glucose yields D–gluconic acid which in phosphorylated form is an important intermediate in carbohydrate metabolism.

2. Aldaric acids: These are dicarboxylic acids derived from aldoses by oxidation of both the aldehydic carbon atoms and the carbon atom bearing the primary hydroxyl group using a strong oxidizing agent like nitric acide. D–Glucose gives D–glucaric acid while D–galactose gives mucic acid on oxidation. Aldartic acids are sometimes useful in the identification of sugars.

3. Alduronic acids: These are sugar acids in which the terminal primary alcoholic group of sugar is oxidized to a carboxyl group, the reducing aldehydic group at the other end remains unaltered. These acids are biologically very important. The uronic acids are components of many polysaccharides.

An important sugar acid is L–ascorbic acid or vitamin C which is the γ –lactone of hexanoic acid having an enediol structure between C–2 and C –3. Ascorbic acid is unstable and readily undergoes oxidation to dehydroascorbic acid. Ascorbic acid is present in large amounts in citrus fruits and tomatoes.

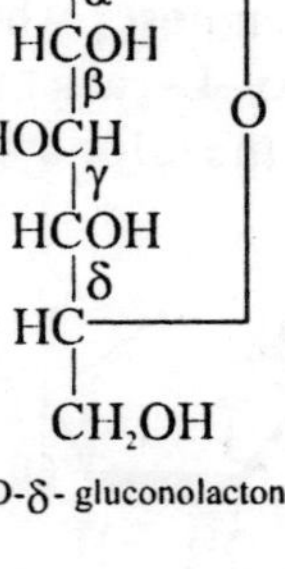

C=O (α) – HCOH (β) – HOCH (γ) – HCOH (δ) – HC – CH_2OH (ring O between C=O and HC)

D-δ- gluconolactone

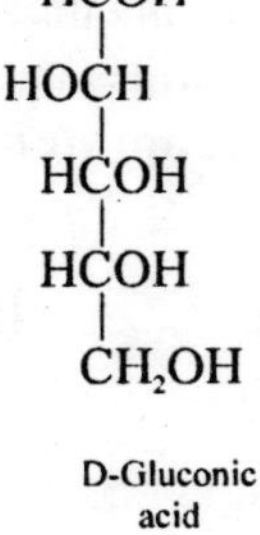

COOH – HCOH – HOCH – HCOH – HCOH – CH_2OH

D-Gluconic acid

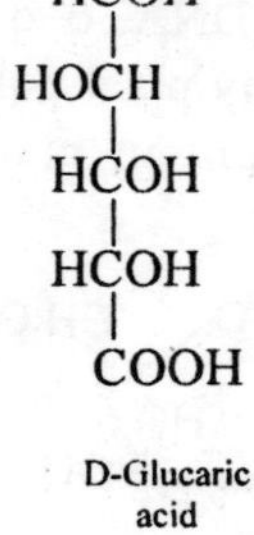

COOH – HCOH – HOCH – HCOH – HCOH – COOH

D-Glucaric acid

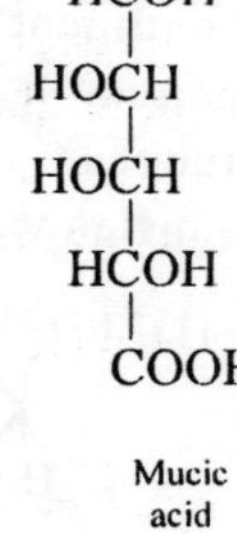

COOH – HCOH – HOCH – HOCH – HCOH – COOH

Mucic acid

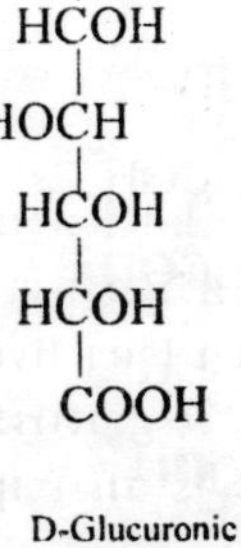

CHO – HCOH – HOCH – HCOH – HCOH – COOH

D-Glucuronic acid

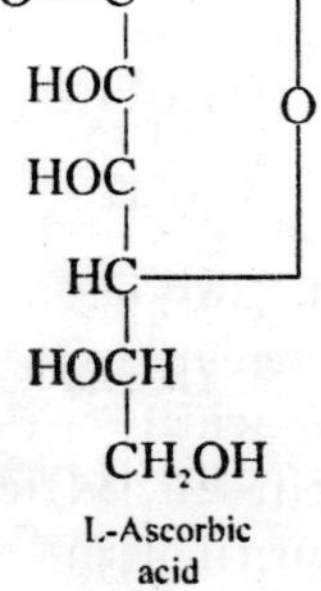

O=C – HOC – HOC – HC – HOCH – CH_2OH (ring O between O=C and HC)

L-Ascorbic acid

O=C – O=C – O=C – HC – HOCH – CH_2OH (ring O between first O=C and HC)

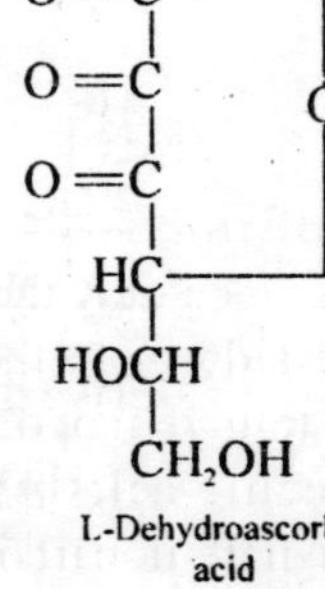

L-Dehydroascorbic acid

Sugar phosphates

Phosphate derivatives of monosaccharides are found in all living cells, in which they serve as important intermediates in carbohydrate metabolism.

Deoxy sugars

If one or more hydroxyl groups of sugars are replaced by hydrogen atoms, the resulting compounds are known as deoxy sugars. The important deoxy sugars are 2–deoxy–D–ribose, a

CH_2OH H H O H HO OH H OPO_3H_2 H OH

α-D-Glucose-1-phosphoric acid

$CH_2OPO_3H_2$ H H O H HO OH H OH H OH

α-D-Glucose-6-phosphoric acid

contituent of DNA, 6–deoxy–L–mannose (L–rhamnose) which is present in many plant polysaccharides and 6–deoxy–L–galactose(L–fucose) present in many glycoproteins like the blood group substances.

$H_2O_3POCH_2$ O $CH_2OPO_3H_2$ H HO H OH OH H

α-D-Fructose-1,6-disphosphoric acid

$H_2O_3POCH_2$ O CH_2OH H HO H OH OH H

α -D-Fructose-6-phosphoric acid

HC=O
CH_2
HCOH
HCOH
CH_2OH

2-Deoxy-D-ribose

HC=O
HOCH
HCOH
HCOH
HOCH
CH_3

L-Fucose

HC=O
HCOH
HCOH
HOCH
HOCH
CH_3

L-Rhamnose

Amino sugars

These are sugars in which a hydroxyl group is replaced by an amino group. The most common sugars are D–glucosamine (2–amino–1–deoxy–D–glucose) and D–galactosamine (2–amino–2–deoxy–galactose) in which the hydroxyl group at carbon atom 2 is replaced by an amino group. D–Glucosamine occurs in many

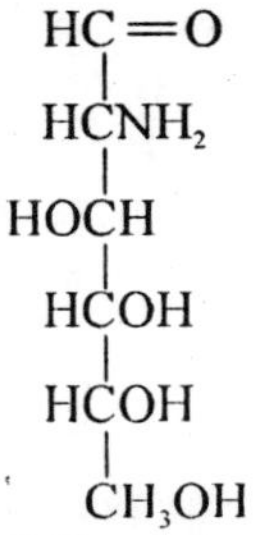

HC=O
HCNH$_2$
HOCH
HCOH
HCOH
CH$_3$OH

D-Glucosamine

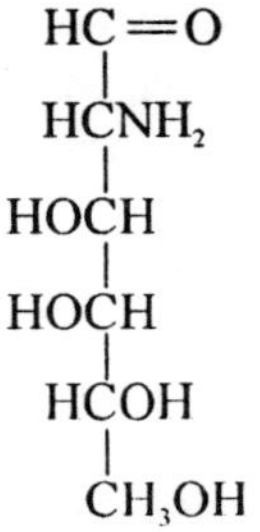

HC=O
HCNH$_2$
HOCH
HOCH
HCOH
CH$_3$OH

D-Galactosamine

HC=O
HC—NH—C(=O)—CH3
HOCH
HCOH
HCOH
CH$_3$OH

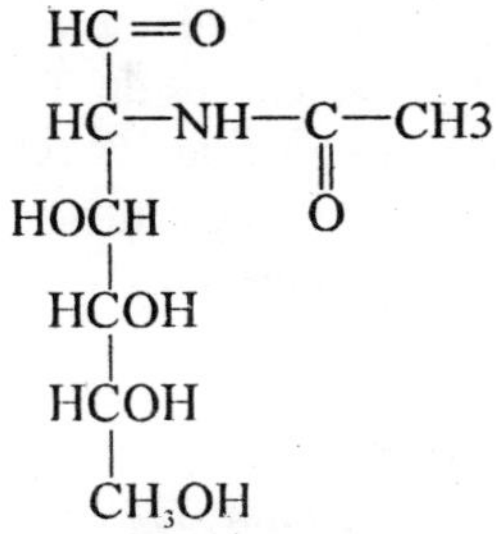

N-Acetyl-D-glucosamine

polysaccharides Galactosamine is a component of glycolipids and of the major polysaccharide of cartilage, chondroitin sulfate.

Two other important derivatives of amino sugars are muramic acid and neuraminic acid. These can be considered as six–carbon amino sugars linked to a three carbon acid. Muramic acid is D–glucosamine to which a lactic acid moiety is attached at C–3. Neuraminic acid is derived from D–mannosamine and pyruvic acid by an aldol condensation. The two compounds are usually acetylated. N–acetylated derivative of neuraminic acid (NANA), known as sialic acid , is a component of glycoproteins present in animal cell membrane.

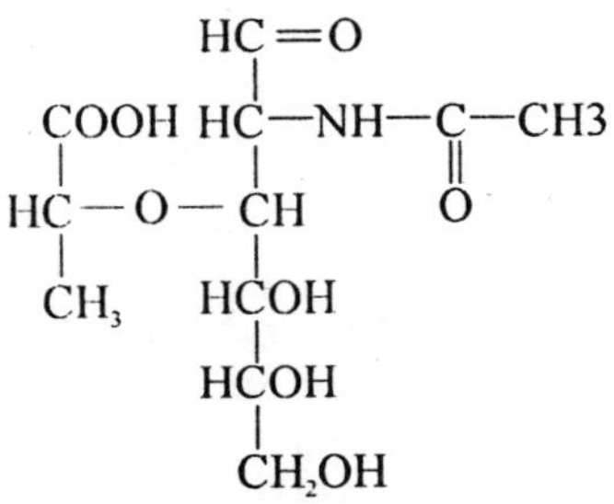

HC=O
COOH HC—NH—C(=O)—CH3
HC(CH$_3$)—O—CH
HCOH
HCOH
CH$_2$OH

N-Acetyllmuramic acid

COOH
C=O
CH$_2$
HCOH
CH$_3$—C(=O)—NH—CH
HOCH
HCOH
HCOH
CH$_2$OH

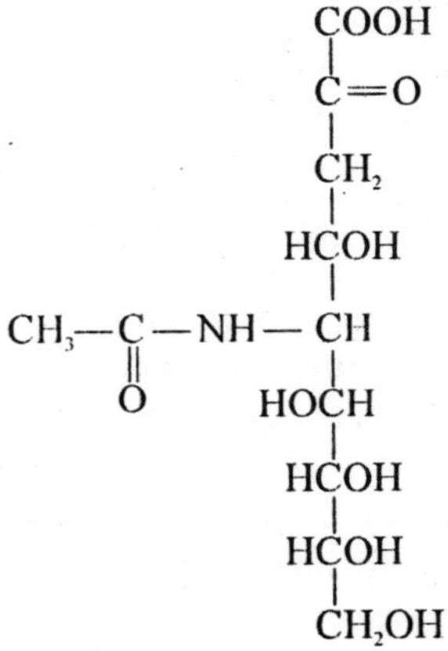

N-Acetylneuraminic acid

Oilgosaccharides

These sugars consist of a short chain of 2 to 8 or 10 monosaccharide units linked by the glycosidic bond (s) with the elimination of water molecule (s). The glycosidic bond is formed most frequently between the anomeric carbon of one sugar residue and a hydroxyl group of the other sugar residue.

Depending on the number of monosaccharide units that are linked, the oilgosaccharides are further classified as disaccharides (two sugar units), trisaccharides (three sugar units), tetrasaccharides (four sugar units), etc. Amongst these, disaccharides are the most important class because of their biological role and relative abundance in natural products.

Disaccharides

The disaccharides may be regarded as glycosides of the form.

```
┌──CH—O—R
│   |
│   CHOH
│   |
O   CHOH
│   |
│   CHOH
│   |
└──CH
    |
    CH2OH
```

where R is a monosaccharide, which either has a potential free aldehydic group (as in maltose and lactose), or has no reducing group (sucrose).

Where R is CH_3, we have methyl glycoside, which has already been discussed. In nature, important glycosides are known in which R is a nonsugar, often of complex structure. For example, phlorhizin (also spelled *phloridzin* and *phlorizin*), found in the bark of Rosaceae, is a combination of glucose and phloretin; amygdalin, present in the seed of the bitter almond, is a combination of glucose and mandelonitrile; digitonin, found in the leaves and seeds of digitalis, is a combination of glucose, galactose, and digitogenin. In the cardiac glycosides (found in certain plants and possessing a characteristic action on the heart), the R is represented by a steroid. (*Glycoside* is the general name for this group of compounds, irrespective of the sugar present; *glucoside* is the more specific name for glycosides that contain glucose as the sugar constituent.)

General properties

1. Those with potentially a free aldehyde or a ketone group can reduce Fehling's solution hence are called reducing disaccharides.

2. The reducing disaccharides have most of the properties of monosaccharides i.e., they can form osazones and show mutarotation, etc.

3. Disaccharides can be hydrolyzed into their constituent monosaccharide units unlike monosaccharides.

4. Some disaccharides may exist in white crystalline solids and are soluble in water and sweet in taste.

5. Disaccharides are not fermented by yeast directly but they are first hydrolyzed to constituent monosaccharides which in turn are fermented.

The most abundant disaccharides in nature are maltose, sucrose and lactose.

Maltose

Maltose yields two molecules of glucose when hydrolyzed. Upon methylation and subsequent hydrolysis, 2,3,4,6–tetramethylglucose are obtained. While these results lead to two possible structures, the following has been selected as more in accord with the facts.

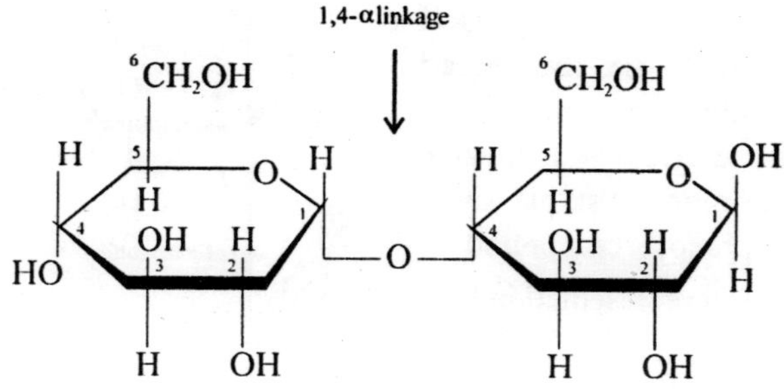

β -Maltose
4.0(α-D-Glucopyranosy-1) β-D-glucopyranose

Lactose

Glucose and galactose are formed upon hydrolysis of lactose.When the lactose is methylated and then hydrolyzed, 2,3,6–trimethylglucose and 2,3,4,6–tetramethylgalactose are obtained, leading to the formula:

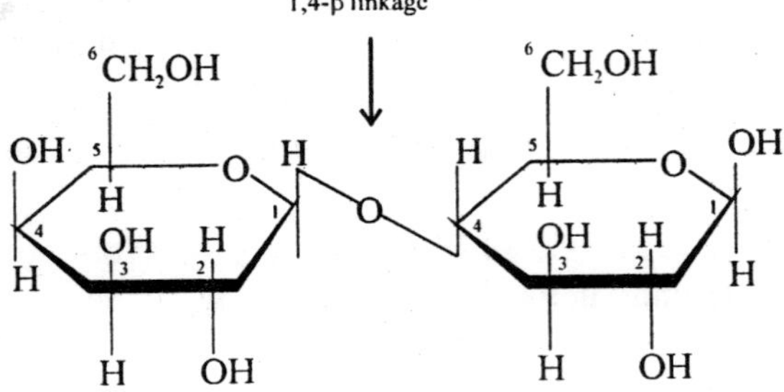

β-Maltose
4.0(β -D-Glucopyranosy-1) β-D-glucopyranose

Both lactose and maltose reduce Benedict's solution, owing to the free OH group at the 1–carbon.

Sucrose

Unlike maltose and lactose, sucrose does not reduce Benedict's solution; it is a nonreducing sugar. When hydrolyzed, it yields glucose and fructose. Methylation and subsequent hydrolysis convert sucrose into a tetramethyl glucose and a tetramethyl fructose.

Sucrose has a pyranose structure for the glucose part, and a furanose structure for the fructose part. It is, therefore, a glucosidofructo-furanoside.

In the free form, fructose itself exists as a stable six–membered pyranose compound, but in the sucrose molecule the fructose exists as the more unstable five–membered furanose compound:

α-D-Fructofuranose

The Haworth representation for sucrose is:

1,2- α,β–linkage

Sucrose
α-D-Glucopyranosy l-β-D-fructofuranose

Note that sucrose has a 1,2 linkage. Note also that the Haworth formula for β–D–fructofuranose has been turned over as well as upside down (with respect to the H and OH groups) in order to show the 1,2 linkage more simply.

Polysaccharides $(C_6H_{10}O_5)_x$

Starch

Starch is found in plants, in layers within a granule (grana) that is surrounded by a thin protein layer. Heating disrupts the granule from which can be isolated two kinds of polysaccharides—amylose and amylopectin.

Amylose. Responsible for the blue color obtained when strach is treated with iodine, amylose is a linear nonbranched polymer of α – D–glucose in a repeating

Glycogen

The polysaccharide glycogen is found in animal tissues, mostly in liver and muscle, and is the storage form of carbohydrate. Like amylopectin, glycogen possesses repeating 1,4–glucose units, interrupted by 1,6–branching points. Glycogen, however, has shorter linear chains than amylopectin, with 12 to 18 glucose residues and more frequent branching. Glycogen represents a mixture of different molecules, since fractions with different solubilities can be obtained from tissue extracts. A model of a glycogen molecule is illustrated in Figure 7–1.

Cellulose

This plant polysaccharide is a repeating glucose polymer and is the most abundant of all the polysaccharides in nature. The glucose are connected by 1,4 β–glucoside bonds, and there are more than 3000 such residues per molecule. The following segment of a cellulose molecule can be viewed as repeating units of the disaccharide cellobiose:

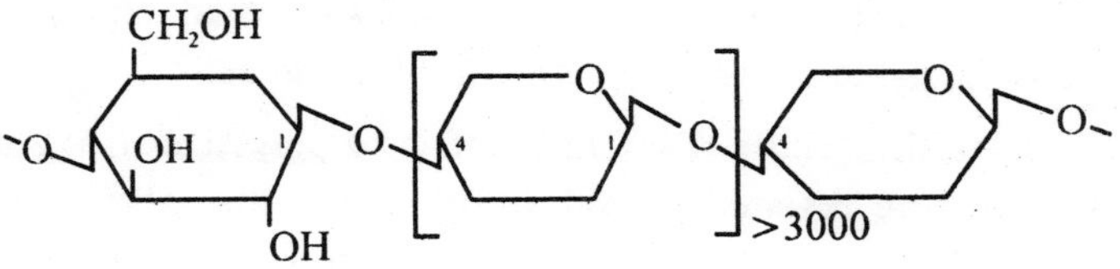

CH_2OH OH 1 O 4 1 O 4 O O O OH >3000

Cellulose

In the woody material of plants, cellulose is cemented together by the complex polymer lignin, which appears to consist of a repeating unit derived from coniferyl alcohol :

-CH=CH·CH$_2$OH
HO
OCH$_3$

Coniferyl alcohol

Acid Mucopolysaccharides

Hyaluronic Acid. Connective tissue contains a polysaccharide that acts as an intercellular cement. Hyaluronic acid is a viscous, high molecular weight (several million) polysaccharide consisting of chains of N–actylglucosamine and glucuronic acid residues. The glycosidic linkages are 1,4 between *N*–acetylglucosamine and glucuronic acid, and 1,3 between glucuronic acid and the glucosamine:

CH$_2$OH CH$_2$OH
COOH O O -O-
HO 4 1 O 4 OH
O 4 OH 1 3
OH NH·COCH$_3$ OH
OH

Hyaluronic acid

The enzyme hyaluronidase causes a depolymerization of the polysaccharide which is accompanied by a decrease in viscosity, as well as a cleavage of the glycosidic bonds between glucosamine–C–1 and glucuronic acid–C–4. The enzyme is widely distributed in microorganisms and mammalian tissues and is believed to play a role in promoting tissue invasion by bacteria of sperm cells.

Chondroitin Sulfate. Cartilage contains a polysaccharide sulfate whose structure resembles hyaluronic acid, with galactosamine sulfate replacing acetylglucosamine. A portion of the chondroitin sulfate chain is shown here:

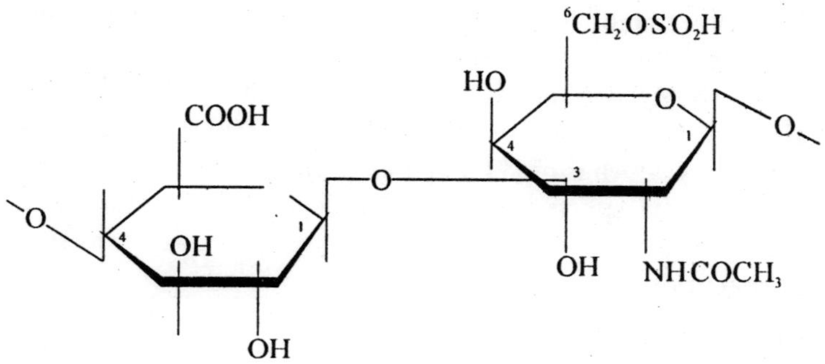

Chondroitin-6-sulphate

Chondroitin sulfate A has the sulfate ester group on the 4–carban, whereas in chondroitin sulfate B, *iduronic acid* replaces glucuronic acid, differing from the latter in its steric configuration at the 5–carbon.

Glycoproteins

Glycoproteins are molecules composed of covalently joined protein and carbohydrates. The carbohydrate is attached to the polypeptide chains of the protein in a series of reactions that are enzymatically catalyzed after the protein component is synthesized.

Glycoproteins in cell and membranes apparently have an important role in the group behavior cells and other biological functions of the membrane. They form a major part of the mucus that is secreted by epithelial cells, where they perform an important role in lubrication and in the protection of tissues lining the body's ducts. Many other proteins secreted from cells into extracellular fluids are glycoproteins. These proteins include hormone proteins found in blood, such as follicle stimulating hormone (FSH),lutenizing hormone (LH), chorianic gonadotropin; and plasma proteins such as immunoglobulins etc. Glycoproteins are also one of the major components of the cell coats of higher organisms.

The carbohydrate per cent within glycoproteins is highly variable. Some glycoproteins such as IgG contain low amounts of carbohydrate (4%). Human ovarian cyst glycoprotein is composed of 70 per cent carbohydrate and human gastric glycoprotein is 82 per cent carbohydrate. Glycoproteins having a very high content of carbohydrate are called protreoglycans.

Metabolism of Carbohydrates

The major anabolic route open for carbohydrates in the human body is **glycogenesis** in which glycogen is synthesized. The catabolic routes are **(i) glycogenolysis (ii) glycolysis (iii) citric acid cycle,** and **(iv) pentose phosphate pathway**. These routes of catabolism liberate energy which is used for various purposes in the body. Certain noncarbohydrate substances lead to biosynthesis of glucose and glycogen, and this process is known as **gluconeogenesis** or **neoglucogenesis.**

Glycogenesis

The process of biosynthesis of glycogen from glucose or other sugars is known as **glycogenesis.** Glycogen is synthesized in practically all the tissues of the body but the major sites are liver and muscles.The aim of storing glycogen is to provide glycosyl units for energy purpose in the muscles, and to maintain blood sugar level within the normal range at the time of fasting so that delicate organs in the body may be protected against harmful effects of *hypoglycemia.* Immediately after taking carbohydrate rich diet, the liver tissue may store glycogen approximately 5–6% of its weight. In adult man weighing 70kg, liver constitutes about 1.8kg. Total glycogen stored in the liver tissue in well fed subjects may vary from 90 to 110g. After a fast of about 12 to18 hrs, liver may become depleted to glycogen. Muscles contain 0.7 to 1.0 per cent glycogen on wet weight basis. Muscles constitute approximately 35kg in an adult individual. Thus, 245 to 350 g glycogen may be stored in the muscles. Muscle glycogen is reduced after *severe exercise*, or when liver glycogen is almost totally depleted.

It has been shown that the reactions leading to synthesis of glycogen in liver, muscles and brain tissues are similar. Glucose is activated by an enzyme **hexokinase** in the presence of ATP and Mg^{++} ions with the formation of glucose –6–PO_4 . The backward reaction is not possible by the same enzyme under physiological conditions. Another enzyme **phospho–glucomutase** change glucose –6–PO_4 to glucose–1–PO_4. This reaction requires the presence of glucose–1, 6–diphosphate as cofactor. In the next stage, glucose–1–PO_4 reacts with uridine triphosphate (UTP) under the influence of the enzyme **uridine diphosphate glucose pyrophosphorylase** (UDPG–*pyrophosphorylase)* to form uridine diphosphate–glucose(UDPG). In this reaction, pyrophosphate is liberated which is finally converted to orthophosphoric acid by another enzyme **pyrophosphatase** making UDPG formation irreversible.UDPG serves as the glucosyl–unit donor in the biosynthesis of glycogen. The actual synthesis of glycogen from glucosyl units of UDPG requires the presence of a small amount of glycogen nucleus in the preformed state (*primer*) to which the glucosyl units from UDPG are attached forming 1, 4–glucosidic linkages (*straight chains*). Such a reaction is catatysed by the enzyme **UDPG–glycogen–transglycosylase,** also known as **glycogen synthetase.** The reaction is stimulated by glucose–6–PO_4 which is believed to bind the enzyme **glycogen synthetase** and stabilize it in the active form. As soon as the polysaccharide chain

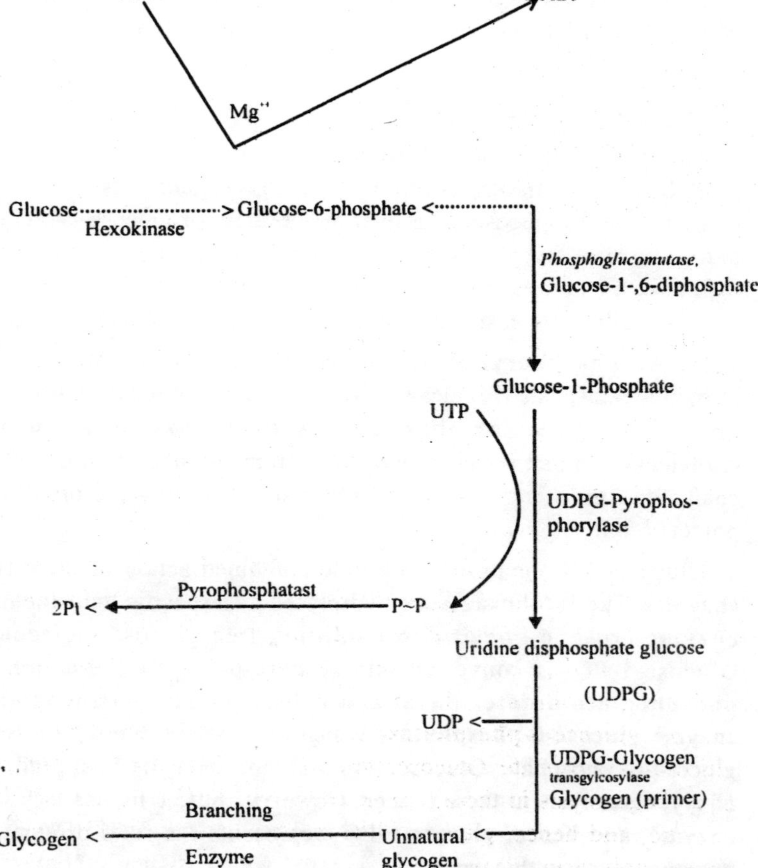

Fig. 8.1 : Biosynthesis of Glycogen in Animal Tissues.

acquires a length of eight glucosyl–units, it is subjected to the action of another enzyme **amylo–1, 6–glucosidase,** also known as **branching enzyme.** This glucosidase cleaves the straight chain fragments and transfers them to the neighbouring chain. Attachment of these fragments occurs forming 1, 6–glucosidic linkages (*branched chains).* **UDPG** further attaches more glucosyl units on these branches. Straight chains are elongated and broken down forming new branches in the same manner. The same process continues till a *tree like structure* of glycogen molecule is synthesized. The molecular weight of glycogen thus synthesized may vary from one to four millions, or even more.

Other sugars such as galactose or fructose may also be converted into glycogen by way of first forming glucose, glucose–6–PO_4,or glucose–1–PO_4.

Glycogenolysis

The process of breakdown of glycogen to glucose or glucose –6–phosphate in the tissues is referred to as **glycogenolysis.** It may be broken down to glucose as in liver and kidney; or glucose–6–PO_4 as in the muscles. The process is enhanced by *hypoglycemia,*or under the influence of certain hyperglycemic hormones. Liver glycogen is metabolically more easily available as compared to muscle glycogen.

Active **phosphorylase** acts upon glycogen in the presence of inorganic phosphate (Pi) cleaving α-1,4-glucosidic linkages from the outer ends of the straight chains. Another enzyme, **a glucan-transferase,** splits trisaccharide residues from one side of the branched chains and transfers these to the other side exposing the branching points (*1, 6 linkages*).

Glucose-1-PO_4 is produced by the combined action of these two enzymes. The 1, 6-linkages are hydrolysed by a specific **debranching enzyme** (α*-1, 6-glucosidase*) splitting free glucose molecules. Glucose-1-PO_4 is converted into glucose-6-PO_4 by the action of **phosphoglucomutase.** Liver and kidney tissues contain another enzyme **glucose-6-phosphatase** which can remove phosphate from glucose-6- phosphate. Glucose, threfore, represents the final products of glycogenolysis in these tissues. However, muscle tissues lack this enzyme, and hence, glucose-6-PO_4 represents the final product of glycogenolysis in this tissue.

A brief reference may be made here of **phosphorylases** involved in hepatic and muscular glycogenolysis. in the liver, phosphoryla exists in an *inactive* form, known as **dephosphophosphorylase** which can be converted into *active* **phosphorylase** (phosphophosphorylase) in the presence of ATP and an enzyme **dephosphophosphorylase kinase.** This enzyme binds phosphate groups to serine in the dephosphophosphorylase molecule. The action of **dephosphophos-phorylase kinase** is promoted by cyclic-AMP (3′5′ *-adenylic acid*). Cyclic-AMP itself is produced from ATP by the action of an enzyme **adenyl cyclase** in the presence of **Mg** $^{++}$ ions.

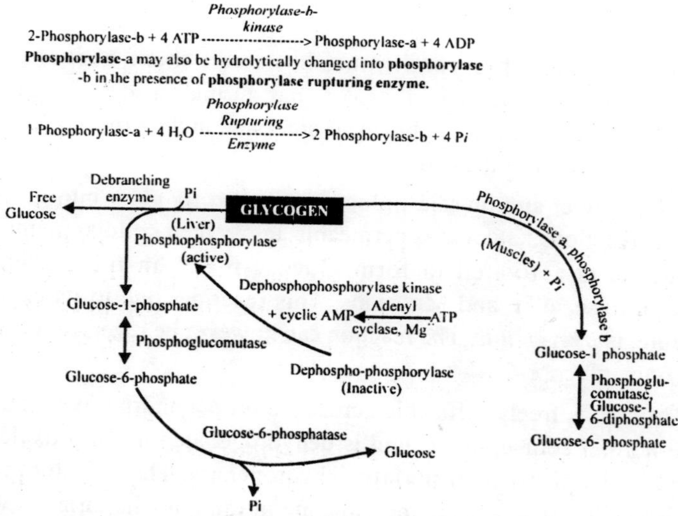

Fig. 8.2 : Mechanism of Glycogenolysis in liver and muscles.

Muscle **phosphorylase** in the rabbit muscles has been shown to exist in two distinct forms, namely **phosphorylase**-a and **phosphorylase** - b. **Phosphorylase**-a (*molecular weight 495,000*) *contains four molecules of pyridoxal phosphate, whereas* ***phosphorylase***-*b* (*molecular weight 242,000*) contains only two molecules of pyridoxal phosphate. **Phosphorylase**-*a* is more active than phosphorylase-*b*. **Phosphorylase**-*b* may be converted into **phosphorylase** -*a* in the presence of ATP and the enzyme **phosphorylase-*b*-kinase.**

$$2 \text{ phosphorylase - b} + 4 \text{ ATP} \xrightarrow{\text{Phosphorylase-b-kinase}} \text{phosphorylase - } a + 4 \text{ ADP}$$

Phosphorylase-*a* may also be hydrolytically changed into **phosphorylase-*b*** in the presence of **phosphorylase rupturing enzyme.**

$$\text{I phosphorylase - }a + 4\,H_2O \xrightarrow[\text{Rupturing}]{\text{phosphorylase}} \text{phosphorylase - }b + 4\,Pi$$

Glycolysis

The biochemical changes occurring in the conversion of glucose or glycogen to pyruvate, and/or lactate are included under **glycolysis.** the reaction sequence in glycolysis is also known as **Embden-Meyerhof Pathway.** Glycolytic reactions do occur in almost all the tissues. The reactions may proceed under anaerobic as well as aerobic conditions, lactic acid does not accumulate and it is changed into pyruvic acid which is further oxidised to CO_2 and H_2O by Krebs cycle.

Under anaerobic conditions, however, there occurs greater accumulation of lactic acid, as is found during muscular contraction. The enzymes involved in glycolysis are found in the *extramitochondrial soluble fraction* of the cell.

In the liver and muscles, glucose can directly enter into glycolytic series reactions. Glucose is permeable across their cellular membranes. Glucose is activated to form glucose-6-PO_4 in the presence of **hexokinase,** ATP and Mg^{++} ions. This reaction, due to its exergonic nature, is *irreversible,* The reaction can however be reversed by another enzyme glucose-6.

Malate is freely diffusible across the mitochondrial membranes. In the normal course, malic acid is dehydrogenated to form oxaloacetic acid in the presence of **malate** dehydrogenase. The cofactor required is NAD. There exists another enzyme in the mitochondrial as well as extramitochondrial fraction, known as malic enzyme which can covert

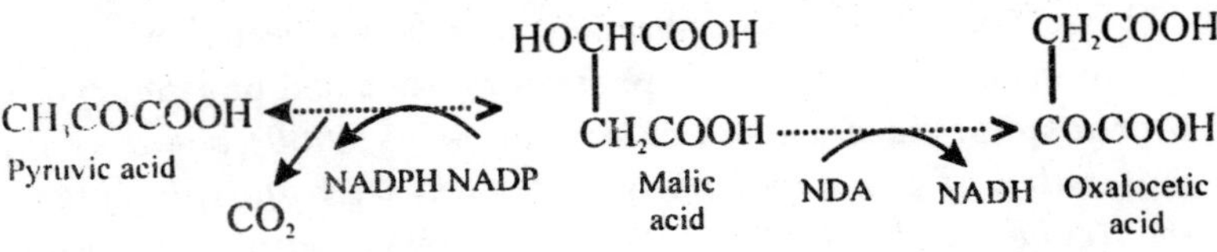

malic acid into pyruvic acid. This enzyme is NADP specific and it also involves decarboxylation. In this way, oxaloacetic acid is regenerated on completion of the cycle.

Energetics of Krebs Cycle - In Krebs cycle, energy is generated at six different sites.

(1) During oxidative-decarboxylation of pyruvic acid, one mole

of NADH is produced per mole of pyruvic acid, converted into acetyl CoA. This on oxidation by electron transport chain leads to synthesis of 3 moles of ATP.

(2) The *second site* is, during conversion of isocitric acid into oxalosuccinic acid when 1 mole of NADH or NADPH is produced which on oxidation produces 3 moles of ATP.

(3) The *third site* is, during conversion of α-ketoglutaric acid (α-KG) into succinyl CoA when one mole of NADH is produced per mole of α-KG consumed which on oxidation leads to production of 3 moles of ATP.

(4) The *fourth site* is during conversion of succinyl CoA into succinic acid which finally produces 1 mole of ATP. This is an example of substrate level phosphorylation.

(5) The *fifth site* of energy generation is, during conversion of succinic acid into fumaric acid which leads to formation of one mole of FAD. H_2 per mole of succinic acid oxidised. This on oxidation by electron transport chain leads to formation of 2 moles of ATP.

(6) The *sixth* and the last site of energy generation is during conversion of malic acid into oxaloacetic acid when I mole of NADH is produced per mole of malic acid converted into oxaloacetic acid. Thus, 3 moles of ATP may be produced at this stage by electron transport chain oxidation.

One mole of pyruvic acid when oxidised to CO_2 and H_2O by Krebs cycle thus leads to the formation of 15 moles of ATP. Since, two moles of pyruvic acid may be obtained per mole of glucose entering into glycolytic reactions, 30 moles of ATP may be synthesized in all.

Under aerobic conditions, complete oxidation of 1 mole of glucose through glycolysis and Krebs cycle may produce 38 moles of ATP.

Gluconeogenesis from Tricarboxylic Acid Cycle Intermediates

The pathway from pyruvate to glucose described above also allows the net synthesis of glucose from various precursors of pyruvate or phosphopyruvate. Chief among them are the tricarboxylic acid cycle intermediates, which may undergo oxidation to malate. Malate then may leave the mitochondria and undergo oxidation to oxaloacetate in the extramitochondrial cytoplasm, where the formation of phospho-enolpyruvate takes place by the action of cytoplasmic phosphoeno-lpyruvate carboxy kinase.

In animal tissues there is an alternative pathway for the generation of phosphoenolpyruvate from α-ketoglutarate and the preceding six-carbon acids of the ticarboxylic acid cycle. The enzymatic steps in this second pathway, which is entirely mitochondrial, are not yet known. Presumably intramitochondrial GTP generated during the oxidation of α - ketoglutarate is employed as phosphate donor in the phosphorylation of oxaloacetate to yield phosphoenolpyruvate. It is possible that phosphoenolpyruvate carboxykinase also participates in this pathway.

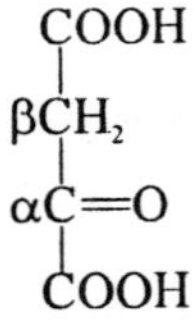

$$
\begin{array}{c}
COOH \\
| \\
\beta CH_2 \\
| \\
\alpha C{=}O \\
| \\
COOH
\end{array}
$$

Oxaloacetic acid

By either mechanism, three carbon atoms of the various tricarboxylic acid cycle intermediates are ultimately convertible into the three carbon atoms of phosphopyruvate. These carbon atoms arise from the α -carboxyl, α -(carbonyl).and *-carbon atoms of oxaloacetate (margin). This pathway has been amply verified by many isotopic tracer experiments carried out on intact animals, as well as on tissue slices or extracts. For example, the isotopic carbon of carboxyl-labeled pyruvate or succinate fed to rats can be recovered as carbon atoms 3 and 4 of the glucose residues isolated from liver glycogen.

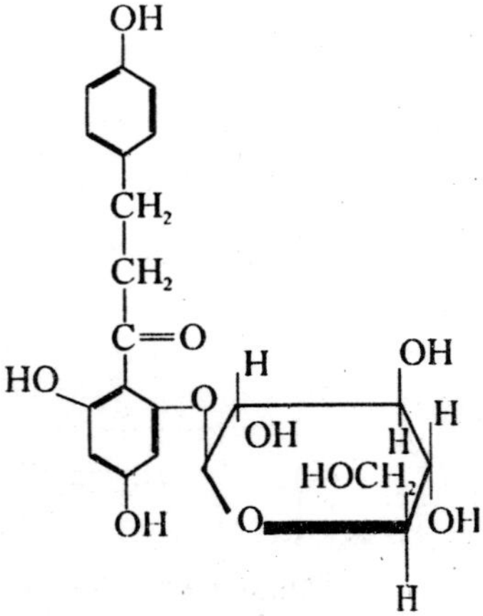

OH, CH₂, CH₂, C=O, HO, O, H, OH, OH, H, H, HOCH₂, OH, O, OH, H

Phloridzin

It has also been established by various experimental approaches that the tricarboxylic acid cycle intermediates give rise to net synthesis of new glucose in vertebrates. In one type of experiment, rats are fasted for 24 hr. or longer, a treatment which reduces the glycogen level in the liver from about 7 percent of the wet weight to 1 percent or less. Feeding of succinate and other cycle intermediates to such fasted rats causes a net increase in the total amount of glycogen. Such a net conversion of tricarboxylic cycle intermediates into glucose is also observed in animals treated with the toxic glycoside *phloridzin* (margin). This poison blocks reabsorption of glucose from the kidney tubule and thus causes blood glucose to be excreted nearly quantitatively into the intermediates to phloridzin-poisoned animals causes excretion of an amount of glucose nearly equivalent to three of the carbon atoms of the intermediate fed.

Gluconeogenesis from Acetyl CoA

It is most important to distinguish between net synthesis of glucose, on one hand, and mere incorporation of an isotopic carbon atom from a labeled precursor into glucose, on the other. For example, when methyl-labeled acetic acid is fed to an animal, the isotopic carbon atom will be incorporated into the glucose residues of liver glycogen, specifically into carbon atoms, 1,2,5 and 6 as some pencil-and -paper work will show. However, in higher animal tissues there is no net formation of new glucose from the two carbon atoms of the acetyl group of acetyl CoA. For one thing, citrate, the six-carbon condensation product of acetyl CoA and oxaloacetate, ultimately undergoes loss of three carbon atoms as CO_2 during its oxidation to phosphoenolpyruvate: thus it can form no more glucose than oxaloacetate can. Moreover, acetyl CoA cannot be directly converted into either pyruvate or succinate in animal tissues. In higher animals, there is no metabolic pathway by which the carbon atoms of fatty acids can be used to form new glucose.

On the other hand, plants and many microorganisms are able to carry out the net synthesis of carbohydrate from fatty acids by way of acetyl CoA, a process made possible by the reactions of the glyoxylate cycle. This cycle permits the net conversion of acetyl CoA to succinate according to the overall reaction.

$$2 \text{ acetyl CoA} + NAD^+ + 2H_2O \longrightarrow \text{succinate} + 2CoA + NADH + H^+$$

Two specific enzymes are required for this pathway, *isocitrate lyase* and *malate synthetase*; they are completely lacking in higher animals. The succinate formed in the glyoxylate pathway yields oxaloacetate, which in turn is the precursor of phosphoenolpyruvate. By this pathway stored fat is converted into glucose by germinating seeds.

Gluconeogenesis from Amino Acids

Some or all of the carbon atoms of the various amino acids derived from proteins are ultimately convertible either into acetyl CoA or into intermediates of the tricarboxylic acid cycle. Those amino acids which can served as precursors of phosphoenolpyruvate, and thus of glucose, are glycogenic amino acids. Examples are glutamic and aspartic acids, which are directly convertible to α -ketoglutarate and oxaloacetate, respectively. There is one amino acid, leucine, which cannot yield net formation of glucose in vertebrates, since all of its carbon atoms are converted into either acetyl CoA or CO_2. However, the carbon atoms of leucine can enter the glucose molecule without net synthesis of the latter. Leucine and other amono acids that yield acetyl CoA on degradation are also capable of forming acetoacetate, particularly in fasting animals. They are thus called ketogenic amino acids. Phenylalanine and tyrosine are examples of amino acids that are both glycogenic and ketogenic, since on degradation they are cleaved to form fumaric acid, which is glycogenic, and acetyl CoA, which is ketogenic.

Table: Fate of Amino Acids

Glycogenic		
	Ala	His
	Arg	Met
	Asp	Pro
	Asn	Ser
	Cys	Thr
	Clu	Trp
	Gln	Val
	Gly	
Ketogenic		
	Leu	
Glcogenic and ketogenic		
	He	
	Lys	
	Phe	
	Tyr	

In plants and many microorganisms no distinction can be made between glycogenic and nonglycogenic amino acids, because all the amino acids may ultimately contribute to the net formation of glucose through the reactions of the tricarboxylic acid and glyoxylate cycles.

Photosynthetic Formation of Hexose by Reduction of Carbon Dioxide

The enzymatic reactions by which light energy is conserved as the phosphate-bond energy of ATP and as reducing power in the form of NADPH in photosynthesizing cells. The ATP and NADPH so generated are then utilized to bring about the reduction of carbon dioxide to form glucose and other reduced products in the dark phase of photosynthesis. Parenthetically, it must be pointed out that the major end products of photosynthesis are hexose residues of cellulose, starch, and other polysaccharides; free glucose per se is not present in significant quantities in most higher plants. We shall use the general term " hexose" to designate all free and combined six-carbon sugar residues formed in photosynthesis.

Although the tissues of higher animals possess the capacity of fixing carbon dioxide as the carboxyl carbon of oxaloacetate and other compounds, the carbon dioxide fixed in such reactions cannot be employed to cause net synthesis of new glucose. For example, when carbon dioxide is fixed in β-carboxyl group of oxaloacetate by the action of pyruvate carboxylase.

$$\text{Pyruvic acid} + CO_2 + \text{ATP} \longrightarrow \text{oxaloacetic acid} + \text{ADP} + \text{Pi}$$

It is ultimately lost again as CO_2 in the subsequent reactions by which three carbon atoms of the oxaloacetate are converted to phosphoe-nolpyruvate and then to glucose, as already outlined. Furthermore, the oxidative decarboxylations of pyruvate to acetyl CoA and of * -ketoglutarate to succinyl CoA are irreversible in animal tissues and thus cannont lead to net formation of glucose from CO_2. From these considerations it is clear that the biosynthetic pathways in green plant cells that lead to net hexose formation from CO_2 must be qualitatively different from the carboxylation reactions taking place in animal tissues.

An important clue to the nature of the pathway from CO_2 to hexose in photosynthesis came from the work of Calvin and his associates. They illuminated green algae in the presence of radioactive carbon dioxide ($^{14}CO_2$) for very short intervals (only a few seconds) and then quickly "killed" the cells, extracted them, and with the aid of chromatographic methods, searched for the earliest radioactive products in which the labeled carbon could be found. One of the compounds which became labeled earliest was 3-phosphoglyceric acid, a known intermediate of glycolysis; the isotope was found predominantly in its carboxyl carbon atom. This carbon atom, which corresponds to the carboxyl carbon atom of pyruvate, does not become labeled in animal tissues in the presence of radioactive CO_2. These findings strongly suggested that the labeled 3-phosphoglycerate is an early intermediate in photosynthesis in algae, a view supported by the fact that 3-phosphoglycerate were then known. After an intensive search, an enzyme that catalyzes incorporation of $^{14}CO_2$ into the carboxyl carbon of 3-phosphoglycerate was found in large amounts in extracts from spinach leaves. This enzyme, called *diphosphoribulose carboxylase or lribulose diphosphate carboxydismutase*, catalyzes the carboxylation and hydrolytic cleavage of ribulose 1,5-diphosphate to form two molecules of 3-phosphoglycerate, one of which bears the isotopic carbon introduced as $^{14}CO_2$. The enzyme has a molecular weight of 550,000 and is extremely abundant in the cell, making up about 15 percent of the total protein of the chloroplast. Electron microscopic observations on spinach chloroplasts have suggested that individual molecules of the enzyme, which have a diameter of about 200 Å, are located on the surface of the thylakoids. Although the enzyme has a rather low turnover number, its very high concentration presumably accounts for the physiological rate of CO_2 uptake. The immediate chemical form in which carbon dioxide is fixed by the enzyme is free CO_2 rather than the HCO_3^- ion.

Fixation of CO_2 by diphosphoribulose carboxylase. The carbon atom of the entering CO_2 is in color.

$CH_2OPO_3^{2-}$
|
CO_2 → C=O
|
HCOH
|
HCOH
|
$CH_2OPO_3^{2-}$

Ribulose 1.5-diphosphate

+

H_2O

↓

$CH_2OPO_3^{2-}$
|
HCOH
|
COO^-

+

COO^-
|
HCOH
|
$CH_2OPO_3^{2-}$

3-Phosphoglycerate

The 3- phosphoglycerate formed by the enzyme can be converted into glucose 6-phosphate by reversal of the glycolytic reactions and the diphosphofructose phosphatase bypass reaction described above.

It is noteworthy that the glyceraldehyde 3-phosphate dehydrogenase of green plants, which is required to reduce 3-phosphoglycerate to glyceraldehyde 3-phosphate, is NADP -linked rather than NAD-linked.

This set of reactions does not by itself account for the fact that all six carbon atoms of hexose are ultimately formed from CO_2 during

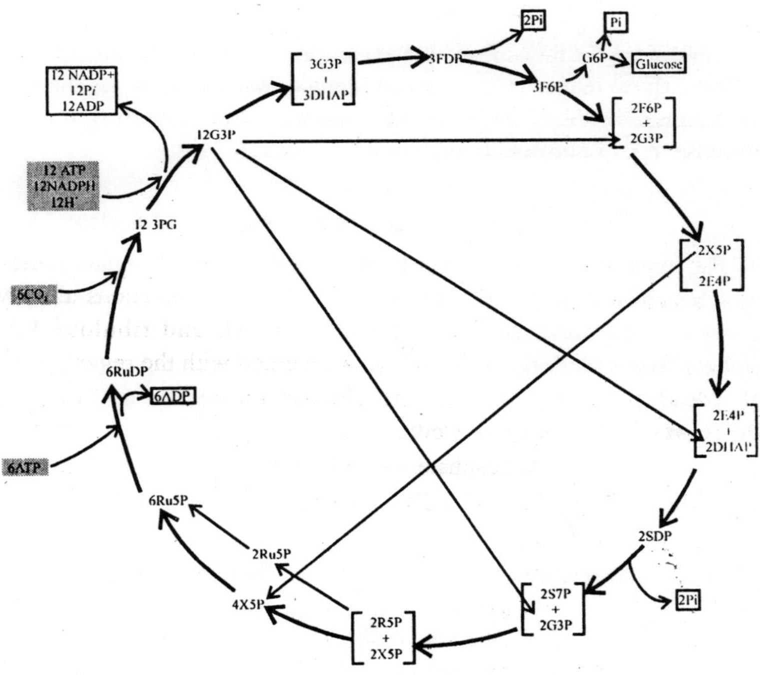

Fig.8.3 : The photosynthetic formation of glucose from CO_2 via the Calvin cycle in spinach leaves. The inputs are shaded in gray and the products in color. Abbreviations are 3 PG = 3-phosphoglyceric acid; G3P= glyceraldehyde 3-phosphate; DHAP = dihydroxyacetone phosphate, FDP = fructose 1,6-diphosphate, F6P=fructose 6-phosphate, G6P = glucose 6-phosphate, E4P= erythrose 4-phosphate, X5P = xylulose 5 - phosphate; SDP = sedoheptulose 1.7- diphosphate; S7P = sedoheptulose 7-phosphate; R5P -ribose 5-phosphate; Ru5P = ribulose 5-phosphate; RuDP = ribulose 1,5-diphosphate.

photosynthesis. To provide such a pathway, a cyclic mechanism for hexose synthesis was proposed by Calvin in which one molecule of ribulose 1,5-diphosphate is regenerated for each molecule of CO_2 reduced. It employs seven of the enzymes of the synthetic glycolytic pathway, three enzymes of the synthetic glycolytic pathway, three enzymes of the phosphogluconate pathway diphosphoribulose caroxylase (described above), and phosphoribulokinase (see below). One way of writing the equation of this complex cycle is as follows:

6 ribulose 1,5-diphosphate + 6 CO_2 + 18 ATP
+ 12 NADPH + 12 H^+ $\longrightarrow$
6 ribulose 1,5-diphosphate + hexose
+ 18 P_1 + 18 ADP + 12 $NADP^+$

Ribulose 1,5-diphosphate is written on both sides of the equation only to show that it is a necessary component which is regenerated at the end of the cycle. The net reaction, after canceling out the ribulose 1,5 -diphosphate, is

$6CO_2$ + 18ATP + 12NADPH + 12 H^+ $\longrightarrow$
hexose + 18 P_i + 18 ADP + 12 $NADP^+$

The separate reactions contributing to this overall equation may now be given in sequence. Reactions (1) to (8) are reactions already given for the formation of glucose from CO_2 and ribulose 1,5-diphosphate; reactions (9) to (15) are concerned with the regeneration of ribulose 1,5-diphosphate. the glucose formed as end product [reaction (8)] is shown in a box.

$6CO_2$ + ribulose 1,5-diphosphate $\longrightarrow$ 12 3-phosphoglycerate [1]

12 3-phosphoglycerate + 12 ATP $\longrightarrow$ 12 1, 3-diphosphoglycerater [2]

12 1,3 - diphosphoglycerate + 12NADPH + $12NADP^+$ $\longrightarrow$
12 glyceraldehyde 3 - phosphate + $12NADP^-$ [3]

5 glyceraldehyde 3 - phosphate $\longrightarrow$
5 - dihydroxyacetone phosphate [4]

3 glyceraldehyde 3 - phosphate 3 dihydroxyacetone phosphate
$\longrightarrow$ 3 fructose 1,6 - diphosphate [5]

3 - fructose 1,6 - diphosphate $\longrightarrow$ 3 fructose 6 - phosphate + 3P*i* [6]

fructose 6 - phosphate $\longrightarrow$ glucose 6 - phosphate [7]

glucose 6 - phosphate $\longrightarrow$ glucose + P*i* [8]

2 fructose 6 - phosphate + 2 glyceraldehyde 3 - phosphate
$\xrightarrow{\text{Transketolase}}$ xylulose 5 - phosphate + 2erythrose 4 - phosphate [9]

2 erythrose 4 - phosphate + 2 dihydroxyacetone phosphate
$\xrightarrow{\text{Aldolase}}$ 2 sedoheptulose 1.7 - diphosphate [10]

2 sedopheptulose 1,7 - diphosphate $\xrightarrow{\text{Phosphatase}}$
2 sedoheptulose 7 - phosphate + 2P*i* [11]

2 sedopheptulose 7 - phosphate + 2 glyceraldehyde 3 - phosphate
$\xrightarrow{\text{Transketolase}}$ 2 ribose 5 - phospyhate +
2 xylulose 5 - phosphate [12]

2 ribose 5 - phosphate $\xrightarrow{\text{Isomerase}}$ 2 ribulose 5 - phosphate [13]

4 xylulose 5 - phosphate $\xrightarrow{\text{Epimerase}}$ 4 ribulose 5 - phosphate [14]

6 ribuloser 5 - phosphate + 6ATP $\xrightarrow{\text{Phosphoribulokinase}}$
6 ribulose 1,5 - diphosphate + 6ADP [15]

Reaction (11) is catalyzed by a phosphatase similar to diphosphof-ructose phosphotase; it is also highly exergonic and is a "pulling" reaction. Reactions (13) and (14) are catalyzed by the pentose isomerase and epimerase described earlier and Reaction (15) by phosphoribulokinase, an enzyme similar in many respects to phosphofr-uctokinase. The latter reaction is also strongly exergonic. the intermediate *sedoheptulose 1, 7-diphosphate* (margin) is similar in its chemical and biochemical properties to its six-carbon analog fructose 1,6 diph-osphate. This complex but therm-odynamically feasible pathway accounts for many of the experimental observations on the formation of hexose from CO_2 during plant photosynthesis; it is now believed to represent the major pathway for this process.

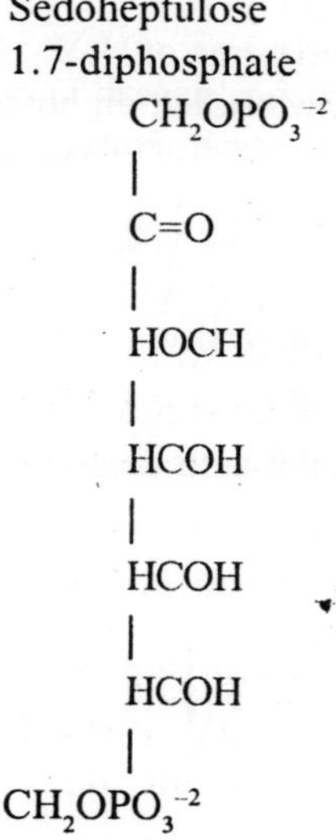

Other Possible Pathways of Hexose Formation

It appears very likely that pathways other than the calvin cycle may exist for the reduction of CO_2 during photosynthesis. In photosynthetic bacteria the oxidative decarboxylations of pyruvate to acetyl CoA and of α-ketoglutarate to succinyl CoA, which are irreversible in ADP-glucose (ADPG) rather than UDPG is the active glucose donor in most plants:

$$ATP + \alpha\text{-D-glucose 1-phosphate} \rightleftharpoons \text{ADP-glucose} + PPi$$

$$\text{ADP-glucose} + (\text{glucose})_n \rightarrow \text{ADP} + (\text{glucose})_{n+1}$$

The first of these reactions is catalyzed by ADP-*glucose pyrophosphorylase*, and allosteric enzyme which is stimulated by 3-phosphoglycerate and fructose 1,6-diphosphate, early products of lthe photosyn-thetic reduction of CO_2. In some plants, ADPG and UDPG are equally active glucose donors.

Conversion of Glucose to Other Monosaccharides

Pathways for the conversion of glucose 6-phosphate into D-fructose 6-phosphate and then into D-mannose 6- phosphate have been described in Chapter 15. Other hexose interconversions proceed via nucleoside diphosphate hexoses formed by the action of pyrophosphorylases. One of the most important is the conversion of D-glucosyl into D-glactosyl residues, which proceeds by enzymatic epimerization of UDP-glucose at carbon atom 4 of the glucose residue, to form uridine diphosphate galactose.

$$\text{UDP-glucose} \rightleftharpoons \text{UDP-galactose}$$

Since the enzyme catalyzing this reaction, *uridine diphosphate glucose epimerase*, has an absolute requirement for NAD, it is believed that epimerization occurs in two separate steps:

$$\text{UDP-glucose} + NAD^+ \rightleftharpoons \text{UDP-4-ketoglucose} + NADH + H^+$$

$$\text{UDP-4-ketoglucose} + NADH + H^+ \rightleftharpoons \text{UDP-galactose} + NAD^+$$

The postulated 4-ketoglucose intermediate (margin), which can accept a pair of hydrogens at the keto group to form eigher epimer, remains tightly bound to the enzyme active site during the catalytic cycle. UDP-galactose formed in this reaction is a required precursor in the synthesis of the disaccharide lactose in the mammary gland.

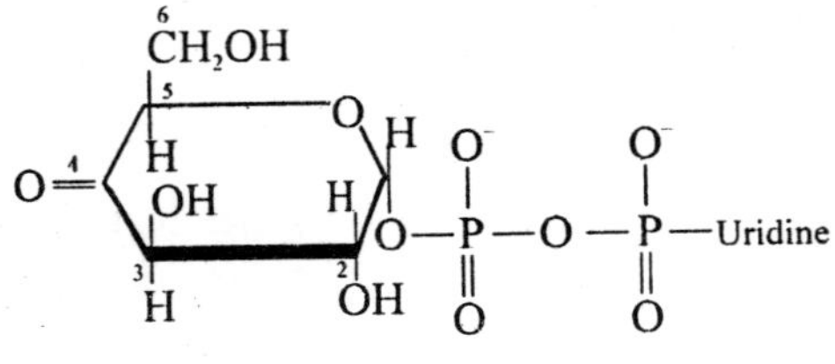

UDP-4-Ketoglucose

Free D-galactose, formed by the enzymatic hydrolysis of lactose in the intestinal tract, is ultimately converted into D-glucose in animal tissues by a series of reactions which has attracted much attention because a genetic defect in one of the steps results in galactosemia, a rare hereditary disorder in human infants. In the liver, D-galactose is first phosphorylated at carbon atom 1 by *galactokinase*, to yield D-galactose 1-phosphate:

$$\text{ATP} + \text{D-galactose} \rightarrow \text{ADP} + \text{D-galactose 1-phosphate.}$$

The resulting galactose 1-phosphate can be converted into UDP-galactose by one of two possible reactions. The first is catalyzed by *phosphogalactose uridyl transferase.*

$$\text{UDP-glucose} + \text{galactose 1-phosphate} \rightleftharpoons \text{UDP-galactose} + \text{glucose 1-phosphate}$$

This enzyme is normally present in high amounts in the liver of infants, but is genetically defective in galactosemic infants. The second enzyme capable of utilizing galactose 1-phosphate is *uridine diphosphate galactose pyrophosphorylase*, which catalyzes the reaction

$$\text{UPT} + \text{galactose 1-phosphate} \rightleftharpoons \text{UDP-galactose} + \text{PP}_\text{I}$$

The latter enzyme is present only in traces in fetal and infant liver, but is much more active in adult liver. The hereditary lack of the first of these two enzymes, the phosphogalactose uridyl transferase, causes infants to lack the capacity to metabolize D-galactose derived from the lactose of milk, since they also lack the second enzyme such infants have an excessively high concentration of D-galactose in the blood and suffer from cataract (opacity) of the lens of the eye as well as mental disorders. This condition can be successfully treated by witholding milk and other sources of galactosemic metabolizes galactose pyrophosphorylase.

UDP-glucose may undergo oxidation to UDP-Dglucuronic acid (margin)

$$\text{UDP-glucose} + 2\ \text{NAD}^+ \rightleftharpoons \text{UDP =glucuronic acid} + 2\ \text{NADH} + 2\ \text{H}^+$$

In this reaction, two oxidation steps occur, one oxidizing the hydroxyl group at carbon atom 6 to the aldehyde and the second oxidizing the aldehyde to a carboxyl group. UDP-glucuronic acid is a precursor of *glucosiduronides* (see below).

Free D-glucuronic acid, which is formed from UDP-glucuronic acid by enzymatic hydrolysis, is a precursor in the biosynthesis of L-ascorbic acid or vitamin C. This process occurs in the liver of most animals with the exception of man, monkeys, the guinea pig,

COOH
H H O H O⁻ O⁻
OH H
HO O—P—O—P—Uridine
H OH O O

Enzymatic synthesis of L-ascorbic acid

UDP-D-glucuronic acid

↓ H_2O → UDP

HC = O
HCOH
HOCH
HCOH
HCOH
COOH

D-Glucuronic acid

↓ NADPH + H^+

CH_2OH	COOH
HCOH	HOCH
HOCH	HOCH
HCOH	HCOH
HCOH	HOCH
COOH	CH_2OH

L-Gulonic acid

↓ H_2O

O=C
HOCH
HOCH O
HC
HOCH
CH_2OH

L-Gulonolactone

↓ -2H

O=C
HOC
‖ O
HOC
HC
HOCH
CH_2OH

L-Ascorbic acid

and the Indian fruit bat. The D-glucuronic acid is first reduced to L-gulonic acid, which lactonizes in the presence of a lactonase to L-gulonolactone, which then is oxidized to L-ascorbic acid, presumably via 3-keto-L-gulonolactone (margin).

Synthesis of Disaccharides and Other Glycosides

Nucleoside diphosphate sugars also function as precursors in the biosynthesis of various disaccharides. In plants, sucrose is formed from glucose and fructose by the following series of reactions:

ATP + glucose → glucose 6-phosphate + ADP

Glucose 6-phosphate ⇌ glucose 1-phosphate

UTP + glucose 1-phosphate → UDP-glucose + PP_i

ATP + fructose → fructose 6-phosphate + ADP

UDP-glucose + fructose 6-phosphate →
UDP + sucrose 6'-phosphate

Sucrose 6'-phosphate + H_2O → sucrose + P_i

Sum: 2ATP + UTP + glucose + fructose →
sucrose + 2 ADP + UDP + PP_i + P_i

If the pyrophosphate formed is hydrolyzed to P_i, then the overall reaction is

2 ATP + UTP + glucose + fructose →
sucrose + 2 ADP + UDP + $3P_i$

Three high-energy phosphate bonds $[\Delta G^{\circ\prime} = -21.9$ kcal] are thus required to form the single glycosidic bond of sucrose [$\Delta G^{\circ\prime}$ for hydrolysis is –6.6 kcal]. The overall reaction of sucrose synthesis from glucose and fructose is therefore quite irreversible. Presumably the strongly exergonic nature of this set of biosynthetic reactions enables the sugar cane to form a juice containing an extremely high concentration of sucrose from very dilute precursors. In some plants, sucrose is formed by an alternative reaction employing fructose rather than fructose 6-phosphate :

UDP-gucose + fructose → UDP + sucrose

Some bacteria contain the enzyme sucrose phosphorylase, which catalyzes the reversible reaction.

Sucrose + P_i ⇌ glucose 1-phosphate + fructose

Although this reaction can be made to form sucrose if the P_i is rapidly removed, under intracellular conditions it normally catalyzes the breakdown of sucrose.

The disaccharide lactose is formed in the mammary gland from D-glucose and UDP-galactose by the action of two enzyme proteins, which together constitute the *lactose synthetase* system. The first, protein A which is found not only in the mammary gland but also in the liver and small intestine, catalyzes the reaction.

$$\text{UDP-galactose} + \text{N-acetyl-D-glucosamine} \rightarrow \text{UDP} + \text{N-acetyllactosamine.}$$

The second, *protein B*, is a proteinlong known as the *α-lactalbumin* of milk; it has not catalytic activity of its own. This protein has quite recently been found to alter the specificity of protein A, regardless of its source, so that it will utilize D-glucose instead of N-acetyl-D-glucosamine as galactose acceptor, causing the A protein to make lactose instead of N-acetyllactosamine:

$$\text{UDP-galactose} + \text{D-glucose} \rightarrow \text{UDP} + \text{lactose}$$

The α -lactalbumin protein is remarkable in another respect: it has a striking homology in its amino acid sequence to the enzyme lysozyme.

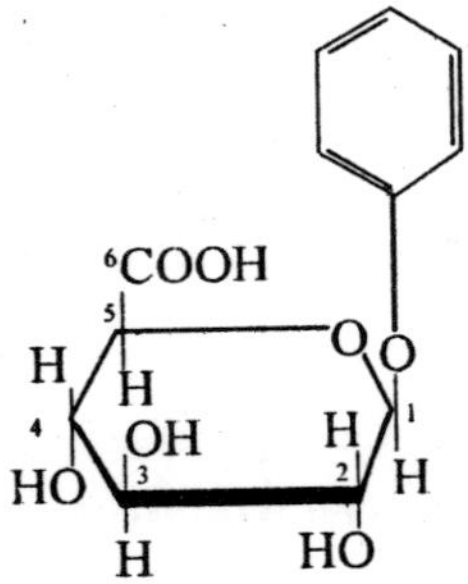

Phenol glucosiduronide

Glucosiduronides (see above) are glycosidic excretory products formed by vertebrates from foreign aromatic or alicyclic alcohols and amines by the action of enzymes in the endoplasmic reticulum of the liver. The general reaction catalyzed is

$$\text{UDP-glucuronic acid} + \text{ROH} \rightarrow \text{R-O-glucosiduronide} + \text{UDP}$$

where ROH is the foreign alcohol. Phenol is excreted by some animals as *phenol glucosiduronide* (margin).

Nucleoside diphosphate sugars a'so may undergo rather complex reduction rections to form nucleoside diphosphate derivatives of deoxy sugars, such as L-fucose and L-rhamnose (Chapter 11), which are important components of lipopolysaccharides of bactereial cell walls. One such reaction is

$$\text{GDP-D-Mannose} + \text{NADPH} + H^+ \rightarrow \text{GDP-L-fucose} + \text{NADP}^+ + H_2O$$

Cori's cycle (Lactic Acid Cycle)

Liver glycogen may get converted into muscle glycogen or vice versa. The inter-relationship is shown by **Cori's** cycle (*lactic acid cycle*). Liver glycogen is broken down into glucose (*by glycogenolysis*) which diffuses into the blood stream and is carried to the muscles where it synthesizes muscle glycogen (*by glycogenesis*). Muscle glycogen breaks into lactic acid (by *glycogenolysis and glycolysis*) which diffuses out of muscles into the blood stream. It is carried to the liver tissue where it can synthesize liver glycogen. It is evident that liver glycogen may be converted into muscle glycogen by intermediate formation of lactic acid. This difference is due to the fact that muscle tissues lack the enzyme **glucose-6-phosphatase,** and hence, formation of free glucose is not possible in the muscles. Glucose-6-phosphate is unable to cross the membrane of muscle cells. Thus, it has got to be converted to lactic acid which is freely permeable. Liver cells contain **glucose-6-phosphatase** which converts glucose-6-phosphate into glucose. Glucose is freely permeable through cellular membranes of liver as well as muscle tissues.

CHAPTER 9

PROTEIN STRUCTURE AND FUNCTION

Proteins

Proteins are the nitrogen containing substances of immense importance (*proteos denotes of primary importance*) to the living beings. These are required by the living beings for the formation of **body tissues, enzymes, hormones, immunoglobulins** and other **intracellular** and **extracellular proteins**. Majority of the proteins are high molecular weight substances having very complex structures. On hydrolysis, proteins give **amino acids**, and some times certain non-protein residue(s) in addition to the amino acids. Usually twenty different amino acids are obtained at the most in the protein hydrolysates. Thus, amino acids are the 'building stones' of the protein molecules. Amino acids in the different protein molecules remain arranged in different but definite sequences which impart a specific type of property in these molecules.

Biological Functions of Proteins

Proteins play crucial roles in virtually all biological processes. The significance and remarkable scope of their functions are exemplified in:

Fig. 9.1: Photomicrograph of a crystal of hexokinase, a key enzyme in the utilization of glucose. [courtesy of Dr. Thomas Steitz and Dr. Mark Yeager.]

1. Enzymatic catalysis. Nearly all chemical reactions in biological systems are catalyzed by specific macromolecules called enzymes. Some of these reactions, such as the hydration of carbon dioxide, are quite simple. Others, such as the replication of an entire chromosome, are highly intricate. Nearly all enzymes exhibit enormous catalytic power. They usually enhance reaction rates by at least a millionfold. Indeed, chemical transformations rarely occur at perceptible rates in vivo in the absence of enzymes. Several thousand enzymes have been characterized, and many of them have been crystallized. The striking fact is that all known enzymes are proteins. Thus, proteins play the unique role of determining the pattern of chemical transformation in biological systems.

2. Transport and storage. Many small molecules and ions are transported by specific proteins. For example, hemoglobin transports oxygen in erythrocytes, whereas myoglobin, a related protein, transports oxygen in muscle. Iron is carried in the plasma of blood by transferrin and is stored in the liver as a complex with ferritin, a different protein.

3. Coordination motion. Proteins are the major component of muscle. Muscle contraction is accomplished by the sliding motion of two kinds of protein filaments. On the microscopic scale, such coordinated motions as the movement of chromosomes in mitosis and the propulsion of sperm by their flagella also are produced by contractile assemblies consisting of proteins.

4. Mechanical support. The high tensile strength of skin and bone is due to the presence of collagen, a fibrous protein.

5. Immune protection. Antibodies are highly specific proteins that recognize and combine with such foreign substances as viruses, bacteria, and cells from other organisms. Proteins thus play a vital role in distinguishing between self and nonself.

6. Generation and transmission of nerve impulses. The response of nerve cells to specific stimuli is mediated by receptor proteins. For example, rhodopsin is the photoreceptor protein in retinal rod cells. Receptor molecules that can be triggered by specific small molecules, such as acetylcholine, are responsible for transmitting nerve impulses at synapses — that is, at junctions between nerve cells.

7. Control of growth and differentiation. Controlled sequential expression of genetic information is essential for the orderly growth and differentiation of cells. Only a small fraction of the genome of a cell is expressed at any one time. In bacteria, repressor proteins are important control elements that silence specific segments of the DNA of a cell. A quite different way in which proteins act in different entiation is exemplified by nerve growth factor, a protein complex that guides the formation of neural networks in higher organisms.

Classification of Proteins

A satisfactory classification of the proteins has not been possible till now as the chemical structure of most of the proteins is not known at present. The following classification is based merely upon solubil·ties and other physical properties of the proteins. On this basis, the proteins have been classifed in three groups, namely **simple proteins, conjugated proteins** and **derived proteins.**

According to structure

Fibrous Proteins. These proteins are associated with cellular elements and often serve the function of supporting specific structures of the cell. Examples are wool, silk fibroin, collagen (connective tissue), myosin (muscle), keratin (hair) and fibrin (blood clot). These proteins are largely insoluble in aqueous media and have high molecular weights that cannot be accurately estimated because of difficulties encountered in their purification. They appear as fibers made up of linear molecules that are arranged roughly parallel to the fiber axis. They are amorphous and some are capable of stretching and contracting. Human fibrin has the molecular dimensions of 38 × 700 Å.

Globular Proteins. In contrast to fibrous proteins, the globular proteins are soluble in aqueous media and can be isolated in the crystalline state. Although not necessarily spherical, they are less asymmetric than fibrous proteins and consist of polypeptides (chains of amino acid residues) that are held together by cross-linked groups or in an aggregated state. Such aggregates are held together in a three-dimensional structure by relatively weak non-covalent bonds.

According to solubility

1. Albumins

These proteins are *soluble* in water and dilute salt solutions, and can be coagulated on heating. These can be precipitated from their solutions upon *full saturation* with ammonium sulphate. Albumins thus salted-out may be purified by the process of dialysis. The coagulated albumins are insoluble in water, dilute acid, alkali and salt solutions. Coagnlation starts above 75°c but this temperature differs for albumins obtained from different sources. The common examples of albumins are **ovalbumin, serum albumin, lactalbumin, myoalbumin**. The plant tissues have also been reported to contain albumins.

2. Globulins

These proteins are insoluble in water and soluble in dilute salt solutions, but can be coagulated on heating. These can be precipitated from aqueous solutions merely by fifty percent saturation with ammonium sulphate. These are found both in the animal and plant tissues. The examples of globulins are **ovoglobulin** (*eggs*), **lactoglobulin** (*milk*), **fibrinogen, alpha, beta and gamma globulins** (*all in serum*), **myosin** and **tropmyosin** (*both in muscles*), **edestin** (*hempseed*) and **excelsin** (*Brazil nut*).

Globulins may be precipitated merely by diluting globulin solutions since these are soluble in dilute salt solutions but insoluble in salt-free water or extremely dilute salt solutions. Globulins may also be precipitated merely by dialysis of the globulin solutions. Alubmins may be separated from globulins by using this technique.

3. Globins

These are neutral type of histones which cannot be precipated by adding ammonium hydroxide as histones can be precipitated. Such proteins can easily ombine with heme forming hemoglobin.

4. Glutelins.

These are insoluble in neutral aqueous solutions but soluble in dilute acid or alkali. Examples are glutenin (wheat) and oryzenin (rice).

5. Prolamines

Prolamines are *water insoluble* but *soluble* in 70 percent ethanol. These proteins are also insoluble in absolute alcohol and other neutral solvents. Such proteins are mainly found in the cereals. The examples of such proteins are **gliadin** of the wheat protein, zein (*corn protein*) and **hordein** (*barely protein*). Gliadin is found to be poor in lysine content. **Zein and hordein** are poor in lysine, tryptophan and valine content. The amide nitrogen content of prolamines is found to be very high.

6. Histones

More basic than most proteins, the histones tend to form complexes with acidic compounds in the cell (nucleic acids). They are soluble in water and insoluble in dilute ammonia solution. Examples are thymus histone (also called nucleohistone because it is found combined with nucleic acids and scombrone (mackerel.).

7. Protamines

Compared with most proteins, these are relatively small molecules. The protamines are soluble in water and basic in character. They are found associated with nucleic acids in the sperm of fish and are often called nucleoprotamines. Examples are salmine (salmon), sturine (sturgeon), clupeine (herring), and cyprinine (carp).

8. Albuminoids

These are characterized by their *marked insolubility* in water and in all other neutural solvents. The **keratins, elastins, collagens** and **keratohyaline** are all the examples of albuminoids. **Keratins** are present in the hairs, nails, horns, hoofs and feathers. The outermost layer of skin is also formed by keratin. **Keratohyaline** is present in the layer of skin. Keratohyaline is later on converted into keratin. These proteins are rich in arginine, lysine and histidine content. These can be hydrolysed only by concentrated acids and alkalies. Although these cannot be usually digested in the human alimentary canal but some of the moths (*cloth moths*) and micro organisms can digest such proteins. **Collagen** is found to exist in the connective tissues and in the bones, and are insoluble in all the neutral solvents. **Elastins** are found to be present in the yellow elastic fibers of the connective tissues. The various ligaments are purely elastins. These can be digested by **pepsin** and **trypsin**. The insolubility of the keratins has been very useful because it prevents dissolutions of body in water, or dilute acids and alkalies.

Conjugated Proteins

These are also known as compound proteins. Conjugated proteins are those proteins which contain in their molecular structure, in addition to protein, some non-protein organic substance(s) called prosthetic group or a metal. These proteins have also been subclassified into the following groups.

(a) Glycoproteins. Those proteins which contain carbohydrate in their molecular structure are included in this group. These include **mucopolysaccharides** and **mucoproteins**. The mucopolysaccharides contain predominantly carbohydrate and less protein, whereas the mucuproteins contain more protein and less carbohydrate. The **mucoproteins** include **ceruloplasmin** which is a blue copper containing protein, transferrin which is a iron transferring protein, **ossemucoprotein** of bones, tendomucoproteins which is a iron transferring protein, osseomucoprotein of bones, **tendomucoproteins** which are found to be present in the tendons, **chondro-mucoproteins** which are found to be present in the cartilage, and **mucin** which is found in the digestive tract.

(b) Lipoproteins. These are those proteins which contain in their molecular structure, in addition to protein, a **lipid substance**. These are found to be soluble in the lipid-solvents. There exists a loose combination between protein and lipid. Such type of proteins occur mainly in the plasma, brain and eggs. In the plasma, these remain bound to mainly globulins. If such combinations contain α -globulin, these are known as the α -lipoproteins, and if β-globulins, then these are called β-lipoproteins.

(c) Nucleoproteins. In addition to nucleohistones and nucleoprotamines, there is a third group of proteins capable of binding nucleic acids. These nucleoproteins are not basic and hold the nucleic acid by secondary valence bonds. They occur in microorganisms and are soluble in isotonic salt solution.

(d) Chromoproteins. These proteins are pigmented owing to the prosthetic nonprotein group. They include hemoglobin (containing heme, an iron-protoporphyrin), ceruloplasmin (copper), hemocyanin (copper), ascorbic acid oxidase (copper) and ferritin (iron).

(e) Dehydrogenases. These are conjugated protein enzymes that possess oxidizing activity because of the presence of prosthetic groups such as NAD^+ (nicotinamide adenine dinucleotide), $NADP^+$ (nincotinamide adenine dinucleotide phosphate), FMN (flavin mononucleotide), and FAD (flavin adenine dinucleotide).

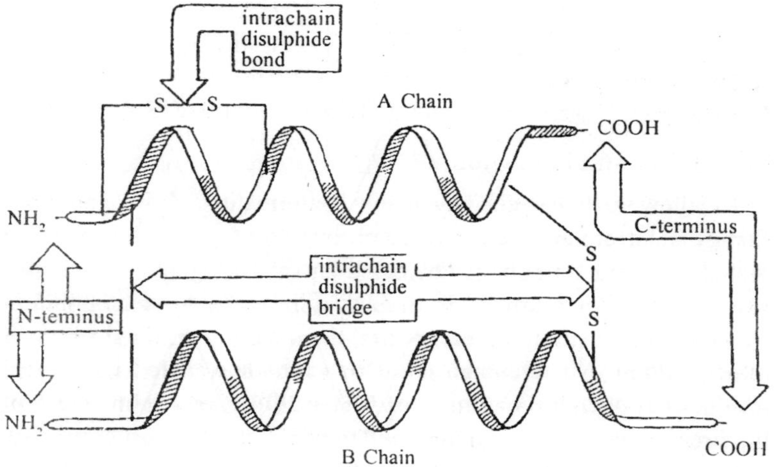

Fig. 9.2: Molecular structure of insulin (After Sheeler and Bianchi, 1987).

(f) Phosphoproteins. These proteins contain **phosphoric acid** in their molecular structure in addition to amino acids. Nucleoproteins and lipoproteins which do also contain phosphoric acid are not included in this group. The important examples of such proteins are **casein** and **vitellin**. Casein is found to be present in the milk, and vitellin is present in the egg-yolk.

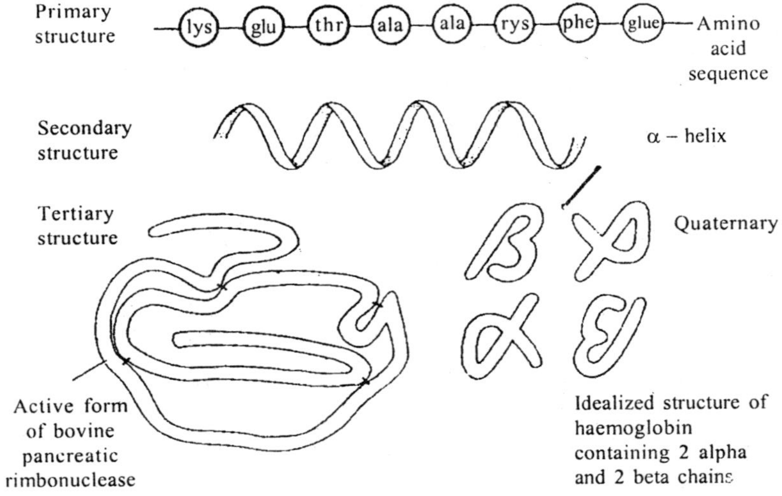

Fig.: 9.3. The four levels of structural organization in protein molecules (after Stansfield, 1969).

(g) Metalloproteins. Such proteins contain **metal (s)** alone which remains attached to the protein. The examples of such protein are **ferritin** (metal portion iron), and **carbonic anhydrase** (Zn).

Myoglobin and hemoglobin

Kendrew in 1950 studied the native conformation of myoglobin, the first globular protein the native conformation of myoglobin, the first globular protein to be studied by the X-ray crystallographic technique. Myoglobin, a heme protein from muscle is similar to hemoglobin in structure and function. Perutz studied the three dimensional structure of hemoglobin. Myoglobin, a single polypeptide with 153 amino acid residues has overall dimensions of 45Å + 30Å + 30Å. About 78% of the structure is in α -helical form and there are no β-pleats. In the case of hemoglobin also more than 70% of the structure is α -helix. This is not however, a characteristic feature of all globular proteins. α -

Chymotrypsin, a proteolytic enzyme has relatively high proportion of β-pleats (45%) and low amount of α -helix (14%). Lysozyme, another globular protein has 40% α -hlelix ad 12% β-pleats. Myoglobin has eight stretches of α -helix interspersed with non-ordered segments (figure 15). The heme group is located in a crevice between two helices. The prosthetic group is on the surface facilitating oxygen binding. Most of the hydrophobic side chains are buried inside and the polar residues are on the surface. As indicated earlier, hemoglobin is made up of two α -chains and two β-chains. The overall conformations of myoglobin, α -and β-chains of hemoglobin are similar. It is probable these three polypeptides are derived from a single ancestral protein by gene mutation. Interaction between α -and β-chains of hemoglobin is very strong whereas, interaction between homologous chains is weak. Thus, the tetrameric protein can easily be dissociated to $\alpha\beta$ dimers. In both α and β chains, some non-polar groups are on the surface which are important in the interaction of the subunits.

Small alterations in the primary structure of a protein can adversely affect its native conformation and function. A well known condition is sickle cell anaemia, a hereditary disorder. About 10% of the American Blacks are heterozygotes (carriers of the defective gene) and 0.4% are homozygotes (affected directly). Hemoglobin content in affected persons is about 50% of the normal value, since they suffer from repeated crisis brought about by physical exercise. They become weak and dizzy and are short of breath. Microscopic examination will show that erythrocytes are abnormal in shape. The cells are long and crescent-like akin to the blade of a sickle. The only alteration in the primary structure is the substitution of valine in place of glutamate in the sixth position in the β-chains. This results in what is known as a 'sticky path' which causes aggregation of hemoglobin molecules when O_2 content is low.

Protein Denaturation

Since the conformation of a protein is solely dependent on week valence forces it can be disrupted by a variety of physical and chemical agents. The process is known as denaturation. The organized structure is lost and the protein assumes a highly disordered form. The protein will lose its biological activity. It becomes more susceptible to proteolytic attack. Many globular proteins are rendered less soluble on denaturation. An interesting instance of a protein losing its biological activity on denaturation is with reference to monellin, a protein 300 times sweeter than socrose. Monellin isolated from an African plant

Dioscorephyllum cumminsis is made up of two polypeptide chains made up of 44 and 55 amino acid residues, non-covalently associated. On denaturation the protein loses its sweet taste. If the denaturation process is not very drastic the process becomes reversible, once the denaturing agent is removed. This phase is known as renaturation.

Moist heat is powerful denaturing agent. Hydrogen bonds are easily disrupted by heat. Extremes of acidity and alkalinity at moderate temperature, disrupt electrostatic interactions and cause conformational changes. Heavy metal ions like Hg^{++} and Pb^{++} can complex with negative charges or sulphydryl groups. The basis of administering egg white immediately to counteract lead poisoning is based on the complexation of egg proteins with the toxic metal ion. Polar organic solvents like ethanol also can cause denaturation, if temperature is not controlled. Trichloroacetic acid and perchloric acid are two common laboratory reagents used for denaturation and precipitation of proteins. Sodium dodecyl sulphate (SDS, $CH_3\,(CH_2)_{10}\,CH_2 - OSO_3^-\,Na^+$) another reagent used in protein analysis is also a powerful denaturing agent . It can dissociate subunits by disrupting non-polar interactions. Urea ($H_2N - CO - NH_2$) and guanidine hydrochloride [$H_2N - C\,(NH_2) = N^+\,H_2Cl^-$] at high concentrations (6.0 – 8.0 M and 2.0 – 3.0 M respectively) act as denaturing agents. Unlike commonly believed, urea does not act by disrupting hydrogen bonding. It acts by interfering with non-polar interactions.

Plasma Proteins

Hemoglobin the major protein in blood (12-15 g/100 ml) is confined to the red cells. Plasma, the cell-free fraction is also rich in proteins (6.3-7.5 g/100 ml). A large array of structurally and functionally different proteins is present in plasma. There are carrier proteins like albumin, transferrin and apolipoproteins. Quantitatively the second important group of proteins after albumin are the immunoglobulins, which are defence proteins. Another group of proteins known as protein inhibitors accounts for 10% of total plasma proteins. A variety of clotting factors which are mostly precursors for proteolytic enzymes involved in blood coagulation are present in relatively low concentrations in plasma. There are circulating protein hormones like insulin and thyroid stimulating hormone in very minute amounts. Enzymes which are mostly nonfunctional in plasma, are also present in traces.

Several techniques are employed for the qualitative and quantitative separation of plasma proteins. A common technique used in clinical laboratories for the separation of serum (rather than plasma to eliminate fibrinogen) is zone electrophoresis on paper, agar gel or cellulose acetate strips at pH 8.6. A typical electrophoretic pattern is shown in figure 17. Five fractions in the order of increasing mobility towards anode namely albumin, α -1, α -2, β and $-\gamma$ globulins are usually obtained by this technique. Each globulin fraction is comprised of more than one protein which can further be separated by sophisticated techniques.

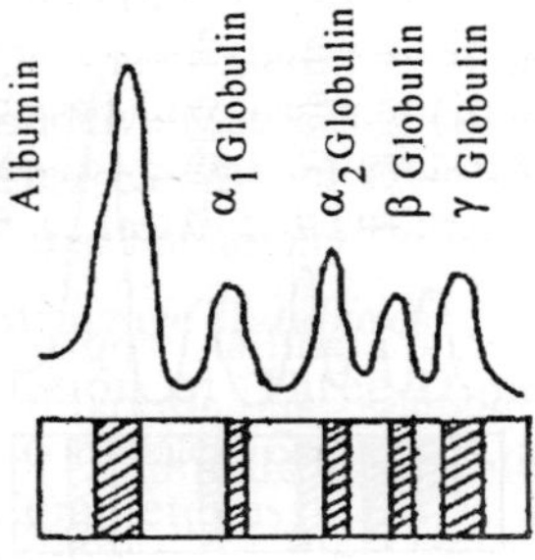

A. After the electrophoretic run the protein bands are stained to visualize the proteins with suitable dyes. Proteins bind tightly to colored dyes through electrostatic and hydrophobic interactions.

B. Semiquantitation of the protein bands by scanning in a densitometer.

Fig 9.4. Electrophoretic pattern of serum proteins

Table 9.1. Plasma proteins

Proteins	Concentration G/100 ml	Mr kDa	Function
Prealbumin	0.03	61	Transport of thyroxine
Albumin	3.2 - 4.2	68	Diverse functions
α_1-Acid glycoprotein	0.10	44	Acute phase protein
α_1-Proteinase inhibitor	0.25	54	Elastase inhibitor
α_2-Macroglobulin	0.20	720	Nonspecific proteinase inhibitor
Haptoglobin	0.10	85	Binds hemoglobin
Ceruloplasmin	0.02	150	Copper transport
Prothrombin	0.02	63	Blood clotting
Transferrin	0.30	85	Transport of iron
Plasminogen	0.05	140	Retraction of fibrin clot
Fibrinogen	0.30	340	Precursor of fibrin clot
γ -Globulins	0.8-1.7	–	Immune reaction

Albumin the major plasma protein, accounts for about 55-60% of the total proteins. It is synthesized exclusively in liver and a healthy liver can synthesize 10-15 g of albumin per day. The life span of an albumin molecule is about 20-25 days. Nearly 40% of plasma calcium is bound to albumin. Albumin is involved in the transport of free fatty acids from adipocyte to liver. Bilirubin, steroid hormones and many water insoluble drugs are also bound and transported by albumin. Along with other plasma proteins albumin acts as a buffer. Albumin has a major role in osmotic regulation and fluid distribution. It accounts for about 80% of the total colloidal osmotic pressure (25 mm Hg) in blood plasma. One g of albumin can hold 18.0 ml of water. Decreases in albumin level will cause fluid accumulation in the interstitial space and soft tissues resulting in edema. Prolonged malnutrition, kidney diseases associated with protein loss in urine, liver cirrhosis and extravagation as in burns will cause this condition. However, in analbuminemia a rare genetic disorder, persons do not develop edema and are apparently healthy. In this condition globumin level increases and arterial blood pressure is lowered to compensate for the absence of albumin.

The predominant component of α_1-globulin (0.3-0.5 g/100 ml) is a glycoprotein known as α_1-proteinase inhibitor which is a powerful inhibitor of neutrophil elastase. Elastase when not regulated can cause tissue breakdown and loss of elasticity in the lung. Two variants of this protein (S and Z) which are far less effective against elastase are known mainly in the Scandinavian countries. Persons with these modified α_1-proteinase inhibitors develop emphysema. α_1-Acid glycoprotein is so called because of its low isoelectric point. The concentration of this protein is significantly increased in plasma in inflammation and other acute disease conditions. Nearly 30% of the α_2-fractions and 40% of β-fraction are accounted for by apolipoproteins, which are described later. The functions of transferrin and haptoglobin which belong to the β and α_2-globulin fractions are also detailed later. Ceruloplasmin a copper containing protein belongs to α_2-globulin fraction. A major protein of this fraction is α_2-macroglobulin which has the capacity to trap and remove from circulation a large number of proteinases and can hence play a regulatory role.

γ-Globulins

This group of proteins which are unique to vertebrates are also called immunoglobulins. Unlike most other plasma proteins which are synthesized in liver, immunoglobulins are made in β-lymphocytes and plasma cells of lymph nodes, bone marrow and spleen. They bind specifically to antigenic sites on other molecules. The antigens, which provide the binding sites, can be proteins, pollysaccharides, nucleic acids or even relatively low molecular weight compounds, which are of foreign origin to the body. Thus, bacterial and viral components can act as antigens and trigger antigen-antibody reactions. Models of such reactions are shown in the figure.

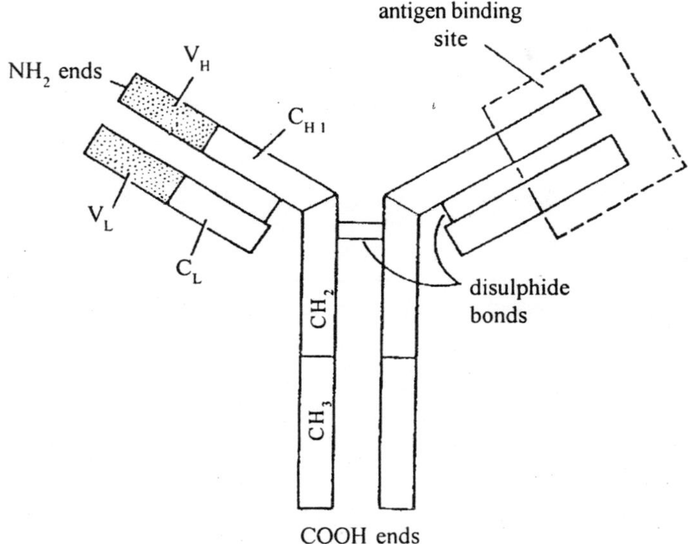

Fig. 9.5: Molecular structure of an immunoglobulin molecule (after Stansfielde, 1986)

Knowledge of immunoglobulins came from the studies on protein in plasma and urine in patients suffering from myelomatosis (immunocytoma tumors). In this condition, γ -globulin level is increased in serum. Characteristic proteins known as Bence-Jones proteins are excreted in large amounts in urine. These proteins precipitate on heating urine samples to 45-50°C but redissolve at higher temperatures. They were identified as low molecular weight proteins (light chains) which are constituent units of immunoglobulins. Two types of light chains known as κ and λ chains are present in immunoglobulins and in humans the former predominates. Each immunoglobulin molecule is a heterotetramer made up of two light chains and two heavy chains. Light chains are simple proteins whereas, the heavy chains are glycoproteins. The light and heavy chains are synthesized as separate molecules and are assesmbled to functional immunoglobulins in the cells of their origin. The immunoglobulins are classified into five groups based on the type of heavy chain.

Table 9.2. Classification of immunoglobulins (Jg)

Group	Heavy Chain	Mr kDa	Native form	Total Mr, kDa	Plasma concentration mg/100 ml	Percent carbohydrate
Ig G	γ	53	$\gamma_2 \kappa_2$	150	800-1200	3.0
Ig A	α	64	$\alpha_2 \kappa_2$	170	150-300	8.0
Ig M	μ	70	$(\mu_2 \kappa_2)_5$	900	50-150	12.0
Ig D	δ	58	$\delta_2 \kappa_2$	160	4	13.0
Ig E	$\in$	75	$\in_2 \kappa_2$	190	0.03	12.0

In general, Ig molecules can be represent as Y-shaped structure. Ig M is a pentameric structure. A polypeptide known as J-chain (Mr 15 kDa) links the monomeric units through disulphide bridges. Ig A also has J- chain and in this instance, aggregation stops at the dimer stage.

Both the heavy chains and light chains consist of distinct globular regions known as domains. These highly organized regions within the individual chains are connected by nonordered polypeptide segments. The heavy chains are linked by disulphide bridges. Further, the heavy chain is also cross-linked by similar bonds with the light chain. One half of the light chain and a quarter of the heavy chain are known as the variable regions. The amino acid sequence considerably varies in this region and are specific for each type of antibody molecule. This is known as the idiotypic variation. The rest of the molecule is known as the constant region in which the amino acid sequences are almost identical in each class of immunoglobulin. The antigden binding site is known as Fab site consisting of light chains and the N-terminal half of the heavy chain. The remaining part of immunoglobulin known as Fc also has important biological functions. For example, the Fc region of immunoglobulin G is responsible for triggering pathways of the immune response that leads to the lysis of unwanted organisms. An example is the complement system that is activated which consists of a series of proteinases.

Ig G is mainly responsible for humoral immunity. It is the only immunoglobulin that can cross through the placenta. Ig A which is present in some secretions is thought to provide surface immunity. Ig A secretion is facilitated by its specific binding to a protein known as secretory component. Ig M is the receptor in the B-lymphocytes and helps in established humoral immunity. Ig D is also found at the surface of B-lymphocytes but its function is not known. Ig E is found in mast cells and basophils. The concentration of Ig E increases in response to allergic reactions.

Disorders of immunoglobins are numerous. Agammaglobulinemia is a rare X-chromosome associated genetic disease affecting only the males. In this condition, there is virtual absence of immunoglobulins in blood plasma. The affected persons are highly susceptible to bacterial infection but react almost normally to viral infections. Hypogammaglobulinemia may be restricted to a single class of immunoglobulins or may involve under-production of all the five types. In myelomas, increased production of a restricted class of immunoglobulins or of a single specific immunoglobulin is seen. Immunoglobulin level in blood is increased generallly in infections. Sometimes the body rejects its own proteins which become antigenic. This results in the autoimmune disorders. Systemic lupus erythomatosis and some types of rheumatoid arthritis are examples. It is probable, buried antigenic sites (epitopes) in endogenous proteins get exposed or the 'self' proteins bind with exogenous triggers and become antigenic in these conditions.

Primary Structure

One of the great achievements of modern chemistry has been the development of techniques for determining the precise sequence of amino acids in a protein. The first protein to have its sequence of amino acids determined was the hormone insulin. Its molecular formula is $C_{254}H_{377}N_{65}O_{75}S_6$. This is small as proteins go. Even so, it took Dr. Frederick Sanger and his colleagues at the University of Cambridge in England 10 years (1944-1954) to work out the exact sequence of amino acids in the protein. Insulin consists of two polypeptide chains containing a total of 51 amino acid residues.

The amino acid cysteine occurs at six positions in the insulin molecules. Wherever two cysteines are close to each other, they can be oxidized (each losing a hydrogen atom). As a result a covalent bond forms between their respective sulfur atoms forming a **disulfide bridge**. In this way two different polypeptides can be drawn into a loop. Sanger

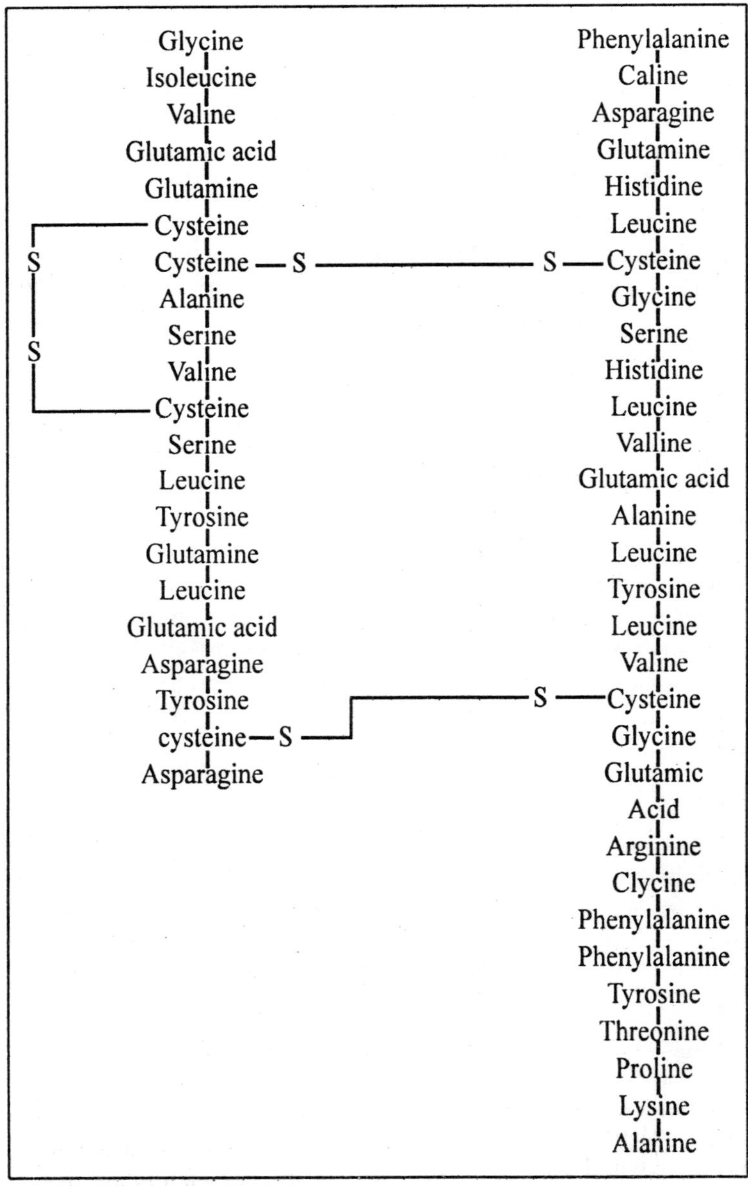

Fig.: 9.6. Sequence of amino acids in the human insulin molecule. The molecule consists of two polypeptide chains held together by two disulfide bridges.

found one *intra* chain and two *inter*chain disulfide bridges in the insulin molecule.

Since the days of Sanger's pioneering work, the techniques of protein sequencing have developed rapidly. It is now possible to have much of the job done automatically by machines. The sequence of amino acids is now known for hundreds of different proteins. The sequence of amino acids in a protein, together with the location of any disulfide bridges, is called the **primary structure** of the protein.

Let us now examine the primary structure of **lysozyme**. Lysozyme is an enzyme found in an egg white, tears, and other secretions. It is responsible for breaking down the polysaccharide walls of many kinds of bacteria, and thus it provides a measure of protection against infection.

The lysozyme in egg white is a simple polypeptide containing 129 amino acid residues. There are four pairs of cysteines, establishing disulfide (S-S) bridges between positions 6 and 127, 30 and 115, 64 and 80, and 76 and 94. The presence of these covalent linkages tells us immediately that we cannot represent the polypeptide as a straight, rigid chain. The chain must fold on itself to allow the pairs of cysteines to be close to each other.

Secondary Structure

To discover the actual configuration of a protein molecules in three-dimensional space, we must turn to another analytical technique, that of x-ray crystallography. A crystal consists of an orderly, stacked arrow of ions or molecules. If a beam of x- rays is passed through a crystal, some of the x- rays will be reflected by atoms in the crystal. Because of the orderly arrangement of the atoms, the reflections will also be orderly. A simple analogy would be the orderly pattern produced by a street light at night when viewed through the grid of a window screen. By examining the pattern of reflections produced as the beam is directed at the crystal from a variety of different angles, it is possible (espically with the help of a computer) to determine the arrangement of atoms in that crystal.

The first proteins to be analyzed by x-ray crystallography were certain structural proteins like the alpha-keratins of wool and fibroin, the protein secreted by the silkworm. These proteins were good choices for study because of their regular, self-repeating structures.

X-ray analysis of alpha-keratins revealed several interesting facts.

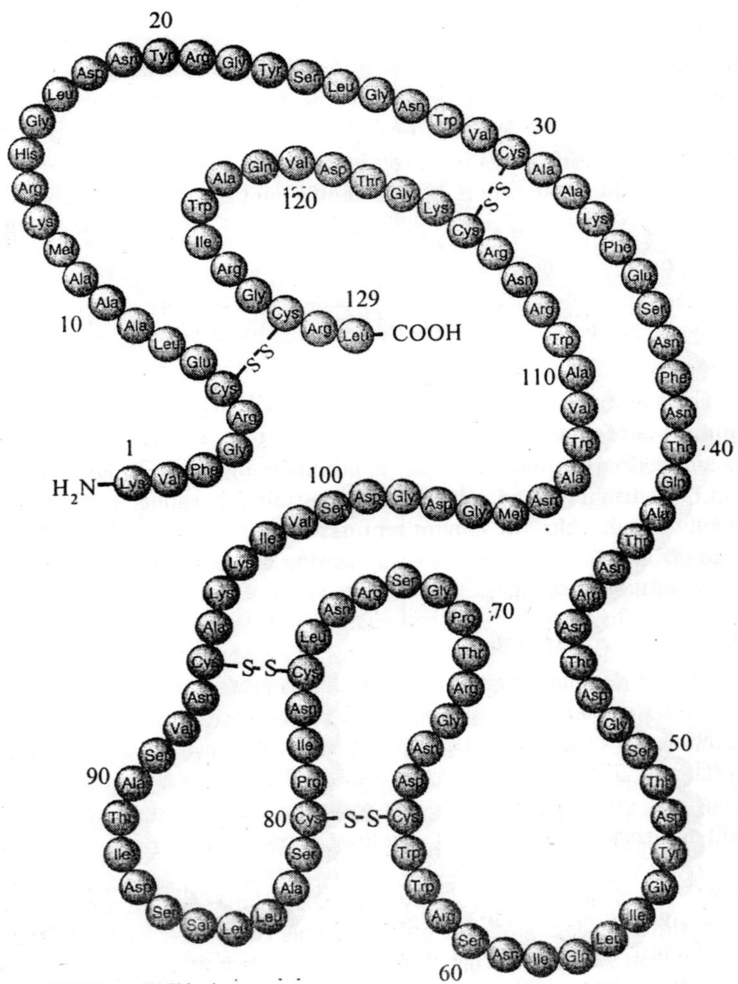

Fig.: 9.7. Primary structure of lysozyme. The molecule consists of a single polypeptide chain of 129 amino acid residues. They are numbered beginning at the amino terminal. There are four intrachain disulfide bridges.

The peptide linkage itself turned out to be very rigid, with all its atoms lying in a plane. The only opportunities for flexibility in a polypeptide occur at the bonds to the alpha carbon—that is, the carbon that carries

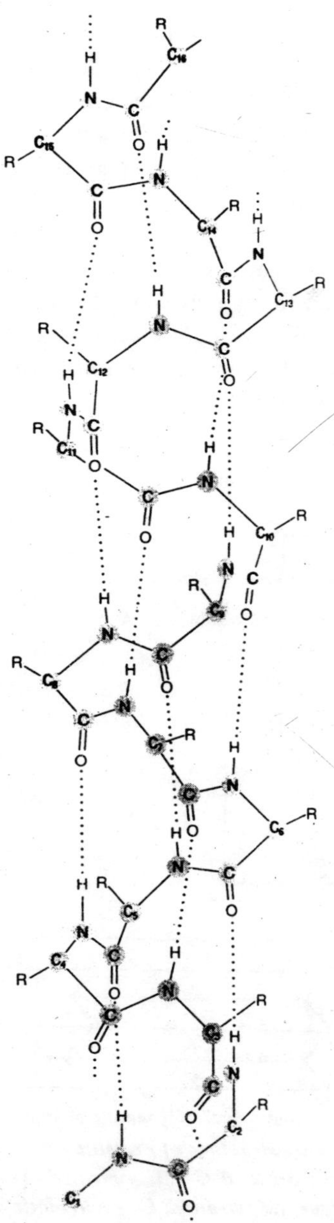

Fig.: 9.8. The alpha helix, a common form of secondary structure turn every 3.6 residues. Note the hydrogen bond that forms between the – C = O group of each peptide bond and the –N–H group of the fourth peptide bond below it in the helix.

the R group. In the alpha-keratins this flexibility is exploited by the formation of a helical twist to the polypeptide chain. Several features of this helix should be noted. (1) The R groups of the amino acid residues all extend to the outsides. (2) The helix makes a complete turn every 3.6 residues. (3) The helix is right-handed; as it recedes, it twists in a clockwise direction. (4) The –C=O (carbonyl) group of each peptide bond extends parallel to the axis of the helix and points directly at the –NH group of the peptide bond four amino acids below it in the helix. A hydrogen bond forms between these groups: –C=O . . . H–N–. This precise three-dimensional arrangements of amino acids is called the **alpha helix**. It is one of the most common examples of **secondary structure** in a protein.

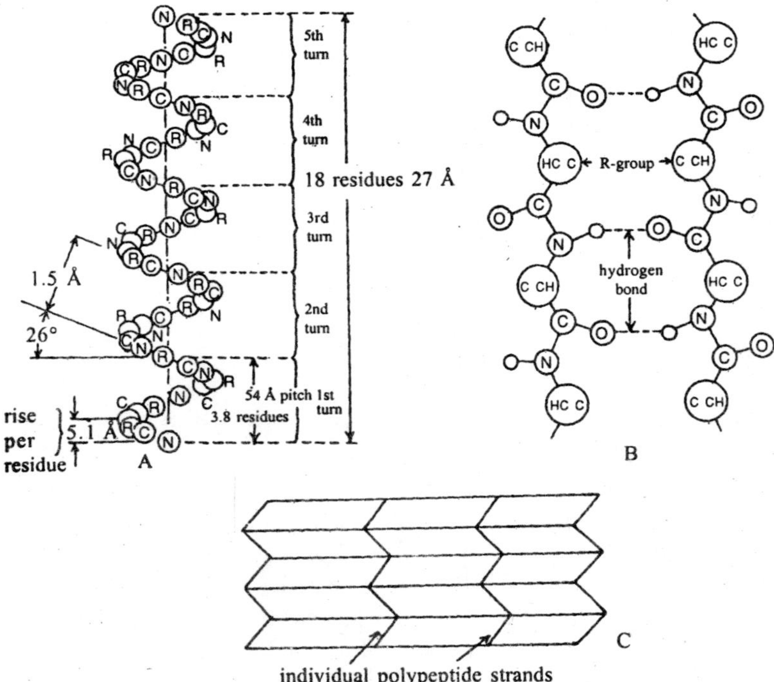

*Fig 9.9. A-C- Secondary structure of proteins : A — arrangement of polypeptide chains in (-helix; **B-C**— β - pleated sheets; **B**— extended β - (antiparallel) polypeptide strands; C— schematic of pleats (top view).*

What makes the alpha helix possible is the large number of hydrogen bonds. The strength of a single hydrogen bond is only about 5% of that of a covalent bond. But the presence of a hydrogen bond between each amino acid in the chain, and the one that is four amino acids away, provides sufficient total bond strength to make the alpha helix a very stable structure. However, it should be possible by breaking these intrachain hydrogen bonds to pull the helix out in much the same way that a spring can be extended. And, in fact, wool fibers can be stretched to about twice their normal length thanks to the molecular properties of the alpha helix.

Fibroin is the protein that silkworms use to spin the threads of their cocoon. It, too, has been subjected to x-ray analysis. In fibroin the polypeptide chains are extended. A number of chains lie parallel to each other, held together by the hydrogen bonds that form between the – C = O and –N–H groups of one chain with the –N–H and – C = O groups of the adjacent chain. In silk it would be better to describe the chains as being antiparallel because adjacent chains run in opposite directions— that is, from N-terminal to C-terminal and vice versa. A series of antiparallel stands lying side by side make up a beta-pleated sheet. It is another important and commonly found example of secondary structure. Because the chains are fully extended, we would not expect to be able to stretch them without breaking covalent bonds. It is for this reason that silk is not extensible. However, the layers of beta-pleated sheet in silk fibers make them very supple.

The alpha helix and the beta-pleated sheet are two of the most common examples of secondary structure, but other orderly arrangements of amino acid residues also occur in polypeptides. All examples of secondary structure are a direct reflection of the particular amino acids and their order that are used in the synthesis of the polypeptide. The chains of fibroin, for example, consist mainly of alternating glycine (hydrophobic) residues. This arrangement enables the chains to rest closes together in the beta-pleated sheet configuration. On the other hand, a chain containing a series of alanine residues will spontaneously fold into an alpha helix with the CH_3 groups of the alanines projecting to the exterior of the helix.

Fig.: 9.10. Collagen triple-helix structure.

Stages, as with cellulose. The fact that the protein is extremely resistant to stretching is an essential part of its functioning. For example in tendons, bone, skin, teeth, and connective tissue. Proteins which entirely in the form of helical coils, such as keratin and collagen, are exceptional.

Tertiary structure

Usually the polypeptide chain bends and folds extensively, forming a precise, compact 'globular' shape. This is the protein's tertiary structure and it is maintained by the interaction of the four types of bond already discussed, namely ionic, hydrogen and disulphide bonds as well as hydrophobic interactions. The latter are quantitatively the most important and occur when the protein folds so as to shield hydrophobic side groups from the aqueous surroundings, at the same time exposing hydrophilic side chains, as described above.

The tertiary structure of a protein can be determined by X-ray crystallography. By early 1959, and after many year's work, John kendrew and Max Perutz had built the first atomic model of myoglobin showing secondary and tertiary structures using this technique. They received the Nobel Prize for their work in 1962:

Primary structure – single polypeptide chain of 153 amino acids, the sequence was elucidated in the early 1960s;

Secondary structure – about 75% of the chain is α -helical (8 helical sections);

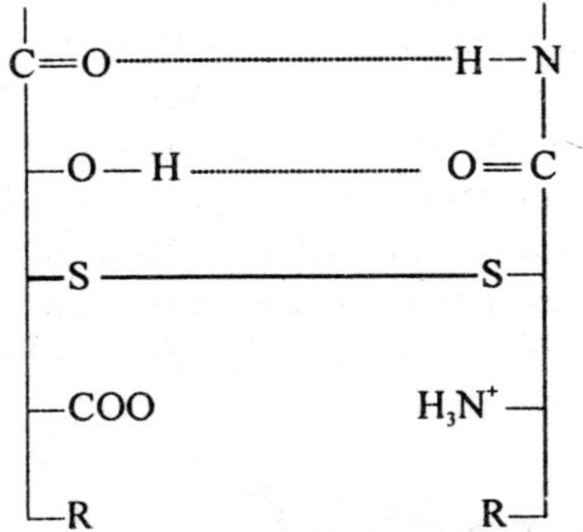

Hydrogen bond between CO and NH group of main amino acid chain

Hydrogen bond between R groups

Disulphide bond formed between cysteine residues

Ionic bond between charged R groups

Hydrophobic interaction of non-popular R groups

Fig 9.11 : Summary of types of bonds stabilising secondary and tertiary structures of proteins. Hydrophobic interactions (associations of non-polar molecules or parts of molecules) to exclude water molecules in the aqueous environment of the cell are particularly important in maintaining structure, as in membranes.

H_2N—Val—Leu—Ser—Glu—Gly—Glu—Trp—Gin—Leu—Val—Leu—His—Val—Tyr—Ala—Lys—Val—
Glu—Ala—Asp—Val—Ala—Gly—His—Gly—Gin—Asp—Ile—Leu—Ile—Arg—Leu—Phe—Lys—
Ser—His—Pro—Glu—Thr—Leu—Glu—Lys—Phe—Asp—Arg—Phe—Lys—His—Leu—Lys—Thr
Glu—Ala—Glu—Met—Lys—Ala—Ser—Glu—Asp—Leu—Lys—Gly—His—His—Glu—Ala—Glu—
Leu—Thr—Ala—Leu—Gly—Ala—Ile—Leu—Lys—Lys—Gly—His—His—Glu—Ala—Glu—
Leu—Lys—Pro—Leu—Ala—Gin—Ser—His—Ala—Thr—Lys—His—Lys—Ile—Pro—Ile—Lys—
Tyr—Leu—Glu—Phe—Ile—Ser—Glu—Ala—Ile—Ile—His—Val—Leu—His—Ser—Arg—His—
Pro—Gly—Asn—Phe—Gly—Ala—Asp—Ala—Gin—Gly—Ala—Met—Asn—Lys—Ala—Leu—Glu—
Leu—Phe—Arg—Lys—Asp—Ile—Ala—Ala—Lys—Tyr—Lys—Glu—Leu—Gly—Tyr—Gin—Gly—COOH

Fig. 9.12. Primary structure of myglobin

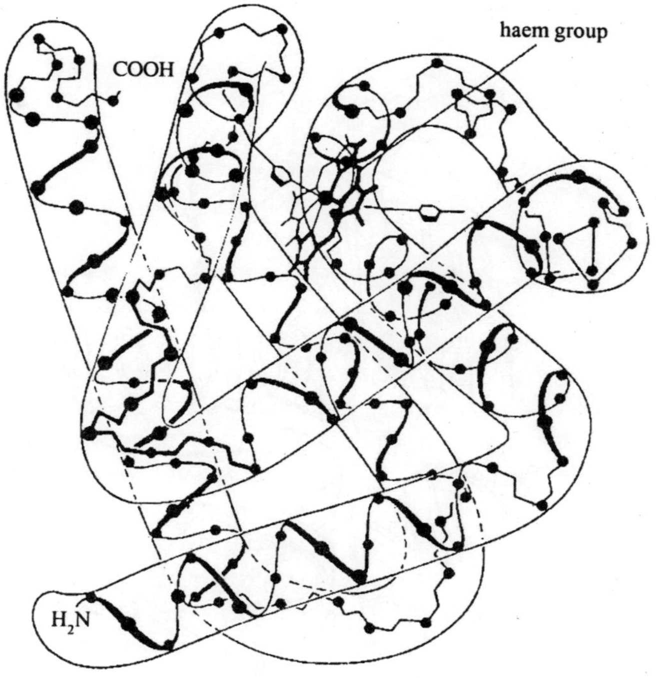

Fig 9.13. Conformation of myoglobin deduced from high resolution X-ray data. There are eight sections of α helix surrounding a haem group wihcih forms a flat disc.

Tertiary structure — non-uniform folding of the α -helical chain into a compact shape;

Prosthetic group — haem group (contains iron).

Myoglobin is formed in muscle where its function is to store oxygen. Oxygen combines with the haem group as in haemoglobin. The haem group gives muscle its red appearance. The elucidation of tertiary structure is still very time-consuming. Use of computers and other techniques to predict tertiary structure, based on knowledge of primary and secondary structures, is a fast-growing area of molecular biology. From this follows the possibility of designing proteins with particular shapes for particular functions, with important applications in industry and medicine.

Quaternary structure

Many highly complex proteins consist of more than one polypeptide chain. The separate chains are held together by hydrophobic

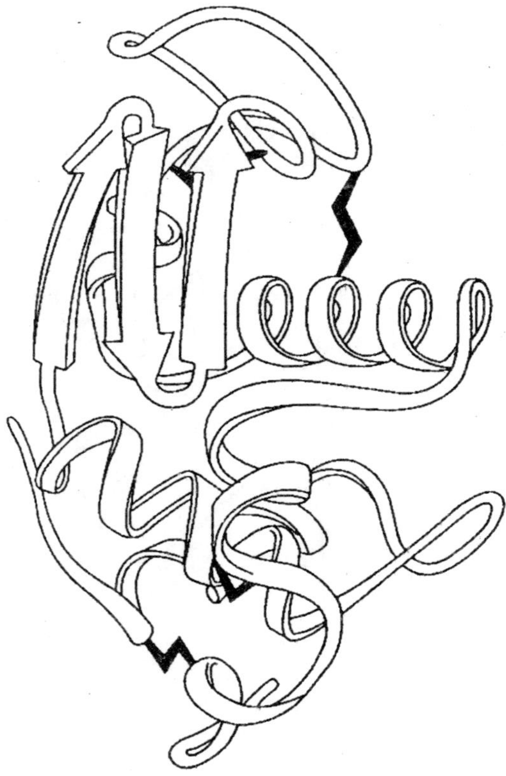

Fig. 9.14. Tertiary structure of lysozyme. The arrows represented a region of β sheet. α helices are shown as regular coils. The rest of the molecule is represented as a ribbon and the four disulphide bridges as black zig-zags.

interactions and hydrogen and ionic bonds. Their precise arrangement is known as the **quaternary structure**. Haemoglobin shows such a structure. It is the red oxygen-carrying pigment found in the red blood cells of vertebrates. It consists of four separate polypeptide chains of two types, namely two α chains and two β chains. These resemble myoglobin in structure. The two α-chains each contain 141 amino acids, while the two β chains each contain 146 amino acids. The complete structure of haemoglobin was worked out by Kendrew and Perutz.

As is typical of globular proteins, its hydrophobic side chains point inwards to the centre of the molecule, and its hydrophilic side chains face outwards, making it soluble in water. A mutation which causes one of the hydrophilic amino acids to be replaced by a hydrophobic amino acid, thereby reducing its solubility, is responsible for the disease sickle cell anaemia.

The protein coats of some viruses, such as the tobacco mosaic virus, are composed of many polypeptide chains arranged in a highly ordered fashion.

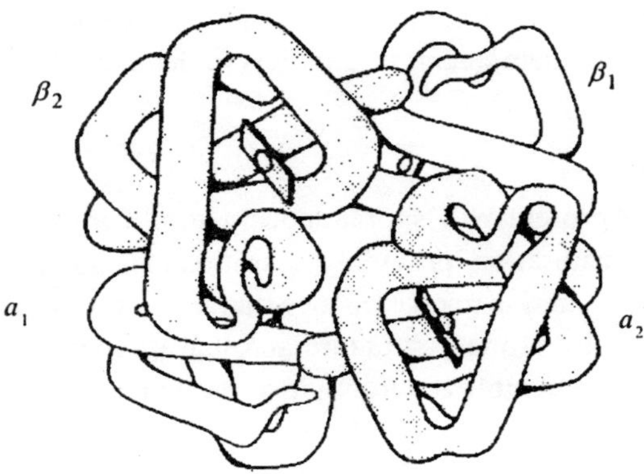

Fig. 9.16. Structure of haemolgobin. The molecule consists of four chains: two α chains and two β chains. Each chain carries a haem to which one molecule of oxygen binds. The assembly of a protein from separate polypeptide chains is an example of quaternary structure.

Denaturation and renaturation of proteins

Denaturation is the loss of the specific three-dimensional shape of a protein molecule. The change may be temporary or permanent, but the amino acid sequence of the protein remains unaffected. If denaturation occurs, the molecule unfolds and can no longer perform its normal biological function. A number of agents may cause denaturation as follows.

Heat or radiation. eg infra-red or ultra-violet light. Kinetic energy is supplied to the protein causing its atoms to vibrate violently, so disrupting the weak hydrogen and ionic bonds. Coagulation of the protein then occurs.

Strong acids and alkalis and high concentrations of salts. Ionic bonds are disrupted and the protein is coagulated. Breakage of peptide bonds may occur if the protein is allowed to remain mixed with the reagent for a long period of time.

Heavy metals. The positively charged ions of heavy metals (cations) form strong bonds with the negatively charged carboxyl groups on the R groups of proteins and often disrupt ionic bonds. They also reduce the protein's electrical polarity (its overall charge) and thus increase its insolubility. This causes the protein to precipitate out of solution.

Organic solvents and detergents. These reagents disrupt hydrophobic interactions and form bonds with hydrophobic (non-polar) groups. This in turn causes the disruption of hydrogen bonding. When alcohol is used as a disinfectant it functions to denature the protein of any bacteria present.

Renaturation

Sometimes a protein will spontaneously refold into its original structure after denaturation, providing conditions are suitable. This is called renaturation, and is good evidence that tertiary structure can be detemined purely by primary structure and that biological structures can spontaneously assemble according to a few general principles.

CHAPTER–10
AMINO ACIDS

Amino Acid Structure

Amino acids

Amino acids are the building blocks of **proteins**. Proteins of all species, from bacteria to humans, are made up from the same set of **20 standard amino acids**. Nineteen of these are α–amino acids with a **primary amino group** $\left(-NH_3^+\right)$ and a **carboxylic acid** (carboxyl; –COOH) group attached to a central carbon atom, which is called the α–carbon atom (C_α) because it is adjacent to the carboxyl group. Also attached to the C_α atom is a hydrogen atom and a **variable side-chain** or 'R' group. The one exception to this general structure is proline, which has a secondary amino group and is really an α **-imino acid**. The names of the amino acids are often abbreviated, either to three letters or to a single letter. Thus, for example, proline is abbreviated to Pro or P.

Stereoisomers

All of the amino acids, except for glycine (Gly or G;), have four different groups arranged **tetrahedrally** around the central C_α atom, and thus can exist in one of two **stereoisomers**. These two stereoisomers (or **enant-iomers**) are **nonsuperimposable, mirror images** and are termed the D and L forms. Stereoisomers are physically and chemically indistinguishable by most techniques, but can be distinguished on the basis of their different optical rotation of plane–polarized light. Molecules are classified as dextrorotatory (D; Greek 'dextro' = right) or levorotatory (L; Greek '*levo*' = left) depending on whether they rotate the plane of plane–polarized light clockwise or anticlockwise. Only the **L–amino acids** are found in proteins. **D-Amino acids** rarely occur in nature, but are found in bacterial cell walls and certain antibiotics.

The 20 standard amino acids

The standard 20 amino acids differ only in the structure of the **side-chain** or 'R' group. They can be subdivided into smaller groupings on the basis of similarities in the properties of their side–chains. They display different **physicochemical properties** depending on the nature of their side–chain. Some are acidic, others are basic. Some have small

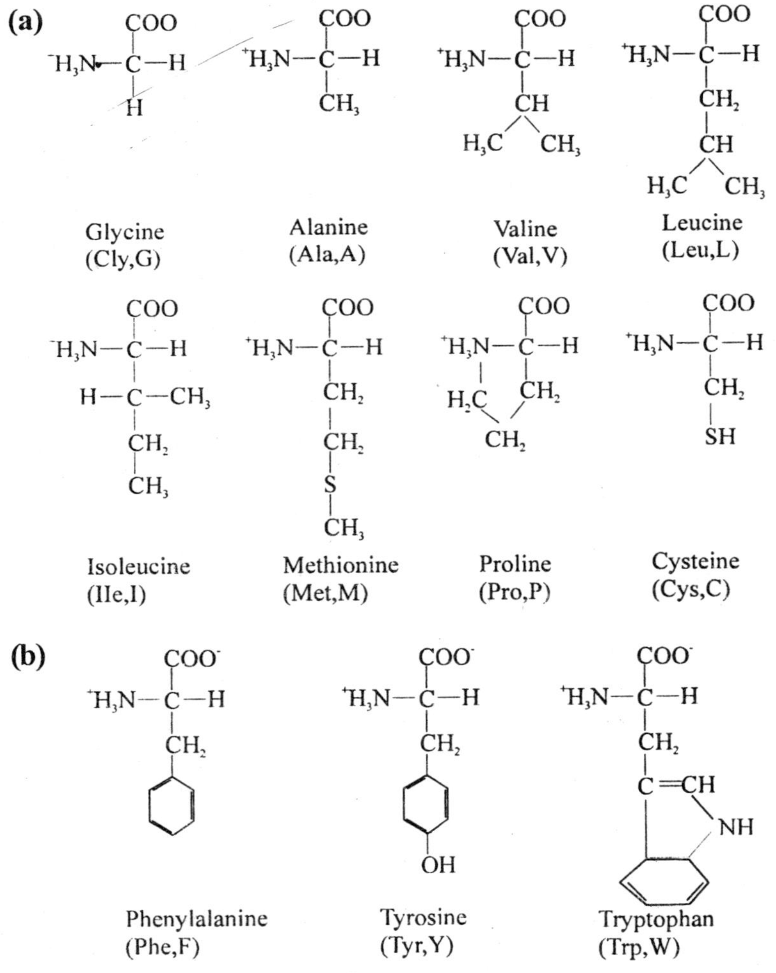

(a)

Glycine (Cly,G) | Alanine (Ala,A) | Valine (Val,V) | Leucine (Leu,L)

Isoleucine (Ile,I) | Methionine (Met,M) | Proline (Pro,P) | Cysteine (Cys,C)

(b)

Phenylalanine (Phe,F) | Tyrosine (Tyr,Y) | Tryptophan (Trp,W)

Fig 10.1 : The 20 standard amino acids ,
(a) Hydrophobic, aliphatic R groups, (b) hydrophobic, aromatic R groups.

side–chains, others large, bulky side–chains. Some have aromatic side–

chains, others are polar. Some confer conformational inflexibility, others can participate either in hydrogen bondings or covalent bonding. Some are chemically reactive.

Hydrophobic, aliphatic amino acids

Glycine (Gly or G), the smallest amino acid with the simplest structure, has a hydrogen atom in the side–chain position, and thus does not exist as a pair of stereoisomers since there are two–identical groups (hydrogen atoms) attached to the C_α atom. The aliphatic side–chains of **alanine** (Ala or A), **valine** (Val or V), **leucine** (Leu or L), **isoleucine** (Ile or I) and **methionine** (Met or M) are chemically unreactive, but hydrophobic in nature. **Proline** (Pro or P) is also hydrophobic but, with its aliphatic side–chain bonded back on to the amino group, it is conformationally rigid. The sulfur–containing side–chain of **cysteine** (Cys or C) is also hydrophobic and is highly reactive, capable of reacting with another cysteine to form a disulfide bond.

(a)

Arginine (Arg,R)	Lysine (Lys, K)	Histidine (His, H)	Aspartate (Asp, D)	Glutamate (Glu, E)
COO^-	COO^-	COO^-	COO^-	COO^-
$^+H_3N—C—H$	$^+H_3N—C—H$	$^+H_3N—C—H$	$^+H_3N—C—H$	$^+H_3N—C—H$
CH_2	CH_2	CH_2	CH_2	CH_2
CH_2	CH_2	C—NH	COO	CH_2
CH_2	CH_2	CH		COO
NH	CH_2	HC—N		
$C{=}NH_2$	NH_3^+	H^+		
NH_2				

(a)

Serine (Ser, S)	Theronine (Thr, T)	Asparagine (Asn, N)	Glutamine (Gin, Q)
COO^-	COO^-	COO^-	COO^-
$^+H_3N—C—H$	$^+H_3N—C—H$	$^+H_3N—C—H$	$^+H_3N—C—H$
CH_2OH	H—C—OH	CH_2	CH_2
	CH_3	C	CH_2
		H_2N O	C
			H_2N O

Fig. 10.2 :. The 20 standard amino acids (a) Polar, charged R group), (b) Polar, uncharged R groups.

Hydrophobic, aromatic amino acids

Phenylalanine (Phe or F), tyrosine (Tyr or Y) and **tryptophan** (Tyr or W) are hydrophobic by virtue of their aromatic rings.

Polar, charged amino acids

The remaining amino acids all have polar, hydrophilic side-chains, some of which are changed at neutral pH. The amino groups on the side-chains of the basic amino acids **arginine** (Arg or R) and **lysine** (Lys or K) (Fig.3a) are protonated and thus positively charged at neutral pH. The side-chain of **histidine** (His or H) can be either positively charged or uncharged at neutral pH. In contrast, at neutral pH the carboxyl groups on the side-chains of the acidic amino acids **aspartic acid** (aspartate: Asp or D) and **glutamic acid** (glutamate; Glu or E) are de-protonated and possess a negative charge.

Polar, uncharged amino acids

The side-chains of **asparagine** (Asn or N) and **glutamine** (Gln or Q) the amide derivatives of Asp and Glu, respectively, are uncharged but can participate in hydrogen bonding. **Serine** (Ser or S) and **threonine** (Thr or T) are polar amino acids due to the reactive hydroxyl group in the side-chain, and can also participate in hydrogen bonding (as can the hydroxyl group of the aromatic amino acid Tyr).

Classification

Amino acids can be classified in two ways based on structure and polarity.

Based on structure, amino acids are grouped into three classes as:

I. **Aliphatic amino acids:** These are straight or open chain amino acids which are further subdivided into four groups as:

   1. **Monoaminomonocarboxylic (neutral) amino acids:** These consist of one amino and one carboxyl groups and hence are neutral to litmus, e.g. glycine, valine, leucine, isoleucine, serine and threonine.

   2. **Monoaminodicardboxylic (acidic) amino acids:** These consist of one amino and two carboxyl groups and hence are acidic to litmus, e.g. aspartic acid and glutamic acid.

   3. **Monocarboxylicdiamino (basic) amino acids:** These consist of one carboxyl and two amino groups and hence are basic to litmus, e.g. lysine, arginine and histidine.

4. **Sulphur-containing amino acids:** These consist of one or more sulphur atoms, e.g. cysteine, cystine and methionine.

II. **Aromatic amino acids:** These contain an aromatic ring in the molecule, e.g. phenylalanine and tyrosine.

III. **Heterocyclic amino acids:** These contain an heterocyclic nucleus in the molecule, e.g. histidine, tryptophan, proline and hydroxyproline.

Structure and range of amino acids

In an amino acid there is a central carbon atom, known as the α carbon atom, to which is always attached an acidic carboxyl group, **–COOH,** a basic amino group, **–NH_2** and a **hydrogen** atom. The fourth position is the only variable part of the molecule. The group there is known as the **R group**. This group gives each amino acid its uniqueness. The simplest amino acid, and therefore the easiest one to learn as an example, is **glycine**, where R is simply hydrogen. When R is –CH_3, the amino acid **alanine** is formed.

Rare amino acids

A small number of rare amino acids occur in the proteins of organisms. They are made from some of the common amino acids. For example, hydroxyproline is made from proline, and is found in the protein collagen; hydroxylysine is made from lysine, and is found in collagen.

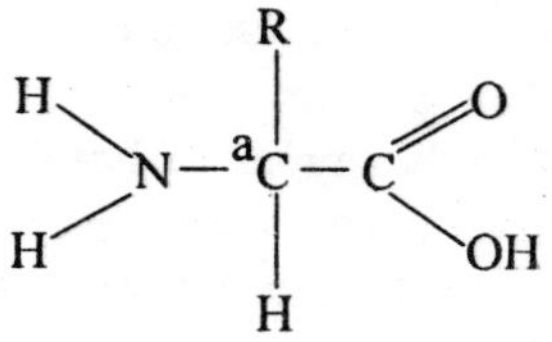

Fig. 10.3 : General formula of an amino acid.

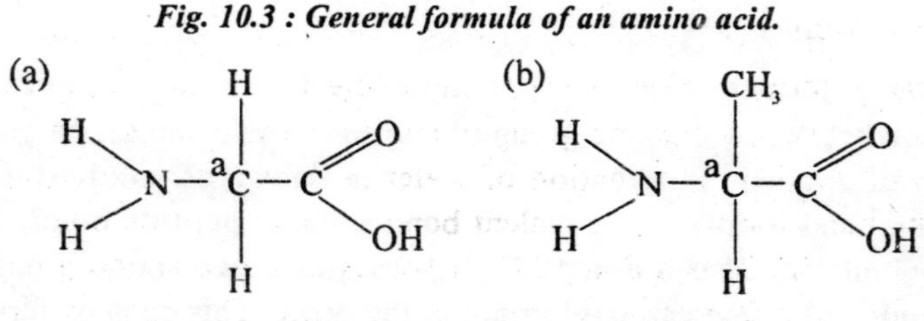

Fig. 10.4 : (a) Glycine. (b) Alanine.

There is no DNA code for the rare amino acids, and they are

made from their parent amino acids after they have been incorporated into a protein.

Non–protein amino acids

Over 150 of these are known to occur, either free or in a combined form in cells, but never in proteins. For example GABA (γ aminobutyric acid) is virtually unique to the nervous system. It is an inhibitory neurotransmitter, important in the brain.

Amino acids are amphoteric

Molecules like amino acids which contain both acid and a basic part are described as **amphoteric.** They are mainly as ions and can carry both a positive charge on the basic part and a negative charge on the acid part. Such ions are described as dipolar and are called zwitterions. This explains the fact that amino acids and proteins can be made to move in an electrical field, as when they are separated by electrophoresis. The charge on the amino acid can be affected by changes in it environment. For example, making a solution more acid would increase the concentration of hydrogen ions, and hence positive charges, and these would tend to cancel out the negative charges on the amino acid.

Bonds used in protein structure

Amino acids combine to form proteins. They are joined together by a type of bond known as a **peptide bond.** Once these bonds have been formed, however, the protein typically folds into a particular shape as a result of four other types of bond, namely ionic bonds, disulphide bonds, hydrogen bonds, and hydrophobic interactions. Studying these bonds us to understand the structure and behaviour of proteins.

Peptide bond

This is formed when a water molecule is eliminated during a reaction between the amino group of one amino acid and the carboxyl group of another. Elimination of water is known as **condensation** and the bond formed is a covalent bond called a **peptide bond.** The compound formed is a **dipeptide**. It possesses a free amino group at one end, and a free carboxyl group at the other. This enables further combination between the dipeptide and other amino acids. If many amino acids are joined together in this way, a **polypeptide** is formed.

R
H
H — N⁺ — C — C =O
H
O⁻
H

$-NH_2$, being a base, possesses a high affinity for H^+ ions

the acidic COOH dissociates, releasing H+ ions which can attach to the basic amino group giving it a prsitive charge

Fig. 10.5 : Neutral zwitterion form of an amino acid.

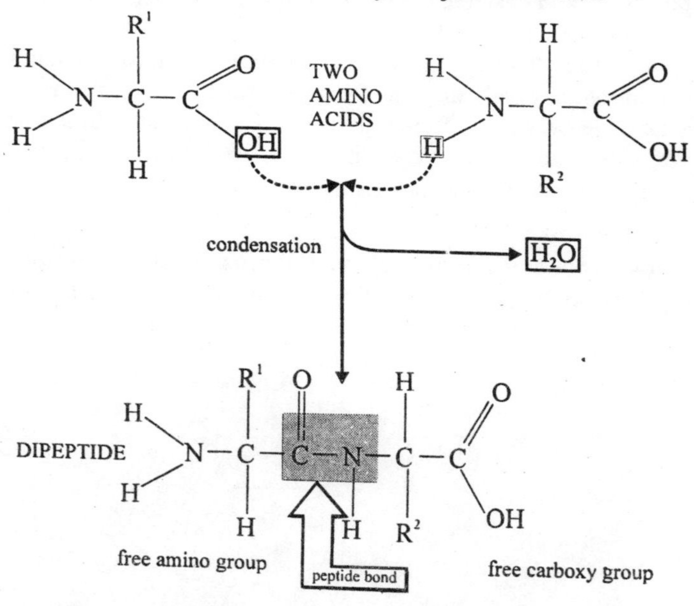

Fig. 10.6 : Formation of a dipeptide by a condensation reaction between two amino acids. Condensation is removal of water.

R^1 O R^3
H
CH N C CH
N C CH N C
H O R^2 H O

*peptide bond

Fig. 10.7 : Part of a polypeptide showing the joining of three amino a

Ionic bond

Acidic and basic R groups exist in an ionised (charged)

certain pHs. Acidic R groups are negatively charged and basic R groups are positively charged. They can therefore be attracted to each other, forming ionic bonds. In an aqueous environment this bond is much weaker than a covalent bond and can be broken by changing the pH of the medium. This helps to explain the disruptive effect that changes in pH can have on protein structure. For example, adding acid to milk makes it curdle because the ionic bonds in casein (milk protein) are broken and the protein ceases to be soluble.

Disulphide bond

The amino acid cysteine contains a sulphydryl group. –SH, in its R group. If two molecules of cysteine line up alongside each other, neighbouring sulphydryl groups can be oxidised and form a disulphide bond. Disulphide bonds may be formed between different chains of amino acids or between different parts of the same chain. In the latter case the disulphide bonds make the molecule fold into a particular shape. They are strong and not easily broken.

Hydrogen bond

Hydrogen bonds have already been discussed in the section on water. When hydrogen is part of an OH or NH group it becomes

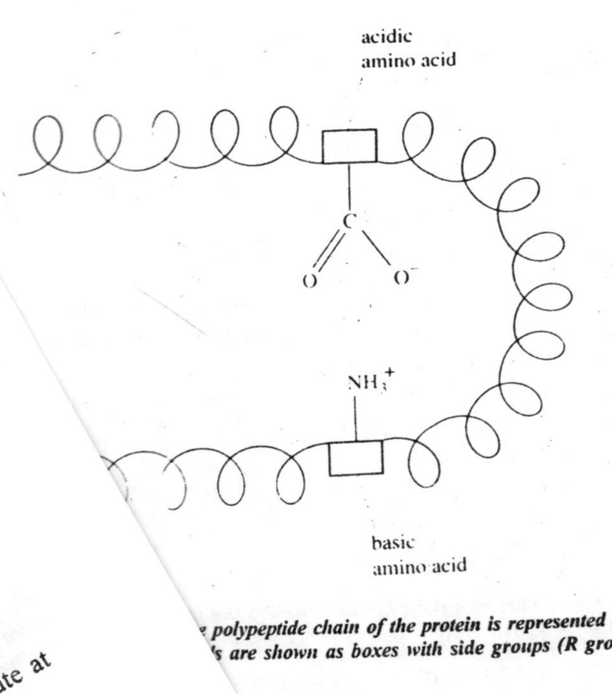

... polypeptide chain of the protein is represented by a ... 's are shown as boxes with side groups (R groups)

slightly positively charged (electropositive). This is because the electrons that are shaped, and which are negatively charged, are attracted more towards the O or N atoms. The hydrogen may then be attracted towards a neighbouring electronegative oxygen or nitrogen atom, such as the O of a C=O group or the N of an NH group. C=O and NH groups occur along the length of plypeptide chains, and they can interact to produce regular shapes such as the α-helix discussed later. The hydrogen bond is weak, but as its occurrence is frequent, the total effect makes a considerable contribution towards molecular stability, as in the structure of the α-helix and silk.

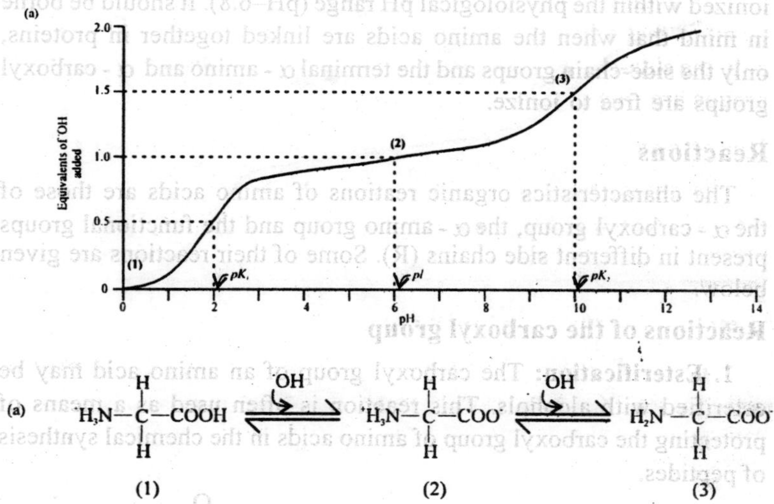

Fig. 10.9 : Ionization of glycine (a) Titration curve of glycine, (b) dissociation of glycine.

Ionization

The 20 standard amino acids have two **acid–base groups;** the α - amino and –carboxyl groups attached to the C_α atom. Those amino acids with an **ionizable side-chain** (Asp, Glu, Arg, Lys, His, Cys, Tyr) have an additional acid-base group. The **titration curve** of Gly is shown in Fig 4a. At low pH (i.e. high hydrogen ion concentration) both the amino group and the carboxyl group are fully protonated so that the amino acid is in the cationic form $H_3N^+ CH_2COOH$. As the amino acid in solution is titrated with increasing amounts of a strong base (e.g. NaOH), it loses two protons, first from the carboxyl group which has the lower pK value (pK = 2.3) (the pK at which this group is half dissociated) and then from the amino group which has the higher pK value (pK = 9.6). The pH at which Gly has no net charge is termed its **isoelectric point, pI**. The α - carboxyl groups of all the 20 standard amino acids have pK values in the range 1.8–2.9, whilst their α - amino acids Asp and Glu have pK values in the range 8.8–10.8. The side-chains of the acidic amino acids Asp and Glu have pK and Lys, have pK values of 3.9 and 4.1 respectively, whereas those of the basic amino acids Arg and Lys, have pK values of 12.5 and 10.8, respectively. Only the side-chain of His, with a pK value of 6.0, is ionized within the physiological pH range (pH–6.8). It should be borne in mind that when the amino acids are linked together in proteins, only the side-chain groups and the terminal α - amino and α - carboxyl groups are free to ionize.

Reactions

The characteristics organic reations of amino acids are those of the α - carboxyl group, the α - amino group and the functional groups present in different side chains (R). Some of their reactions are given below.

Reactions of the carboxyl group

1. Esterification: The carboxyl group of an amino acid may be esterified with alcohols. This reaction is often used as a means of protecting the carboxyl group of amino acids in the chemical synthesis of peptides.

$$H_2N\text{–}\underset{\displaystyle R}{\underset{|}{CH}}\text{–}COOH + C_2H_5OH \rightleftharpoons H_2N\text{–}\underset{\displaystyle R}{\underset{|}{CH}}\text{–}\overset{\displaystyle O}{\overset{||}{C}}\text{–}OC_2H_5 + H_2O$$

2. Reduction: The carboxyl group of an amino acid can be reduced with the potent reducing agent lithium borohydride to yield the corresponding primary alcohol. This reaction is useful in the study of amino sequence.

$$H_2N\text{–}\underset{\displaystyle R}{\underset{|}{CH}}\text{–}COOH \xrightarrow{LiBH_4} H_2N\text{–}\underset{\displaystyle R}{\underset{|}{CH}}\text{–}CH_2OH$$

3. Sorenson's formal titration: The carboxyl group of an amino acid cannot be directly titrated against alkali because of the formation of dipolar ion. In the presence of excess of neutralized formaldehyde, the amino acid solution can be directly titrated against alkali as the amino group which interferes with carboxyl group is converted into a mono– and subsequently into a dimethylol derivatives of an amino acid.

$$HCHO + H_2N\underset{\substack{|\\R}}{C}HCOO^- \longrightarrow HOH_2CHN\underset{\substack{|\\R}}{C}HCOO^-$$

N–Hydroxymethyl derivative

$$HOH_2CHN\underset{\substack{|\\R}}{C}HCOO^- + HCHO \longrightarrow HOH_2C\underset{\substack{|\\CH2OH}}{N}\overset{\substack{R\\|}}{C}H\text{–}COO^-$$

N, N–Bishydroxymethy derivative

Reactions of amino group

1. Reaction with nitrous acid **(Van slyke reaction)**: The amino group of an amino acid will react with the strong oxidizing agent nitrous acid (HNO_2) to liberate N_2. This reaction which is stoichiometric is important in the estimation of α - amino groups in amino acids. The amino acids proline and hydroxyproline do not react. The products are the corresponding α - hydroxy acid and N_2 gas, which can be measured manometrically.

$$R\text{–}\underset{\substack{|\\NH2}}{C}H\text{–}COOH + HNO_2 \longrightarrow R\text{–}\underset{\substack{|\\OH}}{C}H\text{–}COOH + N_2 + H_2O + H^+$$

2. Reaction with ninhydrin: The amino group of amino acids on heating with ninhydrin(oxidizing agent) undergoes oxidation to form ammonia, CO_2 and the aldehyde:

$$R\text{–}\underset{\substack{|\\NH_2}}{C}H\text{–}COOH^+ \text{ oxidized ninhydin} \longrightarrow$$

$$R\text{–}\underset{\substack{||\\O}}{C}H + NH_3 + CO_2 + \text{reduced ninhydrin}$$

A second equivalent of ninhydrin (oxidized) then reacts with the reduced ninhydrin and NH_3 formed in above equation to produce a highly coloured product (Ruheman's purple).

Oxidized ninhydrin + NH_3 + Reduced ninhydrin (hydrindatin) → Purple pigment + $3H_2O$

This reaction is very widely used to estimate amino acids quantitatively in very small amounts. A purple colour is given in the ninhydrin reaction by all the amino acids and peptides having a free α - amino group, whereas, proline and hydroxyproline in which the α - amino group is substitude (imino acids) yield derivative with a characteristic yellow colour.

3. Reaction with 1–fluore–2,4–dinitrobenzene (FDNB): The compound FDNB (Sanger's reagent) reacts with the free amino group on the NH_2–terminal end of a polypeptide as well as the amino groups of free amino acids and produces the bright yellow coloured dinitrophenyl (DNP) – amino acid derivative. This reaction can be used to identify the N–terminal amino acid of a peptide or protein.

4. Reaction with dansyl chloride (5–dimethylamino–naphthalene–1–sulfonyl chloride): The amino group of both free amino acids and peptide chains will react with dansyl chloride to produce a dansyl amino acid derivative which fluoresces under UV light. This reaction may be also used to find our N–terminal amino acid of a peptide or protein.

$$H_2N-CH(R)-CO_2H + \underset{\text{FDNB}}{NO_2-C_6H_3(NO_2)-F} \longrightarrow \underset{\text{DNP-Amino acid}}{NO_2-C_6H_3(NO_2)-NH-CH(R)-CO_2H} + HF$$

5. Reaction with phenylisothiocyanate (Edman's reaction): Phenylisothiocyanate reacts with α - amino group of an amino acid (or peptide) to form the corresponding phenylthiocarbamyl amino acid. In anhydrous acid this compound cyclizes to form a phenylthiohydantion which is stable in acid. This reaction is used both to degrade a polypeptide chain and to identify the N–terminal amino acid in the peptide.

$$C_6H_5-N=C=S + H_2N-\underset{R}{CH}-CO_2H \longrightarrow C_6H_5-NH-\underset{S}{\overset{\|}{C}}-N-\underset{R}{CH}-CO_2H$$

Phenylisothicyanate   Phenylthiocarbamy l Amino acid

$\downarrow H^{'}$

Phenylthiohydantion

Reactions of the R group

Amino acids also show qualitative colour reactions typical of certain functions present in their R groups,e,g. the thioi group of cysteine, the phenolic hydroxyl group of tyrosine and the guanidinium group of arginine.

1. Oxidation of cysteine to cystine:

The thiol group of cysteine is highly susceptible to atmospheric oxygen in the presence of iron salts or by other mild oxidizing agents. The oxidation product is cystine in which the disulfide bond constitutes a covalent bridge between two residues of cystine. Cystine plays an important role in protein structure since its disulfide group serves as a covalent cross link between two polypeptide chains or between two points as a single chain.

$$H_2N-\underset{CH_2-SH}{\overset{COOH}{C}}-H + H-\underset{COOH}{\overset{SH-CH_2}{C}}-NH_2 \xrightarrow{2H^+} H_2N-\underset{CH_2}{\overset{COOH}{C}}-H \quad H_2N-\overset{COOH}{C}-H \quad CH_2-S-S-CH_2$$

L-Cysteine   L-Cysteine   L-Cysteine

2. Reaction of cysteine with Ag^+: The thiol group of cysteine reacts with heavy metal ions such as Hg^{2+} and Ag^+ to form mercaptides.

$$\underset{\text{L-Cysteine}}{H_2N-\overset{\displaystyle COOH}{\underset{\displaystyle \underset{\displaystyle SH}{|}\,CH_2}{\overset{|}{\underset{|}{C}}}}-H} + Ag^+ \xrightarrow{H^+} \underset{\text{Cysteine silver mercaptide}}{H_2N-\overset{\displaystyle COOH}{\underset{\displaystyle CH_2-S-Ag}{\overset{|}{\underset{|}{C}}}}-H}$$

Essential and Nonessential Amino Acids

Autotrophic microorganisms build their carbohydrate, lipids, proteins nucleic acids, and vitamins from N_2, CO_2, and inorganic salts. They are capable of synthesizing all the amino acids and none are essential for growth. Other organisms (as a result of mutations) have lost the capacity to synthesize one or more amino acids, vitamins, etc., and such compounds must be provided in the medium. Such compounds are essential for growth. The rat and man require some amino acids (essential), whereas other are not essential. Table lists the 10 amino acids essential for growth of the

Table. 10.1 : Classifications of Amino Acids with Respect to their Growrh Effects in the Rat

Essential	*Nonessential*
Lysine	Glycine
Troptophan	Alanine
Histidine	Serine
Phenylalanine	Cystine
Leucine	Tyrosine
Isoleucine	Aspartic acid
Threonine	Glutamic acid
Valine	Proline
Arginine	Hydroxyproline
Methionine	Citrulline

Table 10.2 : Minimum and Recommended Intake for Normal man when diet furnishes sufficient nitrogen for synthesis of Nonessential Amino Acids

Amino acid Recommended Daily	*Rqt. (gm.)*	*Minimum Daily Intake(gm.)*
L–Tryptophan	0.25	0.5
L–Phenylalanine	1.10	2.2
L–Lysine	0.80	1.6
L–Threonine	0.50	1.0
L–Valine	0.80	1.6
L–Methionine	1.10	2.2
L–Leucine	1.10	2.2
L–Isoleucine	0.70	1.4

young rat. Table lists the eight amino acids essential for the maintenance of nitrogen equilibrium in young adult men:

In some instances amino acids can furnish energy by being converted to an intermediate (pyruvate), which is changed to glycogen. Ssuch amino acids are called *glycogenic*. It has been estimated that more than one half the protein N of an animal is a potential source of carbohydrate. On the other hand, phenylalanine, tyrosine, leucine, and isoleucine yield products that belong to the ketone bodies, and are called *ketogenic*. In the ketogenic group, phenylalanine, tyrosine, and isoleucine are both glycogenic and ketogenic.

Nitrogen Fixation and Assimilation

Atmospheric nitrogen serves as the source of plant nutrition and, in the form of ammonia, for animal nutririon. Microorganisms in the soil fix nitrogen to form ammonia. Nitrates are the most useful form of nitrogen for crop plants. Among the organisms which carry out nitrogen fixation are the autotrophes *Rhodospirillum rubrum,* a bacterium, *Nostoc muscorum,* a blue–green alga, and *Azotobacter* and *Clostridium pasteuranum,* which are bacterial heterotrophes. Symbiotic nitrogen fixation takes place in nodules on the roots of leguminous plants by *Rhizobium;* neither the roots nor the bacteria can perform this function alone.

In cell–free sustems from microorganisms, pyruvate plays an important and dual role: it provides a source of ATP and it can also act as a reducing agent. As pyruvate is oxidized, the iron–protein *ferredoxin* is reduced, transferring its reducing power to the *nitrogenase* system which catalyzes the conversion of N_2 to ammonia:

$$N_2 + 3H_2 \longrightarrow 2NH_3$$

Little is known of the mechanism of N_2 activation by the enzyme, except for the concept of a redox catalyst which couples ATP hudrolysis and electron transfer in the presence of a divalent metal cation, probably Mg^{++}. Bacterial ferredoxin and flavodoxin, a flavin–containing protein, are capable of acting in the nitrogenase system. The enzyme also catalyzes a wide variety of reduction reactions of substrates, including N_2O, N_3^-, alkynes, HCN, nitriles, and isonitriles. However, the enzyme has a highest affinity for N_2.

Nitrosification refers to the oxidation of ammonia to nitrite. An example of a bacterium that carries on this reaction is *Nitromonas*. *Nitrification* is the oxidation of nitrites to nitrates. An example of a bacterium with this function is *Nitrobacter*. The bacteria involved in these oxidations are autotrophic—oxidize ammonia to nitrites ($\Delta F° =$ –66,500 cal.) and nitrites to nitrates ($\Delta F° = -17{,}500$ cal.) as a source of energy. This energy is utilized for the reduction of CO_2 to carbohydrate and other compounds of carbon.

Higher plants and microorganisms make use of nitrates in the soil by first reducing it to ammonia. *Nitrate reductase* catalyzes the reduction of nitrate to nitrite, and *nitrite reductase* further reduces nitrites to ammonia. The latter reaction may involve as intermediate hyponitrite and hydroxylamine. The enzyme *nitrate reductase* is a pyridine–dependent molybdoflavoprotein, and the *nitrite reductase* is also a pyridine–dependet metalloflavoprotein. In both instances ferredoxin plays a central role as an electron source. The ammonia which is formed can now serve as a source of nitrogen for amino acids in plants and animals.

Fixation of Ammonia as Glutamic Acid and Glutamine

The enzyme *glutamic acid dehydrogenase* catalyzes the synthesis as well as the oxidation of glutamic acid. The equilibrium of the reaction favors glutamic acid synthesis:

$$\alpha\text{ - Ketoglutaric acid} + H^+ + NH_3 \rightleftarrows \text{L–glutamate} + H_2O$$

Whereas the enzyme from plants and most animal tissues makes use of NAD^+, either NAD^+ or $NADP^+$ is utilized by the liver enzyme. One preparation of the enzyme has a molecular weight of 250,000 and is made up of 5 subunits of 50,000 each. These subunits may be held together by Zn^{++}. This reaction may well be the most importanr quantitatively for fixation of ammonia.

The enzyme *glutamine synthetase* catalyzes the fixation of ammonia by glutamic acid in the form of the amide group of glutamine:

$$\text{L–Glutamic acid} + \text{ATP} + NH_3 \rightleftarrows \text{L–glutamine} + \text{ADP} + P_i + H_2O$$

This is an important storage form of ammonia, which may be very toxic in many organisms, as well as aspecific source of ammonia required in many syntheses, e.g., purines and pyrimidines. The enzyme has a molecular weight of 592,000 and consists of 12 subunits of 48,500 each. Two forms of the enzyme have been isolated: *synthetase I,* the most active species, and *synthetase II,* which is much less active and contains covalently bound AMP. The enzyme *adenyl transferase* is responsible for the conversion of I to II:

$$\text{Synthetase I} + \text{ATP} + Mg^{++} \rightleftarrows \text{Synthetase II}$$

This constitutes a system for regulating glutamine synthesis; another enzyme, *glutaminase*, catalyzes the hydrolytic release of ammonia from glutamine. Synthetase II can be deadenylated in a reaction involving as yet unpurified enzymes, α - ketoglutaric acid and a mixture of UTP and ATP. This reaction is inhibited by glutamine.

Amino Acid Nitrogen Interchange; The Nitrogen Pool

Trnasaminases

The presence of enzymes that catalyze the transfer of the amino group from one amino acid to an α - keto acid to form new amino acid and corresponding. The transfer of sulfate fron PAPS to an acceptor —e.g., phenols, chondroitin, dehydroandrosterone, and probably heparin—is catalyzed by specific enzymes.

The anaerobic degradation of cysteine results in the release of H_2S, pyruvic acid, and NH_3:

$$H_2O + \underset{\substack{|\\ COOH \\ \text{Cysteine}}}{\overset{\substack{CH_2SH\\ |}}{CHNH_2}} \longrightarrow \underset{\substack{|\\ COCH \\ \text{Pyruvic acid}}}{\overset{\substack{CH_3\\ |}}{C{=}O}} + H_2S + NH_3$$

Sulfide is removed from the body in the from of $S_2O_3^=$, which is cinverted by small amounts of CN^- to SCN^-:

$$S_2O_3^= + CN^- \rightarrow SO_3^= + SCN^-$$

The enzyme is called rhodaneseor $S_2O_3^=$–transsulfurase.

Cysteine is a precursor of taurine, which is conjugated with cholic acid (a bile acid) to form taurocholic acid, which at neutral pH, is one of the bile salts. The other is glycocholate. This involves the intermediate formation of cholyl CoA:

$$\text{Cholic acid} + \text{CoA} + \text{ATP} + Mg^{++} \rightarrow \text{cholyl CoA}$$

Cholyl CoA + taurine →

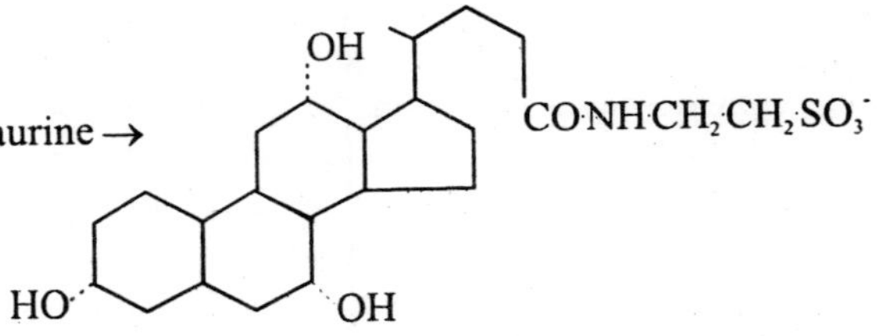

OH

$CO{\cdot}NH{\cdot}CH_2{\cdot}CH_2{\cdot}SO_3^-$

HO OH

Taurocholic acid

Cysteine takes part in reactions with compounds foreign to the animal when these compounds are introduced into the body. An example is the formation of a mercapturic acid (acetyl cysteine) derivative of brombenzene:

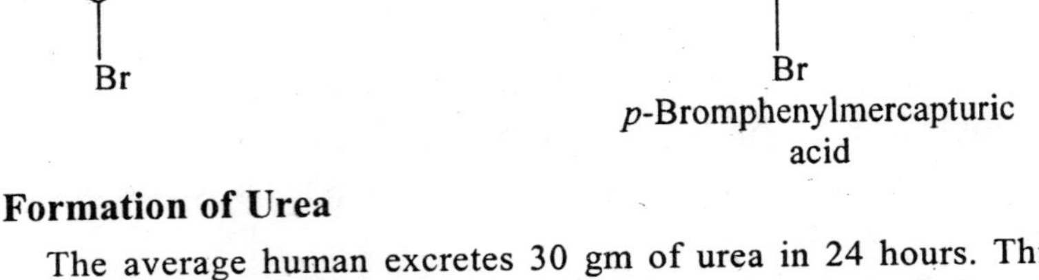

$S{\cdot}CH_2{\cdot}CH{\cdot}COOH$

$NH{\cdot}CO{\cdot}CH_3$

+ Cysteine + "Acetic acid" →

Br Br

p-Bromphenylmercapturic acid

Formation of Urea

The average human excretes 30 gm of urea in 24 hours. This compound has always been regarded as the end product of nitrogen (protein) metabolism. In some species, urea is replaced by uric acid or ammonia .

Early explanations for the formation of urea in the body based on liver perfusion experiments assumed an overall reaction involving ammonia (obtained by the deamination of amino acids) and carbon dioxide present in blood:

$$2NH_3 + CO_2 \rightarrow CO\,(NH_2)_2 + H_2O$$

Table 10.3 : End Products of protein and purine metabolism in various Vertebrates, correlated with the presence of liver Arginase

	End Product of		
	Protein Metabolism	*Purine Metabolism*	*Liver Arginase*
Mammalia	Urea	Allantoin	+
Birds	Uric acid	Uric Acid	–
Reptilia:			
Snakes, lizards	Uric acid	Uric acid	–
Turtles	Urea	Allantoin	+
Amphibia	Urea	Urea	+
Pisces:			
Elasmobranchii (sharks, dogfish, etc.)	Urea	Urea	+
Telesotei (most bony fish)	Ammonia	Urea	+

Later Krebs and Henseleit, utilizing the tissue slice technique, showed that ammouium salts could be converted to urea. Of the many substances tested for their possible stimulating effect on this reaction, ornithine and citrulline showed striking activity. It was discovered that the action of ornithine is a catalytic one; only small quantities were needed, and there was no stoichiometric relation between the quantity of ornithine added and the amount of urea formed.

In an attempt at an explanation, Krebs assumed the formation of an intermediate compound—by the interaction of ornithine, carbon dioxide, and ammonia—which then broke down to form urea. This intermediate compound he showed to be arginine, which in the presence of the enzyme *arginase* forms urea.

In this connection it is important to note that there is a correlation between the ability of a tissue to form urea and its content of arginase. In animals that excrete urea, arginase occurs chiefly in the liver.

It was postulated that citrulline was doubtless an intermediate in the conversion of ornithine to arginine. Citrulline, originally isolated from the seeds of the watermelon, has since been shown to occur in animals.

The theory of Krebs can be summarized as follows:

$$\underset{\text{Ornithine}}{\begin{array}{l}CH_2NH_2\\|\\CH_2\\|\\CH_2\\|\\CH(NH_2)\\|\\COOH\end{array}} \xrightarrow[-H_2O]{CO_2+NH_3} \underset{\text{Citrulline}}{\begin{array}{l}O{=}C{-}NH_2\\|\\CH_2NH\\|\\CH_2\\|\\CH_2\\|\\CH(NH_2)\\|\\COOH\end{array}} \xrightarrow[-H_2O]{NH_3} \underset{\text{Arginine}}{\begin{array}{l}HN{=}C{-}NH_2\\|\\CH_2NH\\|\\CH_2\\|\\CH_2\\|\\CH(NH_2)\\|\\COOH\end{array}} \xrightarrow[\text{(arginase)}]{H_2O} \underset{\text{Ornithine}}{\begin{array}{l}H_2N{=}\underset{\text{Urea}}{\overset{\overset{O}{\|}}{C}}{-}NH_2\\+\\CH_2NH_2\\|\\CH_2\\|\\CH_2\\|\\CH(NH_2)\\|\\COOH\end{array}}$$

What was needed was an explanation of the mechanism of the reactions, the enzymes involved, and the source of energy needed for the production of urea.

The mechanism for the biosynthesis of citrulline from ornithine, carbon dioxide, and ammonia involves two enzymatically catalyzed steps. These are:

$$ATP + CO_2 + NH_3 \rightleftarrows \underset{\text{Carbamyl phosphate}}{NH_2{-}\overset{\overset{O}{\|}}{C}{-}O{-}\overset{\overset{O}{\|}}{\underset{\underset{O^-}{|}}{P}}{-}O^-} + ADP$$

$$\text{Carbamyl phosphate} + \text{ornithine} \rightleftarrows \text{citrulline} + PO_4^{\equiv}$$

In mammalian liver, acetyl glutamic acid or related derivatives of glutamic acid are required as cofactors for the synthesis of carbamyl phosphate.

The mechanism of conversion of citulline to arginine has been clarified by the studies of Ratner, who separated the enzymes required for this conversion and identified the intermediates. The nitrogen donor is not ammonia, but aspartic acid:

$$\underset{\text{Aspartic acid}}{\begin{matrix}COOH\\|\\CH_2\\|\\CH(NH_2)\\|\\COOH\end{matrix}} \xrightarrow[(ATP,\ Mg^{++})]{\text{-Etrulline}} \underset{\text{Argininosuccinic acid}}{\begin{matrix} & & COOH\\ & & |\\HN{=}C{-}NH & {-} & CH\\|\quad & & |\\CH_2NH & & CH_2\\| & & |\\CH_2 & & COOH\\|\\CH_2\\|\\CH(NH_2)\\|\\COOH\end{matrix}} \longrightarrow \underset{\text{Arginine}}{\begin{matrix}HN{=}C{-}NH_2\\|\\CH_2NH\\|\\CH_2\\|\\CH_2\\|\\CH(NH_2)\\|\\COOH\end{matrix}} + \underset{\text{Fumarate}}{\begin{matrix}COOH\\|\\CH\\\|\\HC\\|\\COOH\end{matrix}}$$

The picture of urea synthesis, as it emerges from this work, indicates that the ammonia utilized for urea synthesis does not arise *per se* from oxidative deamination if amino acids. Ammonia is used in the formation of carbamyl phosphate, and some of it is actually supplied in the form of aspartic acid. Of course, aspartic acid can arise in mammalian tissues from ammonia in the following way:

$$NH_3 + \alpha\text{ - ketoglutarate} \rightarrow \text{glutamic acid}$$

$$\text{Glutamic acid} + \text{oxalacetic acid} \rightarrow \alpha\text{ - ketoglutarate} + \text{aspartic acid}$$

The following illustrates the relationship of urea synthesis to the Krebs cycle and also serves to demonstrate the origin and metabolic fate of the amino acids arginie, glutamic acid and aspartic acid. Other amino acids are related to members of the Krebs cycle, e.g., alanine via pyruvic acid, and phenylalanine (and tyrosine) via its oxidative intermediate fumaric acid. The ketogenic amino acids (e.g., phenylalanine) may yield acetoacetic acid which can be oxidized to CO_2 and water, or be used for fatty acid biosynthesis, via acetyl CoA. Aspartic acid is a building block for the pyrimidines and glycine is an important residue in the structures of purices, heme, creatinine, bile salts, etc. Cysteine gives rise to alanine, glutathione, methionine, and choline.

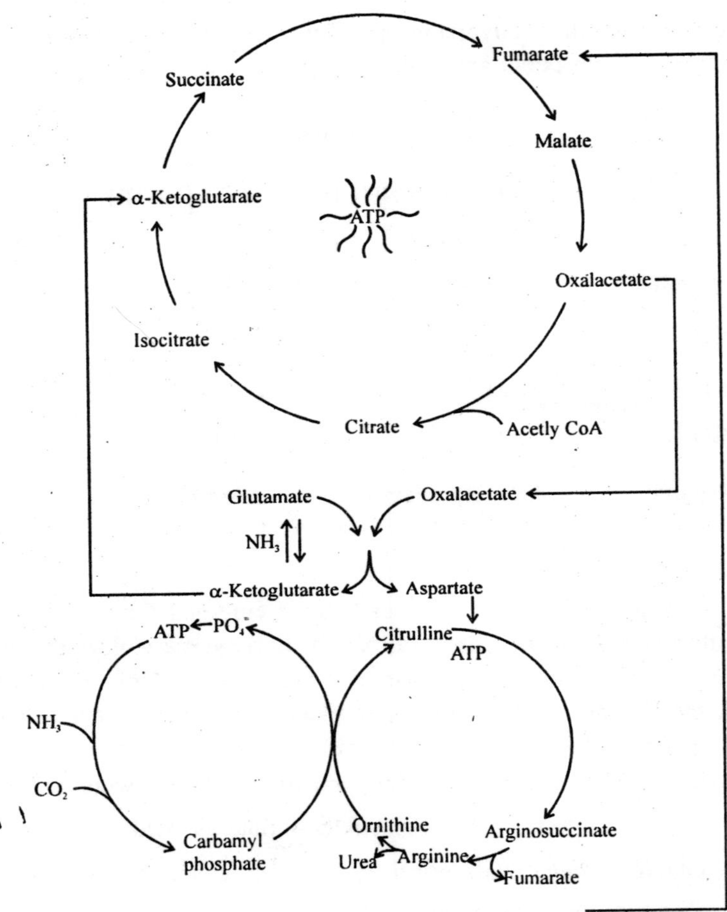

Fig. 10.10 : ***Relationship of urea synthesis to the tricarboxylic acid cycle. Two moles of ATP are required for the synthesis of one mole of urea. (After Ratner* in *McElroy and Glass, eds.:* A symposium on Amino Acid Metabolism, *Johns Hopkins University Press, 1955, p. 249.)***

CHAPTER 11

INDIVIDUAL AMINO ACID METABOLISM

Amino acids are of main importance in the metabolism of all organism because they are the precursors of protein. From nutritional aspects, the amino acids may be classified into three groups, namely **dispensable, indispensable** (*essential*) and **semidispensable** amino acids. Dispensable amino acids are those which can be synthesized at the desired rates in the human body out of the substances ordinarily available in the body. Indispensable amino acids are those which can not be synthesized in the body under ordinary circumstances. Some of the indispensable amino acids may be synthesized under certain special circumstances when special ingredients are made available to the body but this does not happen ordinarily. The semidispensable amino acids are those which under ordinary circumstances are synthesized in only very small quantities which do not commensurate with the body need Examples of each category are given below :

Amino Acids

Indispensable	**Semidispensable**	**Dispensa'**
1. Lysine	1. Arginine	1. Glutar
2. Tryptophan	2. Histidine	2. Aspa'
3. Phenylalaninc	3. Glycine	3. Ala
4. Threonine	4. Cystine	4. Pr
5. Valine	5. Tyrosine	5. '
6. Methionine	6. Serine	
7. Leucine		
8. Isoleucine		

For all practical purposes, it may be assume amino acids are not synthesized in the bo absolutely essential for adequete growth, an **essential anmino acids.** These must be diet in the preformed stage. The nutritive value of a particular protein depends upon the number and quantity of essential amino acids it contains. If a given protein is deficient even in a single essential amino acid, its ability to synthesize body proteins is considerably reduced. Such proteins are merely used by the body as source of energy. For positive nitrogen balance, it is absolutely necessary that all the essential amino acids should be obtained by the body in adequate amounts at the same time. It is not necessary that a single protein should contain all the essential acids. These may be obtained from a **mixture** of several proteins, derived from different sources. Since amino acids are not stored as such in the body, administration of the missing essential amino acids afterwords does not serve the purpose.

Synthesis of Essential Amino Acid

As pointed out already that an amino acid is classed as essential if it must be supplied in the diet for optimal growth of an immature animal ... for the maintenance of nitrogen balance in the mature animal. In ...ontext of this definition histidine may be considered borderline ...tary essential for human beings because it can be synthesized ... salvaged from protein breakdown to a considerable degree. ... since it is essential for infants, this amino acid will be ... the group of nine amino acids, to be described in this

... nonessential amino acids, the biochemical reactions ...ino acids are integrated with the pathways of ... metabolism. The interplay of bidirectional and ... these metabolic relationship may be viewed

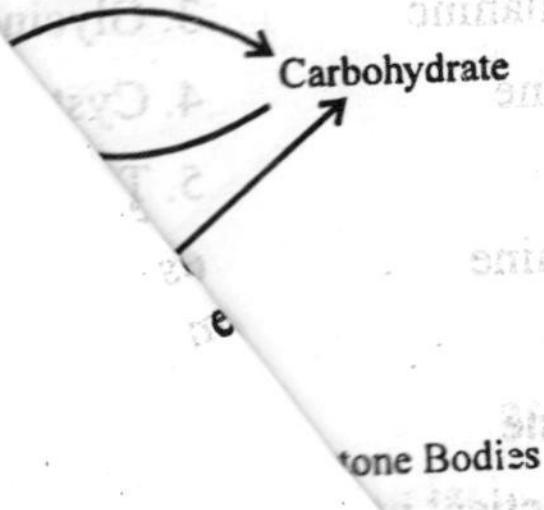

...st recognized in the area ...bolic reactions of both the ... are affected by such genetic

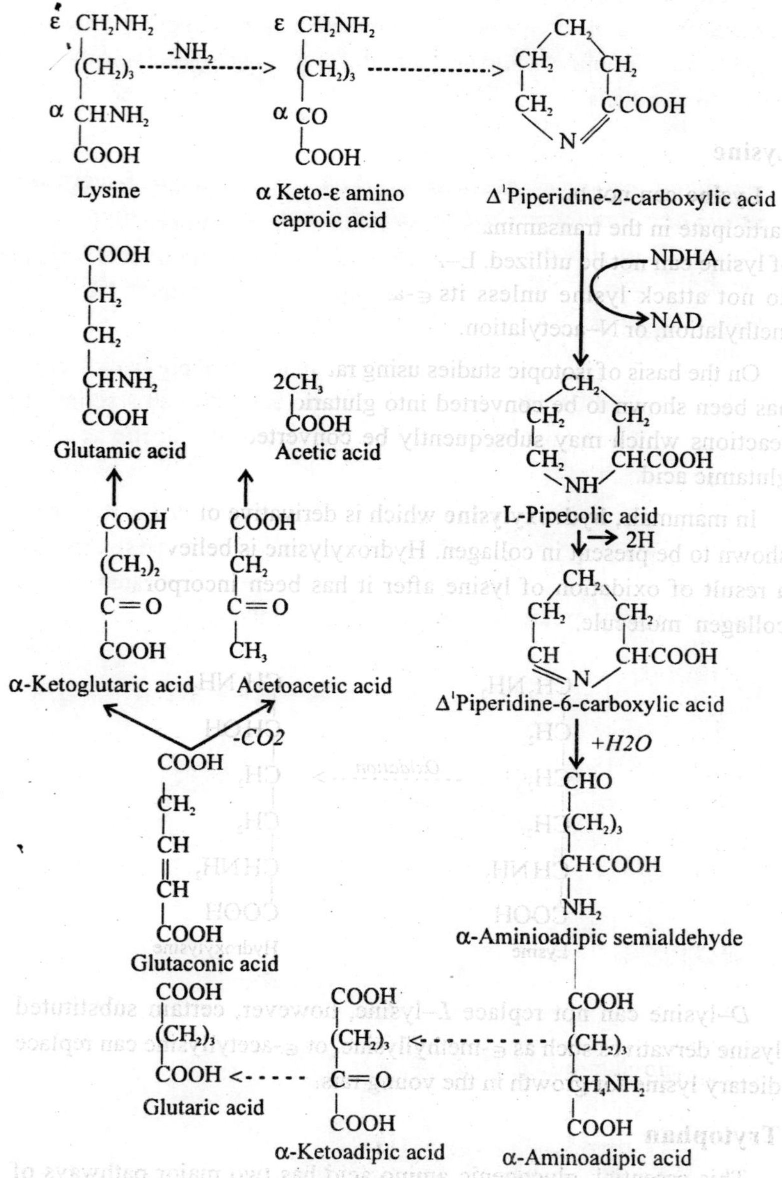

Fig 11.1 Lysine metabolism in the liver tissue.

defects. the inherited errors in metabolism of essential amino acids such as phenylalanine, tryptophan, and the branched chain acids are particularly prominent. We now examine the major catabolic pathways of the essential amino acids and the precursor roles of some of their metabolic products.

Lysine

Lysine can not be synthesized in man or other animals. It can not participate in the transamination reactions in the mammals. D–Isomer of lysine can not be utilized. L–Amino acid oxisases of animals origin do not attack lysine unless its ∈-amino group is protected by N–methylation, or N–acetylation.

On the basis of isotopic studies using rat and guinea–pig liver, lysine has been shown to be converted into glutaric acid through a series of reactions which may subsequently be converted into acetic acid or glutamic acid.

In mammals, **hydroxylysine** which is derivative of lysine has been shown to be present in collagen. Hydroxylysine is believed to arise as a result of oxidation of lysine after it has been incorporated in the collagen molecule.

Lysine		Hydroxylysine
CH_2NH_2		CH_2NH_2
CH_2		$CHOH$
CH_2	*Oxidation* -------->	CH_2
CH_2		CH_2
$CHNH_2$		$CHNH_2$
$COOH$		$COOH$

D–lysine can not replace *L*–lysine, however, certain substituted lysine dervatives such as ∈-methyllysine, or ∈-acetyllysine can replace dietary lysine for growth in the young rats.

Trytophan

This essential, glucogenic amino acid has two major pathways of degradation and gives rise to two biologically active compounds, serotonin and NAD.

(1) Oxygenose, requiring biopterin; decarhboxylase; Swratanin, a Vasoconstrictor

(2) pyrrolase (Cu, heme) → ring opening and exidation → Kynurenine → 2 acketly-CoA, alanine, NAD

bacterial action in lower bowel → Skatole (excreted)

The oxygenase system reponsible for the indroduction of an oxygen atom into the tryptophan ring requires the pteridine derivative, tetrahydrobiopterin:

In the oxygenation of the tryptophan the tetrahydrobiopterin is converted to the dihydroform ($BioH_2$):

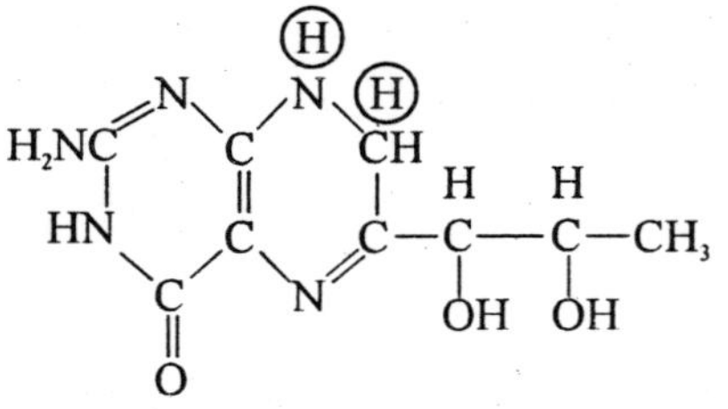

The overall reaction can be summarized schematically as follows:

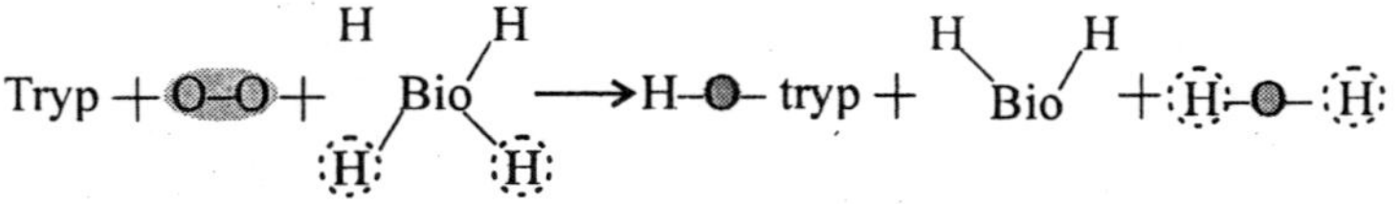

$$\text{Tryp} + \text{O–O} + \text{H}_2\text{Bio} \longrightarrow \text{H–O–tryp} + \text{Bio} + \text{H–O–H}$$

The dihydrobiopterin produced in this reaction can be restored to its tetrahydro form by an NADPH–dependent reduction:

$$\text{BioH}_2 + \text{NADPH} + \text{H}^+ \longrightarrow \text{BioH}_4 + \text{NADP}^-$$

The decarboxylation of the hydroxytryptophan, a reaction prominent in both brain and kidney, yields serotonin. This neurohumoral agent, which is present in brain, platelets, mast cells, and intestinal cells, is a potent vasoconstrictor.

The other major pathway of tryptophan degradation results in its oxidation and ring opening to kynurenine. This pathway, which is prominent in liver, involves a copper–heme enzyme system, termed tryptophan pyrrolase or trytophan 2,3–dioxygenase. Kynurenine can be converted in the liver to quinolinic acid.

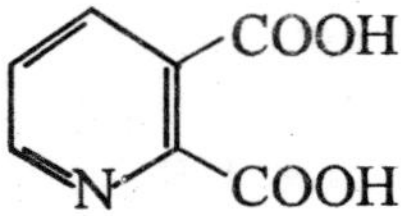

which is an intermediate in the synthesis of nicotinic acid ribonucleotide and NAD. Although tryptophan cannot replace niacin obtained from exogenous sources, it can have a demonstrable "sparing" effect on the dietary requirements for the vitamin. HARTNUP'S DISEASE is a hereditary disorder in which there is a deficiency of the pyrolase. In addition to more serious consequences such as mental retardation, it is predictable that pellagra–like symptoms may be manifest in this disease.

The carbon skeleton of kynurenine can be oxidatively degraded to yield both acetyl–CoA and alanine. Hence, tryptophan is both ketogenic and glcogenic.

The major excretory products of tryptophan metabolism arise from successive deamination and decarboxylation reaction. Indole acetic acid is eliminated in the urine as a glycine conjugate.

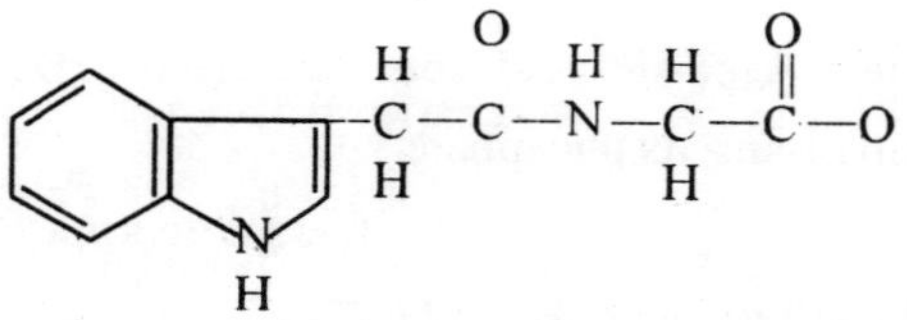

More extensive degradation, resulting in skatole, occurs by bacterial action in the lower bowel.

Threonine

Threonine is a gluconeogenic amino acid. It forms part of most of the proteins.The conversion of threonine of α - oxobytyrate has been

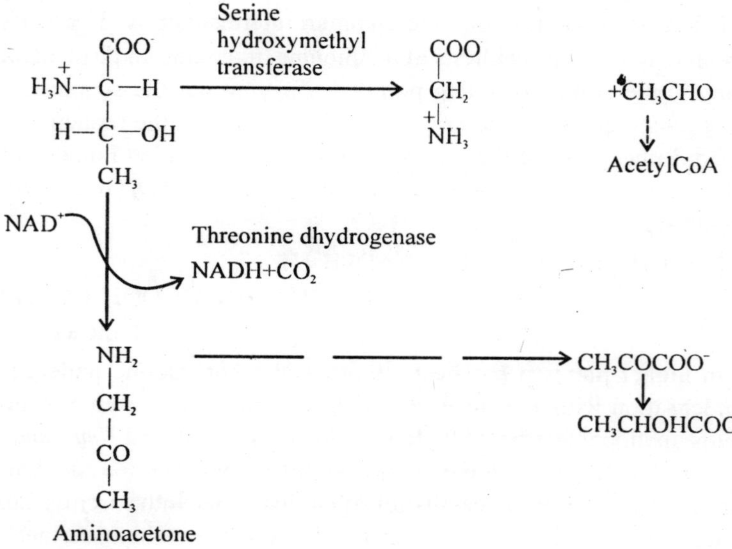

Serine hydroxymethyl transferase

COO^-, $H_3\overset{+}{N}$—C—H, H—C—OH, CH_3 → COO, CH_2, $\overset{+}{N}H_3$ + CH_3CHO → AcetylCoA

NAD^+ → Threonine dhydrogenase → $NADH + CO_2$

NH_2—CH_2—CO—CH_3

Aminoacetone

→ CH_3COCOO^- → $CH_3CHOHCOO^-$

described earlier. The conversion of this compound to propionyl CoA is detailed above. It is believed that alternate catabolic routes for threonine exist. Serine hydroxymethyl transferase is supposed to act on threonine also but in the absence of tetrahydrofolate, to form glycine and acetaldehyde. Aldehyde dehydrogenase can convert acetaldehyde to acetate which can be converted to acetyl CoA. If this sequence of reaction is significant, threonine should be classified as a partly ketogenic amino acid. Threonine dehydrogenase an NAD^+ dependent enzyme, converts threonine to aminoacetone. This is converted to pyruvate and lactate.

Methionine : Aspartic acid gets converted to β - aspartyl semi–aldehyde by involving its phosphate.

$$HOOCCH_2CH(NH_2)COOH + ATP \xrightleftharpoons{\beta\text{-asparto kinase}} H_2O_3POOCCH_2(NH_2)COOH + ADP$$

β - Aspartyl phosphate

$$H_2O_3POOCCH_2CH(NH_2)COOH \xrightleftharpoons{NADPH \rightarrow NADP^+} OHCCH_2CH(NH_2)\,COOH + Pi$$

β - Aspartyl semialdehyde

It has been reported that the common intermediate is β - aspartyl semialdehyde in the synthesis of methionine, threonine and isoleucine. Homoserine is the branching point for carrying out the synthesis of these amino acids. Homeserine gets' synthesized by the reduction of aspartyl phosphate in the presence of the enzyme called homoserine dehydrogenase. The dehydrogenase may use either NADH or NADPH as coenzyme.

$$OHCCH_2CH(NH_2)COOH + NADPH + H^+ \rightleftharpoons HOH_2CCH_2CH(NH_2)COOH + NADP^+$$

Homoserine

In higher plants it has been reported that homoserine undergoes condensation with a cysteine molecule to form homocysteine, which yields methionine after methylation. However *E. coli* and *Neurospora crassa*, involve the complex synthesis of homocysteine. Alcoholic group of homoserine undergoes acylation first either with succinyl CoA or acetyl CoA to form either O–succinyl or O–accetyl–homoserine.

$$HOH_2CCH_2CH(NH_2)COOH + \underset{\displaystyle CH_2COOH}{\underset{|}{CH_2COOH}} \longrightarrow$$

$$HOCC(CH_2)_2\underset{\displaystyle NH_2}{\underset{|}{C}}HCOOH$$

O–Succinyl–homoserine

These products on reacting with cysteine from cystathionine, which then gets hydrolyzed to form homocysteine, pyruvate and ammonia.

$$HOOC(CH_2)_2COO(CH_2)_2\underset{\displaystyle NH_2}{\underset{|}{C}}HCOOH \xrightarrow{HOOCCH(NH_2)CH_2SH}$$

$$HOOC\underset{\displaystyle NH_2}{\underset{|}{C}}HCH_2S(CH_2)_2CH(NH_2)COOH \xrightarrow{H_2O}$$

$$HSCH_2CH_2\underset{\displaystyle NH_2}{\underset{|}{C}}HCOOH + CH_3COCOOH + NH_3$$

Homocysteine

As in higher plants, homocysteine gets methylated at —SH end and the product of the reaction has been found to be methionine. The methyl donor has been found to be another amino acid, serine.

$$HSCH_2CH_2CH(NH_2)COOH \xrightarrow{\text{Serine}} CH_3SCH_2CH_2CH(NH_2)COOH$$

Methionine

Leucine

Leucine can undergo transamination reaction forming α - **ketoisocaproic acid** which is oxidatively decarboxylated in the presence of NAD and CoA.SH to form **isovaleryl coenzyme A** which is an important precursor for synthesis of branched–chain fatty acids. Isovaleryl coenzyme A on further oxidation is converted into β - **methylcrotonyl coenzyme A.** Carboxylation of this product gives β - **methylglutaconyl coenzyme A.** A molecule of water is then added across the double bond forming β - **hyd-roxy–** β - **methylglutaryl coenzyme A** (*HMG–CoA*). This is an important compound which is involved in the biosynthesis of cholesterol and ketone bodies. HMG–CoA is finally degraded into acetoacetic acid and acetyl coenzyme A.

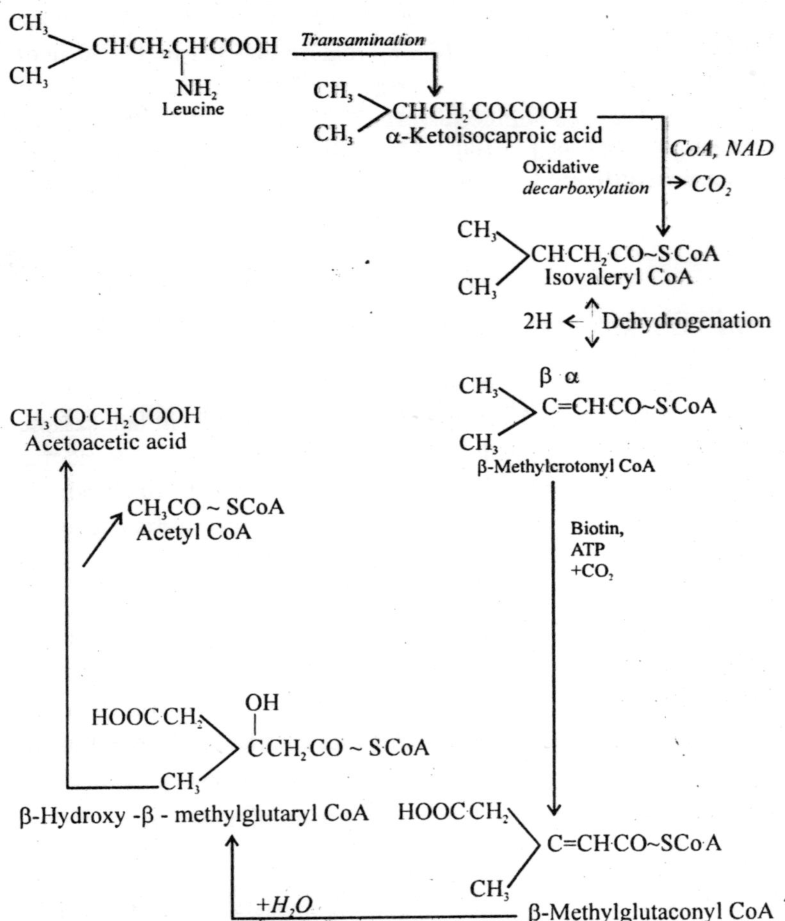

Fig. 11.2 :. Metabolism of Leucine

Leucine is a **ketogenic amino** acid.

A block in the metabolism of leucine is an inborn error of metabolism and in infants causes a disease known as *maple syrup urine* disease. In this disease, oxidative decarboxylation of α - keto–150–carpic acid is blocked, and hence, it is excreted in the urine in excessive amounts.

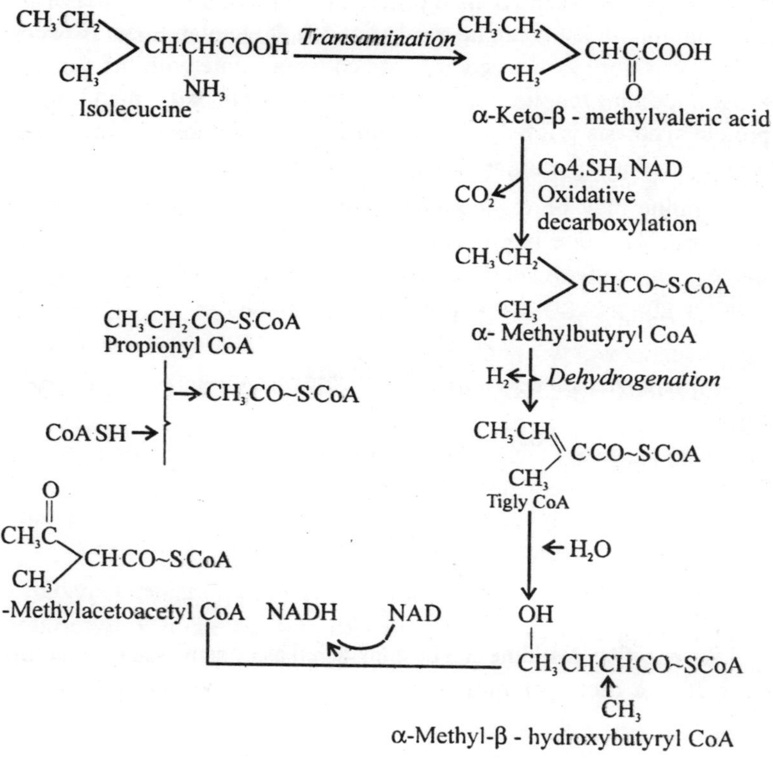

Fig. 11.3 : Metabolism of Isoleucine.

Metabolism of Semidispensable and Dispensable Amino Acids

The dispensable and semidispensable amino acids are not in any way lesser important in comparison to the indispensable (*essential*) amino acids. These are also similarly required for protein biosynthesis and formation of other biogenic compounds. The only difference being, body has adopted itself for the biosynthesis of dispensable, or semidispensable amino acids but has as yet failed to adopt itself for synthesis of indispensable amino acids. Such amino acids are also designated by some authors as **synthesizable** amino acids.

Semidispensable Amino Acids:

Arginine

Arginine is essential in the diet of *only young ones* but is not essential for adults. It can be synthesized in the human body. However in rats, it is not synthesized at a rate commensurate with the demand. Most important function of this amino acid in the body in addition to protein synthesis is it helps in the formation of urea from ammonia and is converted into ornithine to fix up more ammnonia.

Arginine may be regenerated from ornithine through a series of reactions. Arginine is gluconeogenic in nature. It is converted into glucose in *diabetes mellitus*. Besides this, arginine can donate an amidine group to glycine and synthesize creatine.

$NH_2(NH=)C{\cdot}NH{\cdot}(CH_2)_3{\cdot}CH(NH_2){\cdot}COOH$ (Arginine) —Arginase→ $CH_2(NH_2)(CH_2)_2CH(NH_2){\cdot}COOH$ (Orinithine) + $(NH_2)_2C{=}O$ (Urea)

Histidine

Like arginine, histidine can be considered as a "semiessential" amino acid. It is a dietary requirement for the young animal but can be

$HC{=}C{-}CH_2{-}CH(NH_3){-}COO^-$ (imidazole ring: HN^+, C–H, NH) —histidose→ $HC{=}C{-}C(H){=}C(H){-}COO^-$ (ring: HN^+ 3, C 2, NH 1) Uroconic Acid —ring opening→ →

$^-COO-(CH_2)_2-CH(-COO^-)-NH-CH{=}NH$ Formiminoglutamate (FIGLU)

Original C-2 of his → CH

Tetrahydrofalate (FH_4) → formimino-FH_4

→ glutamate → a-ketoglutarate → → → PEP → → glucose

synthesized at a rate adequate to maintain nitrogen balance in the adult. In E.*coli* and *Salmonell,* histidine is synthesized from phosphoribosyl–l–pyrophosphate (PRPP), ATP, and glutamine. Its glycogenic character is evident from the given reactions.

A genetic disease, HISTIDINEMIA, manifests itself by a low activity of histidase. In normal metabolism the degradative pathway initiated by histidase results in the formation of a glutamate derivative, formiminoglutamate (FIGLU).The formiminomoiety can be transferred enzymatically to tetrahydrofolate to form N^5–formiminoterrahydrofolate, an important ONE–CARBON COMPOUND. In a folate deficiency, FIGLU will be excreted because there is inadequate tetrahydrofolate to "trap" the formimino group.

In most cells of lung, liver, and gastric mucosa, histidine undergoes an enzymatic decarboxylation.

```
            +
           NH2                                              +NH3
HC=C —CH2—C —COO⁻        HCO3⁻          HC=C—CH2—CH
 /   \       H              ↑             /   \          H
HN⁻  NH              ———————————→       HN⁺   NH
  \\ /                                     \\ /
   C                                        C
   H                                        H
                                         Histamine
```

to form histamine, a powerful vasodilator. This amine is liberated in traumatic shock and is localized in areas of inflammation.

Tyrosine

Since phenylalanine may be converted into **tyrosine** irreversibly, tyrosine is not an indispensable amino acid. One of the major pathways of phenylalanine metabolism is through its prior conversion to tyrosine, hence, the fate of tyrosine is essentially the same as described under phenylalanine metabolism.

D–Tyrosine can not be metabolized by human beings. **Inability** to utilize tyrosine is observed in the inherited diseases such as *tyrosinosis, alkaptonuria* and *phenylketonuria.*

In *tyrosinosis* the keto–acid of tyrosine., *p*–hydroxyphenylpyruvic acid is excreted in considerable amounts in the urine. However, such patients can utilize homogentisic acid. *Tyrosinosis* is rather a very rare disease. For metabolic details of tyrosine, see phenylalanine metabolism.

Phenylalanine and Tyrosine

Humans and most all other animal organisms cannot sythesize the benzene ring. Hence, phenylalanine is a dietary essential. Tyrosine is synthesized from phenylalanine. Other than to be utilized for protein synthesis in normal metabolism, the only known fate of phenylalanine is its conversion to tyrosine. Much of the dietary requirement for phenylalanine is due, therefore, to the need for tyrosine.

Synthesis of Tyrosine

The synthesis of tyrosine from phenylalanine again illustrates a type of synthetic reaction in which molecular oxygen is a direct reactant. The hydroxylation of phenylalanine, like that of tryptophan, requires tetrahydrobiopterin:

$$C_6H_5-CH_2-\underset{H}{\overset{^{+}NH_3}{C}}-COO + \text{tetrahydrobiopterin} + O{=}O \xrightarrow[\text{hydroxylase}]{\text{phenylalanine}}$$

$$NADPH+H^+ \rightarrow NADP^+$$

$$HO-C_6H_4-CH_2-\underset{H}{\overset{^{+}NH_3}{C}}-COO + \text{dihydrobiopterin} + HO{-}H$$

This reaction is the beginning of the metabolism of phenylalanine in normal individuals. In about 1 of 10,000 to 20,000 newborns, there is a genetic defect because of which phenylalanine hydroxylase is missing or deficient. As a consequence, phenylalanine cannot be converted to tyrosine at a normal rate in these individuals and a second pathway of phenylalanine metabolism, which is normally insignificant,comes into play:

$$C_6H_5-\underset{H}{\overset{H}{C}}-\underset{H}{\overset{\overset{+}{N}H_3}{C}}-COO^- \xrightarrow[2H^+]{\overset{+}{N}H_4} C_6H_5-\underset{H}{\overset{H}{C}}-\overset{O}{\overset{\|}{C}}-COO^- \xrightleftharpoons{H^+ + NADH \quad NAD^+} C_6H_5-\underset{H}{\overset{H}{C}}-\underset{H}{\overset{OH}{C}}-COO^-$$

Phenylpyuvate

Phenyllacatate

(blocked) ↓ tyrosine

There is now an increased concentration of phenylalanine in all body fluids and phenylpyruvate and phenyllactate are excreted in high levels. This disease is called **phenylketonuria.** Early diagnosis of this tragic condition is absolutely essential. A common therapy is a diet low in phenylalanine.

Alkaptonuria

The first step in the catabolism of tyrosine is the loss of its α – amino group to form the keto acid.

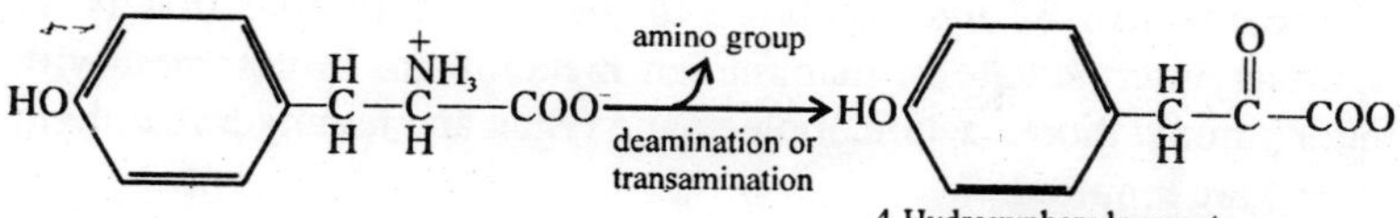

$$HO-C_6H_4-\underset{H}{\overset{H}{C}}-\underset{H}{\overset{\overset{+}{NH_3}}{C}}-COO^- \xrightarrow[\text{deamination or transamination}]{\text{amino group}} HO-C_6H_4-\underset{H}{\overset{H}{C}}-\overset{O}{\overset{\|}{C}}-COO$$

4-Hydroxyphenylpyruvate

The next reaction is a complicated oxidation rearrangement in which another hydroxyl group is added to the ring. the carboxyl group is lost as CO_2, and the positions of the phenyl hydroxyls are changed relative to the side chain. The product of this reaction is homogentisic acid:

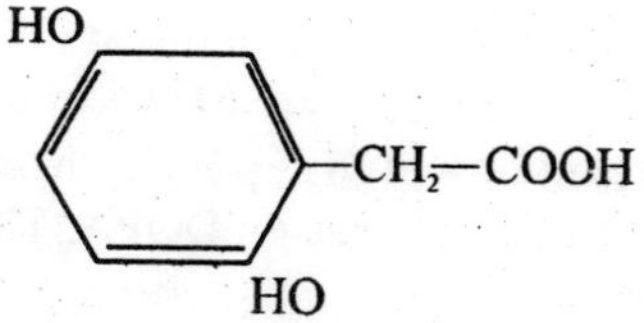

$$(HO)_2C_6H_3-CH_2-COOH$$

In a subsequent oxidation reaction, the homogentisic ring is opened, giving an eight –carbon compound:

$$\text{Homogentisic acid} \xrightarrow[\text{oxidose}]{\text{homogentisic}} OOC-\underset{H}{C}=\overset{H}{C}-\underset{O}{\underset{\|}{C}}-CH_2-\underset{O}{\underset{\|}{C}}-CH_2-COO$$

Fumarylacetoacetate

This compound can be hydrolyzed to fumarate and acetoacetate:

$$OOC-\underset{H}{C}=\overset{H}{C}-COO+CH_3-\underset{O}{\underset{\|}{C}}-CH_2-COO$$

Phenylalanine and tyrosine are, therefore, both glucogenic and ketogenic.

ALKAPTONURIA is an inherited disorder caused by the absence of homogentisic oxidase, a condition in which homogentisate accumulates and is excreted in the urine. Upon standing, the urine of these individuals turns dark because the homogentisate is further oxidized to a melaninlike compound. Alkaptonuria is generally a benign condition.

Tyrosine as a Precursor of Melanin, Catecholamines, and Thyroxine

Tyrosine can be hydroxylated to yield a derivative that is an important intermediate in mammalian metabolism. In common with other hydroxylation reaction, molecular oxygen and tetrahydrobiopterin ($BioH_4$) are required:

$$HO-C_6H_4-CH_2-\underset{H}{\overset{\overset{+}{N}H_3}{C}}-COO^- \xrightarrow[\text{tyrosine hydroxylase, or tyrosinase}]{BioH_4,\ O_2,\ Cu} (HO)_2C_6H_3-CH_2-\underset{H}{\overset{\overset{+}{N}H_3}{C}}-COO^-$$

L-3,4-Dihydroxyphenylalanine (DOPA)

MELANIN is the brownish–black pigment of skin that is produced in the melanocytes in the basal layer of the epidermis. Melanin is a group of high molecular weight polyhydroxyphenyl polymers produced by the action of tyrosinase on DOPA. The oxidation involves two quinone intermediates:

$$\text{DOPA} \xrightarrow[\text{tyrosinase}]{(O)} \text{DOPA Quinone} \rightarrow \xrightarrow{(O)} \rightarrow \text{Indole 5,6-quinone} \xrightarrow{(O)} \rightarrow \text{Melanin}$$

DOPA Quinone: O=, O= ring –CH_2–HC–COO^-, $\overset{+}{N}H_3$

Indole 5,6-quinone: O=, O= ring –CH=CH–N(H)

A deficiency in the tyrosinase system of the melanocytes results in the condition called albinism.

Another important precursor function of **DOPA** is its role in the synthesis of the catecholamines in the adrenal medulla and sympathetic never terminals.

$$\text{DOPA} \xrightarrow[\text{decarboxylase}]{\text{DOPA}} HCO_3 + (HO)_2C_6H_3-\underset{H}{\overset{H}{C}}-\underset{H}{\overset{H}{C}}-\overset{+}{N}H_3$$

Dopamine

O_2 ↓ Dopamine β–hydroxylase (cu,ascorbate)

$$(HO)_2C_6H_3-\underset{H}{\overset{OH}{C}}-CH_2-\overset{+}{N}H_3 \xrightarrow[\text{S-adenosylhomocysteine}]{\text{S-adenosylmethionine}}$$

Norepinephrine

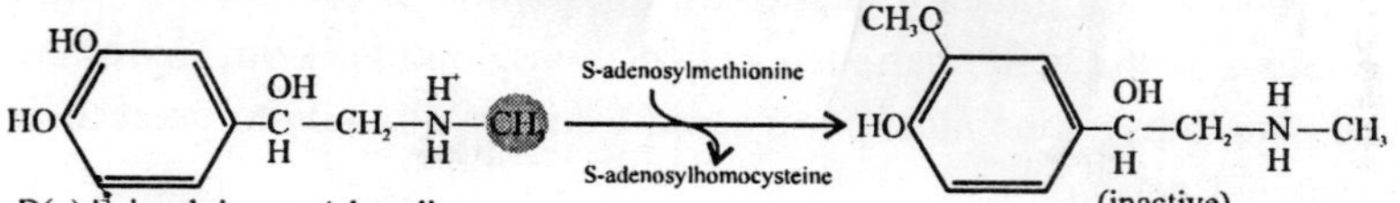

HO, HO-C_6H_3-C(OH)(H)-CH_2-N(H)(H)-CH_3 —(S-adenosylmethionine → S-adenosylhomocysteine)→ CH_3O, HO-C_6H_3-C(OH)(H)-CH_2-N(H)(H)-CH_3

D(–) Epinephrine, or Adrenalin (inactive)

Dopamine β - hydroxylase is a Cu–containing mixed–function oxidase. It does not require the tetrahydrobiopterin as in the case of the prior hydroxylation reactions involving the phenyl ring.

Two hormones of the thyroid gland, thyroxine and triiodothyronine, are iodinated derivatives of tyrosine. The overall pathways for the synthesis of these related compounds can be represented as follows:

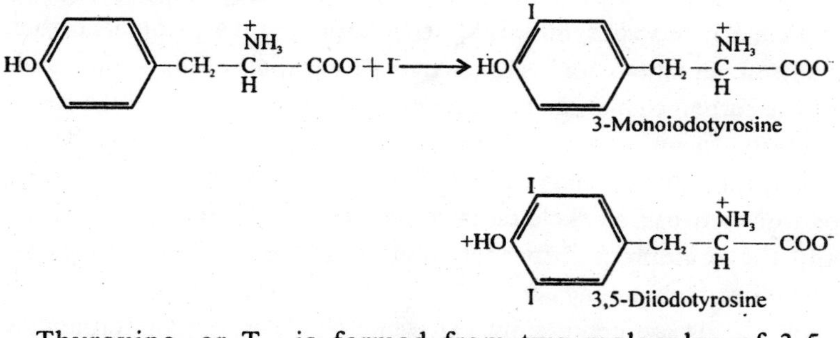

$$HO-C_6H_4-CH_2-\overset{\overset{+}{NH_3}}{\underset{H}{C}}-COO^- + I^- \longrightarrow HO-C_6H_3(I)-CH_2-\overset{\overset{+}{NH_3}}{\underset{H}{C}}-COO^-$$

3-Monoiodotyrosine

$$+HO-C_6H_2(I)_2-CH_2-\overset{\overset{+}{NH_3}}{\underset{H}{C}}-COO^-$$

3,5-Diiodotyrosine

Thyroxine, or T_4, is formed from two molecules of 3,5–diidotyrosine:

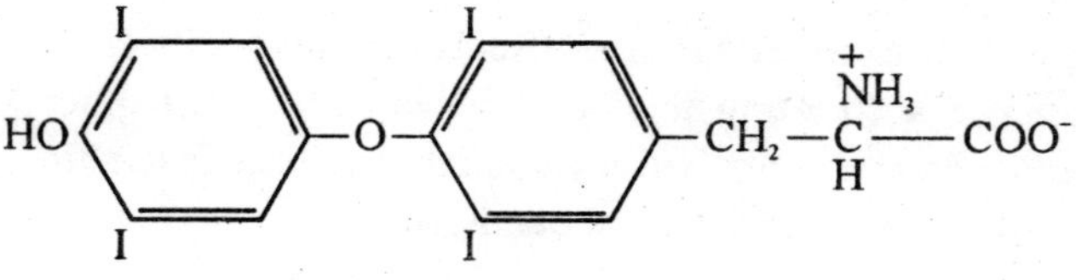

$$HO-C_6H_2(I)_2-O-C_6H_2(I)_2-CH_2-\overset{\overset{+}{NH_3}}{\underset{H}{C}}-COO^-$$

Thyroxine

Triiodothyronine is the product of the reaction of 3–monoiodotyrosine and 3.5–diiodothyronine:

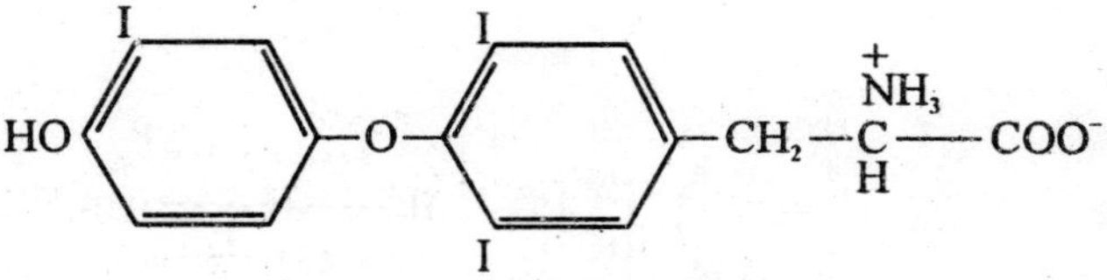

$$HO-C_6H_3(I)-O-C_6H_2(I)_2-CH_2-\overset{\overset{+}{NH_3}}{\underset{H}{C}}-COO^-$$

Triiodothyronine, or T_3

T_3 and T_4 are linked to a specific to globulin of the thyroid gland colloid and the combination is called thyroglobulin. Both thyroxine and triiodothyronine must be released from the globulin to exert their hormonal action.

Synthesis of Non Essential Amino Acids

Whether they are dietary essentials or not, all amino acids are required for protein synthesis, all serve as precursors of many compounds that are essential for life, and all can serve as sources of metabolic fuel. Now examine the major biochemical reactions that permit the pathways of amino acid metabolism to be integrated with the pathways of carbohydrate aand lipid metabolism. Toward this end, it is helpful to categorize amino acids as glucogenic or ketogenic. If there can be a net synthesis of phosphoenolpyruvate from its carbon skeleton, an amino acid is glucogenic. Hence, all of the amino acids whose carbon chains can be converted into intermediates of the Krebs cycle or to pyruvate, will be glucogenic. Most amino acids are glucogenic. Some amino acids are degraded in such a manner that a part of their carbon skeleton is convertible to phosphoenol pyruvate and the remainder into acetoacetate and other ketone bodies. Therefore, these amino acids are both glucogenic and ketogenic. Some amino acids are completely ketogenic because their intermediate oxidation products can be accounted for entirely as acetyl–CoA and its ketone body derivatives.

Let us now examine the amino acids that are not dietary essentials for humans, accounting for their biosynthesis, their glucogenic or ketogenic character, and their roles as intermediates in the biosynthesis of metabolically essential compounds.

Serine

Serine is a nonessential amino acid because it can be synthesized from glycine and also from 3–phosphoglyceric acid, an intermediate of glycolysis

$$\begin{array}{c} COO^- \\ | \\ HCOH \\ | \\ HC{-}O\text{Ⓟ} \\ H \end{array} \xrightarrow{\nearrow 2H} \begin{array}{c} COO^- \\ | \\ C{=}O \\ | \\ HC{-}O\text{Ⓟ} \\ H \end{array} \xrightarrow[\text{glutamate}]{\text{transamination}} \begin{array}{c} COO^- \\ | \quad + \\ HC{-}NH_3 \\ | \\ HC{-}O{-}\text{Ⓟ} \\ H \end{array} \xrightarrow{\nearrow P_i} \begin{array}{c} COO^- \\ | \quad + \\ HC{-}NH_3 \\ | \\ HC{-}OH \\ H \end{array}$$

Role in Cysteine Synthesis. The carbon chain of serine can become the carbon chain of the amino acid cysteine:

S-Adenosylmethionine → (CH_3 ↑, adenosine ↓) → $SH-(CH_2)_2-HC(\overset{+}{N}H_3)-COO^-$ Homocysteine

Homocysteine + $HO-CH_2-HC(\overset{+}{N}H_3)-COO^-$ → $H_2C-S-CH_2-HC(\overset{+}{N}H_3)-COO^-$ / $H_2C-HC(\overset{+}{N}H_3)-COO^-$

Cystathionine

Cystathionine + HOH → α-Aminobutyric acid + $HS-CH_2-HC(\overset{+}{N}H_3)-COO^-$

Cysteine

In humans the formation and splitting of cystathionine are irreversible reactions. Methionine is an essential amine acid. It cannot be made from cysteine.

Function in Complex Lipid Synthesis. One of the major precursor functions of serine is its role in the biosynthesis of the complex lipids. Serine itself can be a component of the complex lipids such as the following:

$$\begin{array}{l} \text{Fatty acyl}-O-CH_2 \\ \text{Fatty acyl}-O-CH \\ \qquad\qquad\quad H_2C-O-\overset{O}{\overset{\|}{P}}(O^-)-O-CH_2-\overset{H}{C}(COO^-)-\overset{+}{N}H_3 \end{array}$$

Phosphatidlserine

In addition, serine can lose its COOH group to form aminoethanol:

$$HOCH_2-\underset{H}{C}(\overset{+}{N}H_3)-COO^- \xrightarrow{HOH} HO-CH_2-CH_2(\overset{+}{N}H_3) + HCO_3^-$$

By successive addition of methyl groups to the nitrogen by S–

adenosylmethionine, the ethanolamine is converted to choline:

$$CH_3-\overset{\displaystyle CH_3}{\underset{\displaystyle CH_3}{\overset{|}{\underset{|}{N^+}}}}-CH_2-CH_2OH$$

All of these reactions can occur with phosphatidyl derivatives.

$$\text{Phosphatidylserine} \xrightarrow[CO_2]{} \text{phosphatidylethanolamine} \xrightarrow{\cdot CH_3} \xrightarrow{\cdot CH_3} \xrightarrow{\cdot CH_3} \text{phosphatidylcholine}$$

Cysteine

It will be recalled that this amino acid can be synthesized from a reaction between serine and homocysteine.

Cysteine is glucogenic because it can be converted to phosphoenolpyruvate:

$$\text{Cys} + \alpha\text{-KG} \longrightarrow \text{Glu} + HS-CH_2-C(=O)-COO^-$$

$$\underset{\beta\text{-Mercaptopyruvic acid}}{HS-CH_2-C(=O)-COO^-} \xrightarrow[\text{Sulfide}]{} \text{pyruvate} \longrightarrow \text{phosphoenolpyruvate}$$

Oxidation of the —SH Group. The oxidizability of the cysteine sulfhydryl group gives the amino acid considerable metabolic versatility and also makes it an important determinant of protein structure:

$$2\text{Cys—SH} \longrightarrow \text{Cys—S—S—Cys} + 2\text{H}$$

Thus, both inter–and intrapeptide chain—S—S—linkages can be formed:

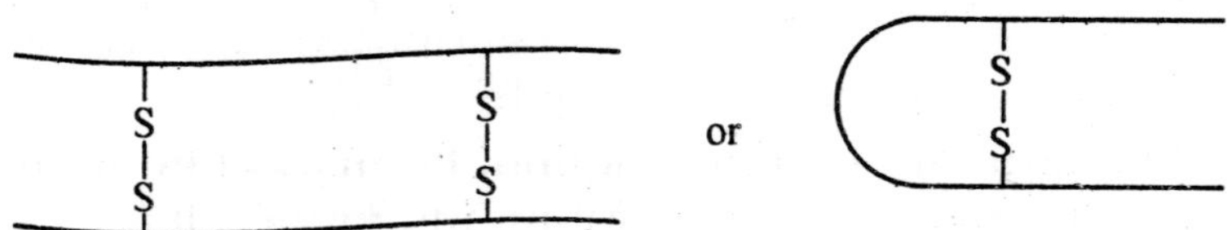

The S of cysteins can also be oxidized beyond the S—S level:

$$\begin{array}{c} HS \\ | \\ CH_2 \\ | \\ HC-\overset{+}{N}H_3 \\ | \\ COO^- \end{array} \rightarrow \rightarrow \begin{array}{c} SO_3H \\ | \\ CH_2 \\ | \\ HC-\overset{+}{N}H_3 \\ | \\ COO^- \\ | \\ SO_4^{2-} \end{array} \text{Cysteic Acid or Cysteine Sulfinic Acid} \xrightarrow{\nearrow HCO_3^-} \begin{array}{c} SO_3H \\ | \\ CH_2 \\ | \\ HC-\overset{+}{N}H_3 \\ H \end{array}$$

Taurine

SO_4^{2-} → active sulfate or excreted

Formation of Bile Salts. Bile salts are formed as follows:

Cholic acid (H_3C, R, C—O, =O) + Tourine → H_3C, R, C(=O)—N(H)—CH_2—CH_2SO_3

lipophilic portion | hydrophilic part

Cholic acid → Tourochalate

Taurocholate is an important bile salt that acts as a detergent in the emulsification process associated with fat digestion and absorption.

Aspartate

That this glucogenic amino acid is a dietary nonessential is evident from the following synthetic sequence:

Glucose ← PEP + GDP + CO_2

Glucose → pyruvate

pyruvate + CO_2 + ATP → axaloacetate + ADP + P_i

axaloacetate —GTP→ PEP + GDP + CO_2

axaloacetate + Glu ⇌ Asp + a-KG

It should be remembered that the oxaloacetate also serves as the "trigger" for the tricarboxylic acid cycle.

Participation in Synthesis of Urea, Purines, and Pyrimidines. Aspartate participates in three vital synthetic reactions. It contributes one of the nitrogens in the synthesis of urea, serving as a substrate for the synthesis of arginosuccinate:

$$^{-}OOC-CH_2=\underset{H}{\overset{\overset{+}{NH}}{C}}-COO^{-} + H_2N-\overset{O}{\overset{\|}{C}}-\overset{H}{N}-(CH_2)_3-\underset{\underset{+}{NH_3}}{\overset{H}{C}}-COO^{-} \rightarrow$$

Citrulline

$$\begin{array}{c} ^{-}OOC-CH_2-\overset{H}{C}-COO^{-} \\ | \\ NH \\ | \\ HN=C-\overset{H}{N}-(CH_2)_3-\underset{\underset{+}{NH_3}}{\overset{H}{C}}-COO^{-} \end{array}$$

Aspartate is also one of the nitrogen donors in purine synthesis.

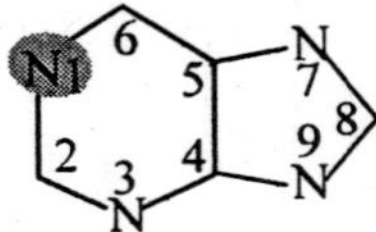

Finally, both its carbon chain and nitrogen become a major part of the pyrimidine ring system.

Glutamate

As pointed out previously, this amino acid is formed from α - ketoglu-tarate by either transamination or reductive amination. Glutamate is glucogenic because it can be converted to phosphoenolpyruvate:

Glutamate → α - ketoglutarate → succinyl–CoA → succinate → fumarate → malate → oxaloacetate → phosphoenolpyruvate(PEP) → → → glucose

Role in Biosynthesis of Urea, Glutathione, and γ - Aminobutyric Acid. Glutamate plays a significant role in biosynthetic reactions and as a precursor of other biological compounds. As pointed out earlier,

$$Glu-\alpha\text{-}NH_2 \rightarrow \alpha\text{-}KG$$

$$NH_4^{+} \xrightarrow{HCO_3^{-} \quad ATP \quad HOH} \text{carbamoyl phospltate}$$

it is one of the prime amino group donors in transamination. One of the nitrogens of urea is provided by glutamate.

This amino acid is also one of the three constituent acids of glutathione:

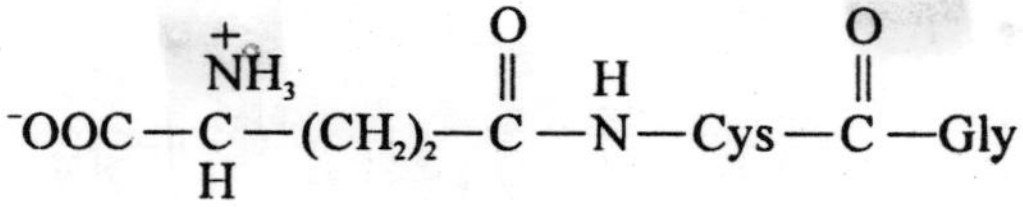

$$^{-}OOC-\overset{\overset{+}{NH_3}}{\underset{H}{C}}-(CH_2)_2-\overset{\overset{O}{\|}}{C}-\overset{H}{N}-Cys-\overset{\overset{O}{\|}}{C}-Gly$$

A product of glutamate, γ - aminobutyric acid (GABA), is essential for brain function because it retards transmission of nerve impulses.GABA is produced by the loss of the number one carboxyl group of glutamate:

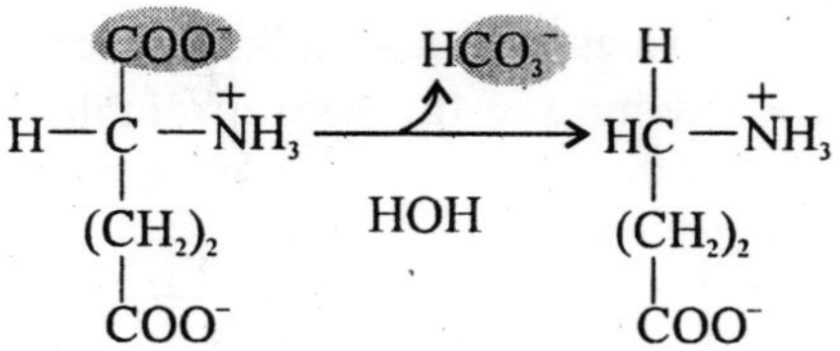

$$H-\overset{\overset{COO^{-}}{|}}{\underset{\underset{\underset{COO^{-}}{|}}{(CH_2)_2}}{C}}-\overset{+}{N}H_3 \xrightarrow[HOH]{HCO_3^{-}} H\overset{\overset{H}{|}}{\underset{\underset{\underset{COO^{-}}{|}}{(CH_2)_2}}{C}}-\overset{+}{N}H_3$$

It may be noted that 75 percent of the total free amino acids of brain can be accounted for by aspartate, glutamate, and their derivatives.

Amide Group of Glutamine as a Nitrogen Source in Biosynthetic Reactions. When converted to glutamine, the amide group on C–5 of glutamate becomes a prime source of nitrogen in many biosynthetic reactions. N^3 and N^9 of the purine ring system and the amino nitrogen at position 2 of guanylic acid are derived from glutamine:

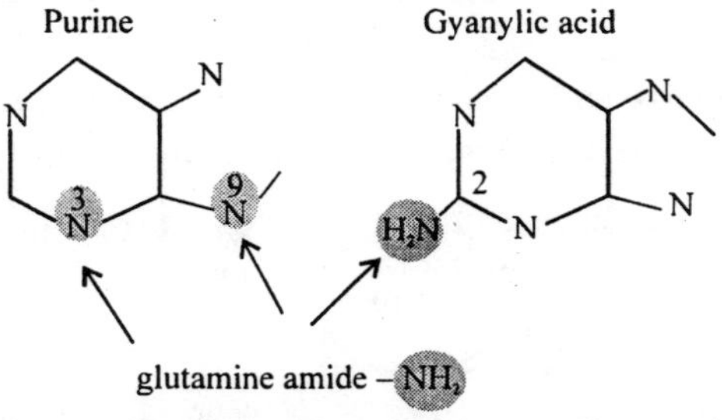

Purine

Gyanylic acid

N N 3N 9N

N N 2 H_2N N N

glutamine amide $-NH_2$

In the cytoplasm of eukaryotic cells is a carbamoyl phosphate synthetase that needs glutamine as a nitrogen donor. This carbamoyl phosphate synthetase provides a carbamoyl group for pyrimidine synthesis.

$$\underset{(-NH_2)}{\text{Glutamine}} + HCO_3^- + 2ATP + HOH \longrightarrow H_2N-\underset{\underset{O}{\|}}{C}-O\text{Ⓟ} + 2ADP + P + \text{glutamate}$$

Glutamine also donates its amide nitrogen in the synthesis of amino sugars:

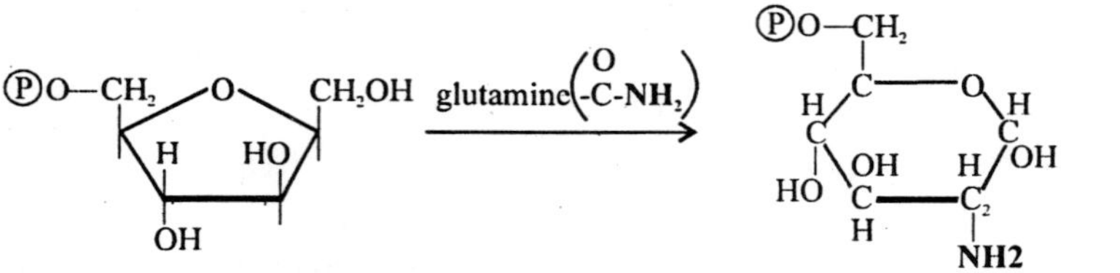

ⓅO—CH_2 O CH_2OH H HO OH — glutamine($\overset{O}{\|}$C-NH_2) → ⓅO—CH_2 C O H H C C OH HO C H OH C H NH_2

Fructose-6-phosphate → Glucosamine-6-phosphate

Glutamine participates in a cross–linking reaction of fibrin in blood clotting process, resulting in the final insoluble polymeric form, or HARD CLOT:

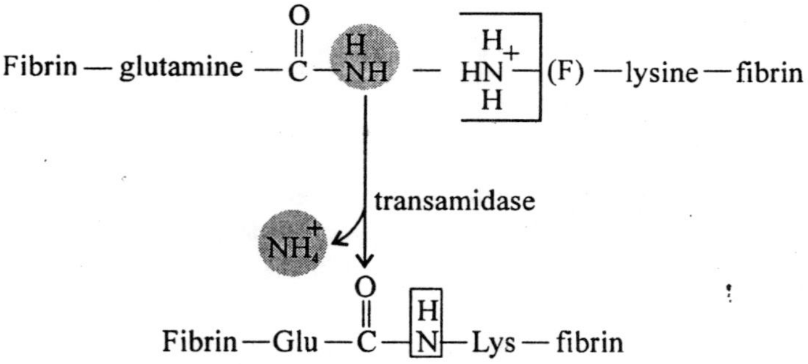

$$\text{Fibrin — glutamine} - \overset{\overset{O}{\|}}{C} - \overset{H}{NH} \quad - \quad [H\overset{H}{\underset{H}{N^+}}] (F) \text{— lysine — fibrin}$$

$$\downarrow \text{transamidase} \quad (\nearrow NH_4^+)$$

$$\text{Fibrin—Glu} - \overset{\overset{O}{\|}}{C} - [\overset{H}{N}] \text{—Lys — fibrin}$$

Alanine

The synthesis and metabolism of the carbon chain of alanine is apparent from the following metabolic sequence:

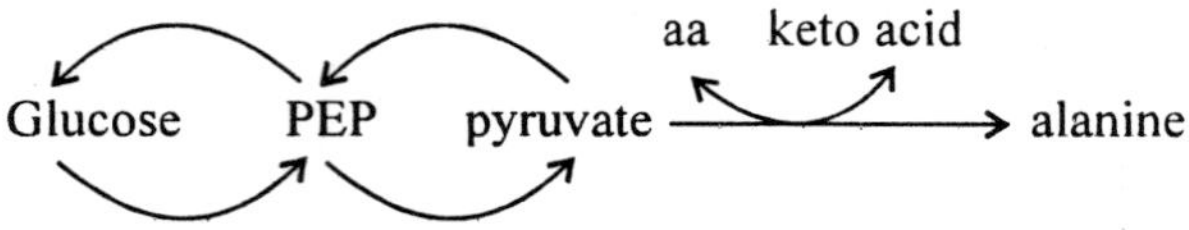

Glucose ⇄ PEP ⇄ pyruvate —(aa → keto acid)→ alanine

Alanine–Glucose Cycle. The Cori cycle represents a metabolic link between liver and muscle. The liver furnishes glucose, produced from lactate by gluconeogenesis, to skeletal muscle. In the muscle lactate is regenerated from pyruvate in skeletal muscle, so can alanine be produced from glucose by glycolysis. Just as lacate is formed reductively from pyruvate by transam,ination from a number of amino acids. Because of its interconvertibility with pyruvate by transamination, alanine can serve in the same manner as lactate in a recycling mechanism between skeletal muscle and the liver :–

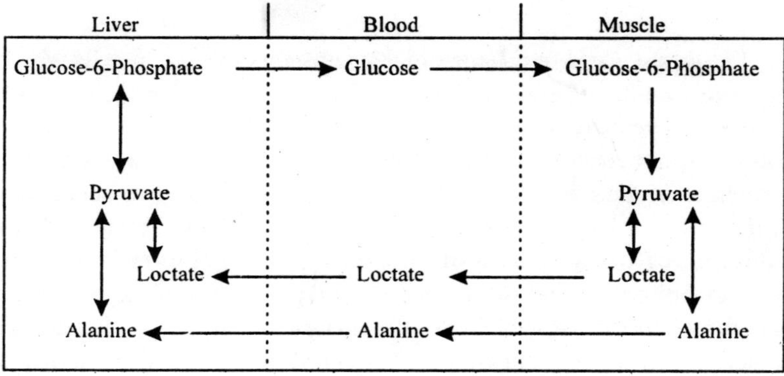

CHAPTER 12
LIPIDS

Lipids are water–insoluble organic substances found in cells which are extractable by nonpopular solvents such as chloroform, either, and benzene. There are several major classes and subclasses of lipids, most of which may occur in different molecular species, depending on the structure of their fatty acid components. Lipids appear to serve four general functions: (1) as structural components of membranes, (2) as intracellular storage depots of metabolic fuel, (3) as a transport form of metabolic fuel, and (4) as protective components of the cell walls of many bacteria, of the leaves of higher plants, of the exoskeleton of insects, and of the skin of vertebrates. Some substances classified among the lipids have intense biological activity; they include some vitamins and their precursors, as well as a number of hormones.

Until only a few years ago, lipid biochemistry was considered an uninteresting and hopelessly complex field. However, the perfection of new chromatographic methods for high–resolution separation and analysis of lipids has been an important factor in opening this field to more penetrating investigation. It appears probable, from the tempo of recent research, that we are now on the eve of important discoveries regarding the role of lipids in the structure of cell membrances.

Occurrence

Fats and oils are widely distributed in nature in both plant and animal tissues. They occur in relatively high concentration in seeds of certain plants (oilseeds) where they function to supply food for use of the growing seedlings. Animals store deposits of fats in their adipose tissues; these stored fats constitute a reserve which can be used as the source of energy.

Physiological Functions

The general and important functions of some classes of lipids in biological systems are as follows:

1. Major sources of metabolic energy in animals, insects, birds and high lipid containing seeds.
2. Basic structural components of cell membrances.
3. As a protective water-proof coating on the surface of cuticle of leaves of fruits of plants, feathers of birds and as insect secretions.
4. As cell-surface components concerned in cell recognition, species specificity and tissue immunity.
5. Intense biological activity – some have profound biological activity; they include some of the vitamins and hormones.
6. Fats stored subcutaneously in warm-blooded animals serve as insulation against as unfavourable environment and also fatty tissues around vital organs give protection against mechanical injuries, and
7. As activators of enzymes – for example, phosphatidylcholine micelles for activation of microsomal enzymes.

General properties

Although the properties vary from one class to other, some of the general properties of lipids are:

1. Soluble in nonpolar solvents but only sparinglyl soluble in water.
2. Greasy or fat–like in nature and show translucent properties.
3. Polar lipids are amphipathic (Greek, amphi, double) i.e. , one end of a lipid molecule, the head, is polar or ionic and therefore, hydrophilic; the other end, the trail (hydrocarbon) is nonpolar and therefore hydrophobic.
4. Most lipids contain fatty acids. The glyceride esters of saturated fatty acids are usually liquids at room temperature, and
5. Fats and oils containing unsaturated fatty acids slowly become rancid when exposed to light, heat, moisture and air.

Classification

Although there are different ways of classification, a useful classific-ation of lipids by Bloor is as follows:

1. Simple lipids: These are esters of fatty acids with alcohols; saponifiable; include the most abundant of all lipids such as fats and oils or triglycerides and the less abundant waxes.

2. Compound or complex lipids: These are esters of fatty acids containing othe groups in addition to alcohol and fatty acids; saponifiable; include phosphoglycerides and sphingolipids.

Table 10.1 Some naturally occurring fatty acids.

Carbon atoms	Structure	Systematic name	Common name	Melting point [°C]
Saturated fatty acids				
12	$CH_3(CH_2)_{10}COOH$	n-Dodecanonic	Lauric acid	44.2
14	$CH_3(CH_2)_{12}COOH$	n-Tetradecanoic	Myristic	53.9
16	$CH_3(CH_2)_{14}COOH$	n-Hexadecanoic	Palmitic	63.1
18	$CH_3(CH_2)_{16}COOH$	n-Octadecanoic	Stearic	69.6
20	$CH_2(CH_2)_{18}COOH$	n-Eicosanoic	Arachidic	76.5
24	$CH_2(CH_2)_{22}COOH$	n-Tetracosanoic	Lignoceric	86.0
Unsaturated fatty acids				
16	$CH_3(CH_2)_5CH{=}CH(CH_2)_7COOH$		Palmitoleic	–0.5
18	$CH_3(CH_2)_7CH{=}CH(CH_2)_7COOH$		Oleic	13.4
18	$CH_3(CH_2)_4CH{=}CHCH_2CH{=}CH(CH_2)_7COOH$		Linoleic	–5
18	$CH_3CH_2CH{=}CHCH_2CH{=}CHCH_2CH{=}CH(CH_2)_7COOH$		Linoleic	–11
20	$CH_3(CH_2)_4CH{=}CHCH_2CH{=}CHCH_2CH$ $CHCH_2CH{=}CH(CH_2)_3COOH$		Arachidonic	–49.5
Some unusual fatty acids				
18	$CH_3(CH_2)_5CH{=}CH(CH_2)_9COOH$ (trans)		trans-Vaccenic acid	44
19	$CH_3(CH_2)_5HC{-}CH(CH_2)_9COOH$ (ring bridged by CH_2)		Lactobaillic	
19	$CH_3(CH_2)_7CH(CH_2)_8COOH$		Tuberculostearic	
24	$CH_3(CH_2)_{21}CHCOOH$ (with OH on CH)		Cerebronic	

3. Derived lipids: These are derived from the hydrolysis of above two classes of lipids; nonsaponifiable (except fatty acids); include fatty acids, sterols, terpenes and fat– soluble vitamins.

Fatty Acids

Although fatty acids occur in the free state in only trace amounts in most cells and tissues, we shall discuss them in some detail since they are the building blocks of several classes of lipids, including the neutral fats, phosphoglycerides, glycolipids, cholesterol esters, and some waxes. Over 70 different fatty acids have been isolated from various cells and tissues. All possess a long hydrocarbon chain and a terminal carboxyl group (margin). The chain may be saturated, or it may have one or more double bonds; a few fatty acids contain triple bonds. Fatty acids differ primarily in chain length and in the number and position of their unsaturated bonds. The following Table gives the structures of some important naturally occuring saturated and unsaturated fatty acids.

Some generalizations may be made on the fatty acids present in lipids of higher plants and animals. Nearly all have an even number of carbon atoms and have chains that are between 14 and 22 carbon atoms long; those having 16 and 18 carbons are by far the most abundant. In general, unsaturated fatty acids predominate over the saturated type, particularly in the neutral fats and in cells of poikilothermic organisms living at lower temperatures. Unsaturated fatty acids have lower melting points than saturated fatty acids; most neutral fats rich in unsaturated fatty acids are liquid down to 5°C or lower. In most of the unsaturated fatty acids in higher organisms, there is a double bond between carbon atoms 9 and 10; additional double bonds usually occur betwen C_{10} and the methyl-terminal end of the chain. In fatty acids containing two or more double bonds, the double bonds are never found in conjugation (that is, –CH= CH–CH=CH–) but are separated by one methylene group (that is, $-CH=CH-CH_2-CH=CH-$). The double bonds of nearly all the naturally occurring unsaturated fatty acids are in the *cis* geometrical configuration. The most abundant unsaturated fatty acids in higher organisms are oleic, linoleic, linolenic, and arachidonic acids.

In general, bacteria contain fewer and simpler types of lipids than the cells of higher organisms. The fatty acids of E. coli lipids consist of C_{12} to C_{18} saturated acids (some of which contain a substituted methyl group or a cyclopropyl group) and C_{16} or C_{18} monounsaturated acids. Fatty acids with more than one double bond have not been found in bacteria.

Nomenclature Of Fatty Acids

A brief comment on the nomenclature of fatty acids is appropriate before dealing with fatty acid metabolism. The systematic name for a fatty acid is derived from the name of its parent hydrocarbon by the substitution of *oic* for the final *e*. For example, the C_{18} saturated fatty acid is called octadecenoic acid because the parent hydrocarbon is octadecane. A C_{18} fatty acid with one double bond is called octadecenoic acid; with two double bonds, octadeca*dienoic* acid; and with three double bonds, octadeca*trienoic* acid. The symbol 18: 0 denotes a C_{18} fatty acid with no double bonds, whereas 18:2 signifies that there are two double bonds. Fatty acid carbon atoms are numbered starting at the carboxyl terminus:

$$\underset{\omega}{H_3C}-(CH_2)_n-\overset{3}{\underset{\beta}{C}}-\overset{2}{\underset{\alpha}{C}}-\overset{1}{C}\begin{matrix}{\scriptstyle /\!/} O \\ \diagdown OH\end{matrix}$$

Carbon atom 2 and 3 are often referred to as α and β, respectively. The methyl carbon atom at the distal end of the chain is called the ω carbon. The position of a double bond is represented by the symbol Δ followed by a superscript number. For example, *cis*-Δ^9 means that there is a *cis* double bond between carbon atoms 9 and 10; *trans*-Δ^2 means that there is a *trans* double bond between carbon

Fig. 12.1 : L ***Photomicrograph of a fat cell. A large globule of fat is surrounded by a thin rim of cytoplasm and a bulging nucleus. (Courtesy of Dr. Pedro Cuatrecasas.)***

atoms 2 and 3. Fatty acids are ionized at physiological pH, and so it is appropriate to refer to them according to their carboxylate form: for example, palmitate or hexadecanoate.

Properties of Fatty Acids

Saturated and unsaturated fatty acids differ significantly in structural configuration. In saturated fatty acids, the hydrocarbon tails can exist in an infinite number of conformations because each single bond in the backbone has complete freedom of rotation. Actually, however, the extended form, which is the minimum energy form, is the most probable configuration. Unsaturated fatty acids, on the other hand, have a rigid kink in their hydrocarbon chains contributed by the nonrotating double bond (s). The *cis* configuration of the double bonds in naturally occurring fatty acids produces a bend of about 30° in the aliphatic chain. whereas the *trans* forms have nearly the same conformation as that of the saturated chains. The *cis* forms are less stable than the *trans* forms; they may be converted into the latter by heating with certain catalysts. In this way oleic acid can be readily converted to its trans isomer elaidic acid, which has a much higher melting point. There is only one naturally occurring trans fatty acid of any abundance, namely, vaccenic acid. In fatty acids with multiple double bonds, their cis configuration causes the hydrocarbon chain to become kinked and shortened. Presumably, these structural features of unsaturated fatty acid chains are biologically significant, particularly in membranes.

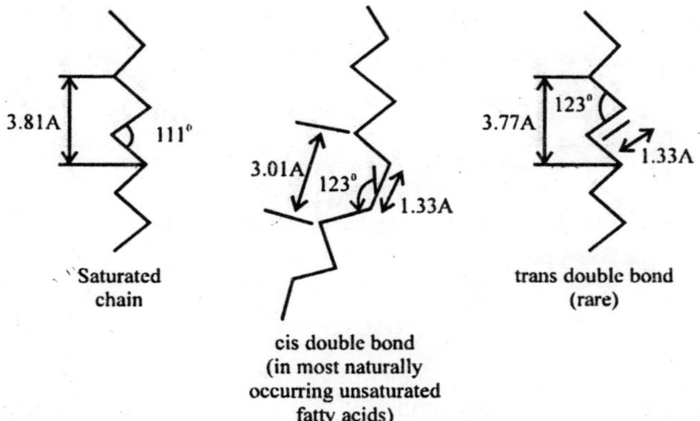

Fig. 12.2 : Configuration of double bonds. At the left are shown space-filling models of saturated and unsaturated fatty acids (anionic forms).

Long-chain fatty acids (C_{16} to C_{18}) are essentially insoluble in water; however, their Na^+ and K^+ salts, called soaps, form micelles in water that are stabilized by hydrophobic interactions.

Naturally occurring fatty acids, whether saturated or unsaturated, show no absorption of light in the visible and near-ultraviolet range. However, those unsaturated fatty acids having more than one double bond may be isomerized by heating with KOH into fatty acids in with the double bonds are in conjugation (that is, —CH=CH—CH=CH—). Since conjugated double bonds have characteristic light absorption in the zone 230 to 260 nm, polyunsaturated fatty acids may be quantitated by spectrophotometry after isomerization.

Unsaturated fatty acids undergo addition reactions at their double bonds. For example, they readily add halogens such as iodine or chlorine. Quantitative titration with halogens thus yields information on the relative number of double bonds in a given sample of fatty acids or lipid.

Triglycerides or acylglycerols

These are the most abundant group of naturally occurring lipids. Chemically they are the esters of fatty acids with the trihydric alcohol, glycerol. One, two or all the three of the hydroxyl groups of glycerol can be esterified to give rise to mono-, di- and triglycerides, respectively. While the mono- and diglycerides are important intermediates in the metabolism of lipids, most of the natural lipids are triglycerides. In the glycerides, fatty acids are linked to glycerol by an ester bond between the hydroxyl groups of glycerol and the carboxyl groups of the fatty acids.

$$\begin{array}{l} CH_2-OH \\ | \\ CH-OH \\ | \\ CH_2-OH \end{array} + \begin{array}{c} R_1-COOH \\ + \\ R_2-COOH \\ + \\ R_3-COOH \end{array} \underset{+3H_2O}{\overset{-3H_2O}{\rightleftharpoons}} \begin{array}{l} CH_2-O-\underset{\underset{O}{\|}}{C}-R_1 \\ | \\ CH-O-\underset{\underset{O}{\|}}{C}-R_2 \\ | \\ CH_2-O-\underset{\underset{O}{\|}}{C}-R_3 \end{array}$$

Glycerol — Fatty acids — A triglyceride (A fat or an oil)

(Where R_1, R_2 and R_3 represent the fatty acid residues which may be same or different).

According to the identity and position of three fatty acids esterified to glycerol, triglycerides can be classified as below:

1. Simple triglycerides: These contain single kind of fatty acids in all the positions, e.g. tristearin, tripalmitin and triolein.

2. Mixed triglycerides: These contain two or more kinds of fatty acids e.g. β-oleo-α, α[1]-stearopalmitin or 1-stearo-2-oleo-3-palmitin.

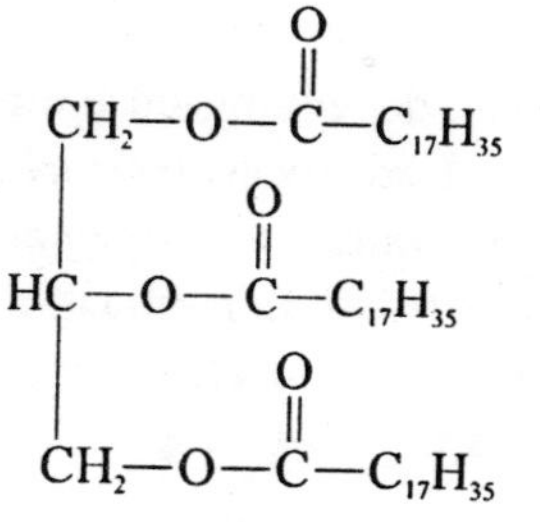

$$\begin{array}{l} CH_2-O-\overset{O}{\overset{\|}{C}}-C_{17}H_{35} \\ | \\ HC-O-\overset{O}{\overset{\|}{C}}-C_{17}H_{35} \\ | \\ CH_2-O-\overset{O}{\overset{\|}{C}}-C_{17}H_{35} \end{array}$$

Tristearin
(A simple triglyceride)

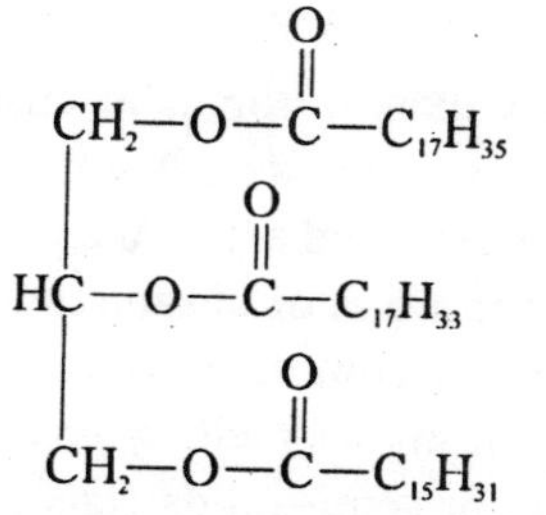

$$\begin{array}{l} CH_2-O-\overset{O}{\overset{\|}{C}}-C_{17}H_{35} \\ | \\ HC-O-\overset{O}{\overset{\|}{C}}-C_{17}H_{33} \\ | \\ CH_2-O-\overset{O}{\overset{\|}{C}}-C_{15}H_{31} \end{array}$$

β-Oleo- α,α' - stearopalmitin
(A mixed tryglyceride)

Fig. 12.3 : Simple and mixed triglycerides.

Most of the triglycerides which occur in nature are mixed triglycerides. All fats and oils from both animal and plant origin are triglycerides. Those obtained from animal fats contain a higher percentage of saturated fatty acids while those of plant origin are rich in unsaturated fatty acids.

Biological importance

Triglycerides are the main storage form of energy in higher animals, migratory birds and oilseeds. They have the highest calorific value (about 9.0 kcal/g) in contrast to carbohydrates and proteins (about 4.0 kcal/g each). This is because they are more highly reduced than the latter two. Fats also serve as thermal insulators slowing the loss of heat through the skin and regulate body temperatures and afford protection to vital organs.

General properties

Some of the general properties of triglycerides can be summarized as below:

1. Specific gravity: It is lower than that of water.

2. Greasy or Fat–like and show translucent properties.

3. Solubility: All triglycerides are insoluble in water and do not tend by themselves to form highly dispersed micelles. They are soluble in nonpolar solvents.

4. Melting point: It is determined by their fatty acid composition. In general the melting increases with the increasing carbon chain–length of the saturated fatty acid components whereas it decreases with increase in degree of unsaturation.

5. They exist in the solid or liquid form, depending on the nature of the constituent fatty acids. Most plant triglycerides have low mlelting points and are liquids at room temperature since they contain a large proportion of unsaturated fatty acids (oils). In contrast, animal triglycerides contain a higher proportion of saturated fatty acids resulting in higher melting points and thus at room temperature they are solids or semi–solids (fats).

6. Optical activity: Although glycerol itself is optically inactive, carbon atom 2 becomes asymmetric whenever the fatty acid substituents on carbon atoms 1 and 3 are different.

7. Autoxidation: Fats and oils rich in unsaturated fatty acids slowly oxidize when exposed to atmospheric oxygen, light, heat and moisture and develop off-flavour and off-odour.

Chemical reactions

1. Hydrolysis: All triglycerides on hydrolysis yield three molecules of fatty acids and one molecule of glycerol when boiled with acids

$$\begin{array}{l} CH_2-O-\underset{\underset{O}{\|}}{C}-R_1 \\ | \\ CH-O-\underset{\underset{O}{\|}}{C}-R_2 \\ | \\ CH_2-O-\underset{\underset{O}{\|}}{C}-R_3 \end{array} \xrightarrow{+3H_2O} \begin{array}{l} CH_2-OH \\ | \\ CH-OH \\ | \\ CH_2-OH \end{array} + \begin{array}{c} R_1-COOH \\ + \\ R_2-COOH \\ + \\ R_3-COOH \end{array}$$

A triglyceride — Glycerol — Free fatty acids

or bases or by the action of enzymes called lipases.

2. Saponification: Hydrolysis of tryglycerides or fats and oils by alkali is called saponification. The free fatty acids formed by the hydrolysis of triglycerides react with excess of alkali to form metallic salts called soaps. Sodium and potassium salts of fatty acids are soluble in water (soft soap) but other metal salts such as calcium, magnesium and barium are water–insoluble (hard soaps).

$$\begin{array}{l} CH_2-O-\overset{\overset{\displaystyle O}{\|}}{C}-R' \\ | \\ HC\;-O-\overset{\overset{\displaystyle O}{\|}}{C}-R' \\ | \\ CH_2-O-\overset{\overset{\displaystyle O}{\|}}{C}-R'' \end{array} \xrightarrow[\substack{H_2O,\ OH^- \\ (40\text{-}120^{\circ}C)}]{3KOH} \begin{array}{l} CH_2OH \\ | \\ HC-OH \\ | \\ CH_2OH \end{array} + \begin{array}{l} R-COO^-K^+ \\ \quad + \\ R'-COO^-K^+ \\ \quad + \\ R''-COO^-K^+ \end{array}$$

A triglyceride — Glycerol — Potassium salt of fatty acid (soap)

3. Oxidation: Many fats and oils when stored for long time, often become rancid - develop off-flavour and off-odour. One type of rancidity, called hydrolytic rancidity is caused by the growth of microorganisms which secrete lipases and split triglycerides into mono-and diglycerides, glucerol and fatty acids. If fatty acids of low molecular weight are released, they impart unpleasant taste and odour. This kind of rancidity which commonly occurs in butter can be reduced by refrigeration (butter is stored at low temperature), by exclusion of water or destroying the microorganisms.

The other type of rancidity called the oxidative rancidity occurs due to the autoxidation of the unsaturated fatty acids at their double bonds yielding short-chain acids and aldehydes having rancid taste and odours. This can be prevented by the addition of compounds like vitamin E (antioxidant)

In case of oxidative rancidity, oxygen adds to the olefinic bonds of unsaturated fatty acids to produce either cleavage or polymerization. The slow oxidation of unsaturated fatty acids in edible fats is associated with a cleavage type of reaction.

Phospholipids

The phospholipids include the *phosphoglycerides* (lecithins, cephalins, and plasmalogens), the *phosphoinositides, and the sphingolipids* (sphingomyelins).

Phosphoglycerides

Lecithins. The term describes a group of compounds that are glycerl esters of two fatty acid molecules and esterified at its third C atom with phosphoric acid, which in turn is bound by ester linkage to a nitrogenous base— choline:

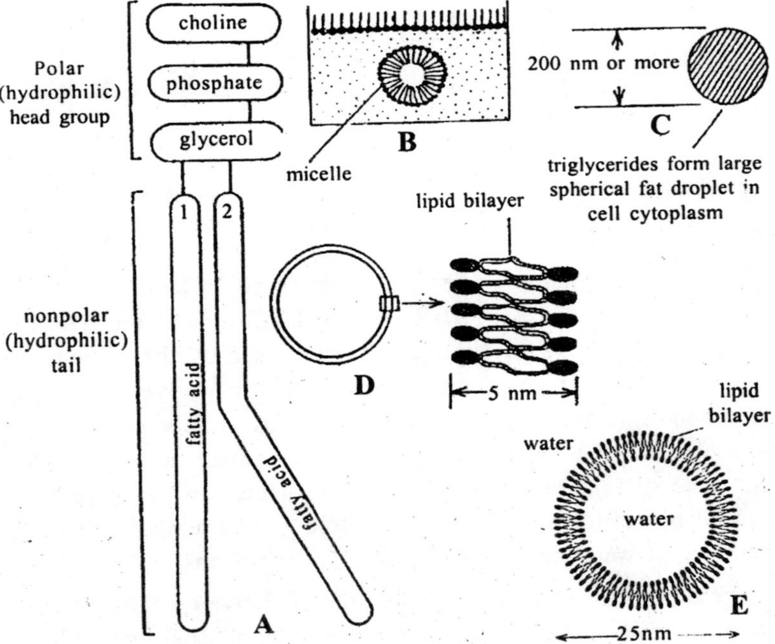

Fig. 12.4 : ***Formation of various types of lipid aggregates. A— Schematic representation of a phospholipid molecule; B—Formation of micelle and monolayer film; C—Formation of a fat droplet by triglycerides; D—Formation of self– sealing lipid bilayer (e.g., liposome); E— Cross section of a liposome (after Alberts et al., 1989).***

Although theoretically the phosphorylcholine group may also be attached to the β - carbon of glycerol, it has been shown that the lecithins in nature are of the α - variety. Also since the 2-carbon of glycerol is asymmetric, there are two possible isomers, although the lecithins found in nature appear to be of the L variety.

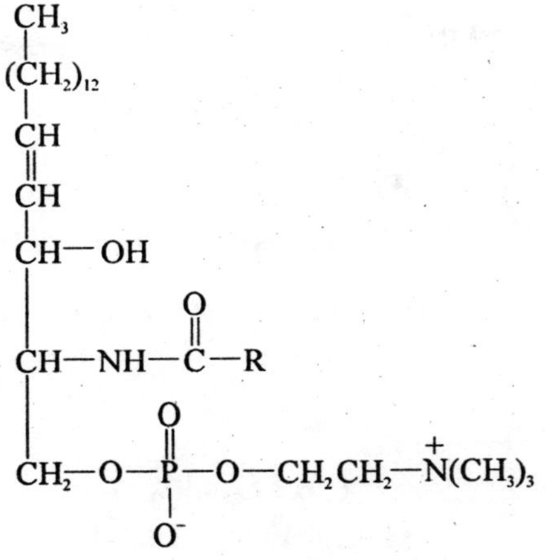

$$CH_3-(CH_2)_{12}-CH=CH-CH(OH)-CH(NH-\overset{O}{\overset{\|}{C}}-R)-CH_2-O-\overset{O}{\overset{\|}{\underset{O^-}{\underset{|}{P}}}}-O-CH_2CH_2-\overset{+}{N}(CH_3)_3$$

Fig. 12.5 : General chemical formula of sphingomyelin

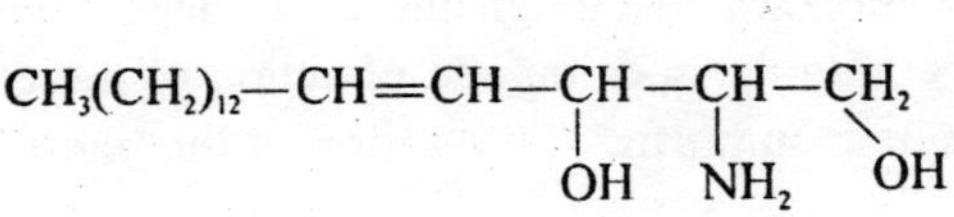

$$CH_3(CH_2)_{12}-CH=CH-\underset{OH}{\underset{|}{CH}}-\underset{NH_2}{\underset{|}{CH}}-CH_2-OH$$

Fig. 12.6 : ***Chemical formula of sphingosine.***

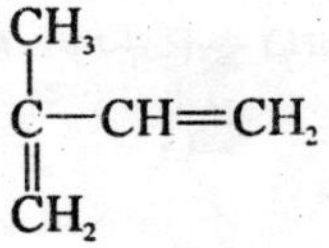

$$CH_2=\underset{}{\overset{CH_3}{\overset{|}{C}}}-CH=CH_2$$

Fig. 12.7 : ***Isoprene.***

Many different lecithins are possible because of the wide variety of fatty acids that may be attached to the glycerol moiety. Lecithins are found in relatively large quantities in egg yolk and liver and constitute most of the phospholipid fraction of dog or human plasma.

Lecithin is a waxy colourless solid, that turns yellow and then brown on exposure to air and light. It is extremely hygroscopic and can be prepared in crystalline form at very low temperatures. It is insoluble in acetone or methyl acetate but is dispersed in an aqueous medium if bile salts are present.

```
                    O
                    ‖
          H2C—O—C—R
  O        |
  ‖        |
R'—C—O—CH
           |        O
           |        ‖
          H2C—O—P—O—CH2·CH2·N(CH3)3
                    |              +
                    O-
```

L-α- Lecithin
(phosphatidyl choline)

Phosphatidases. These enzymes act on the lecithin molecule or on compounds similar to, or derived from, lecithin and hydrolyze specific ester linkages. Although there has been some question in the past concerning the specific site of action of these enzymes, the following diagram summarizes the action of the phosphatidases:

```
                     (B) O
                      ↓  ‖
  O (B)   H2C—O ↕ C—R
  ‖  ↓     |      ↑
R'—C ↕ O—C—H  (L) O            (C)
   ↑       |       ‖             ↓
  (A)     H2C—O ↕ P—O—CH2·CH2·N(CH3)3
                 ↑ |               +
                (D) O-
```

The letters A, B, C, and D refer to the respective phosphatidases whereas L refers to lysophosphatidase, which acts on lysolecithin to yield 1 mole of fatty acid. This is in contrast to B, which splits 2 moles of fatty acid from 1 mole of lecithin. Formulas for the various hydrolytic products include.

Fatty acids of lecithin. The fatty acids found in lecithin include both saturated (stearic and palmitic) and unsaturated acids (oleic, linoleic, linolenic, and arachidonic). Although there is some variability in composition among lecithins, there appears to be a pattern with

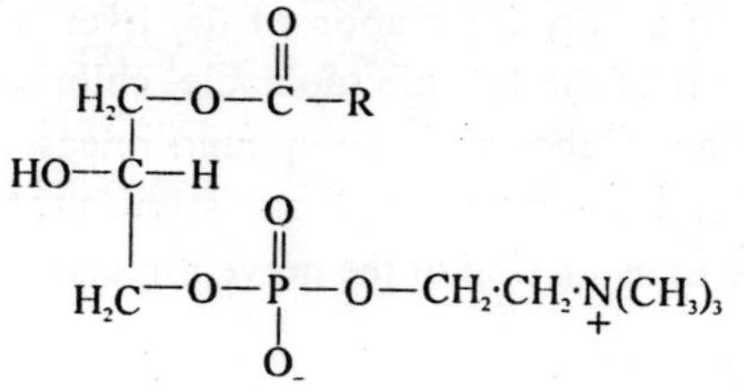

$$\begin{array}{l} H_2C-O-\overset{\overset{O}{\|}}{C}-R \\ HO-\overset{|}{C}-H \\ H_2C-O-\overset{\overset{O}{\|}}{\underset{\underset{O^-}{|}}{P}}-O-CH_2\cdot CH_2\cdot \underset{+}{N}(CH_3)_3 \end{array}$$

L-α - Lysolecithin

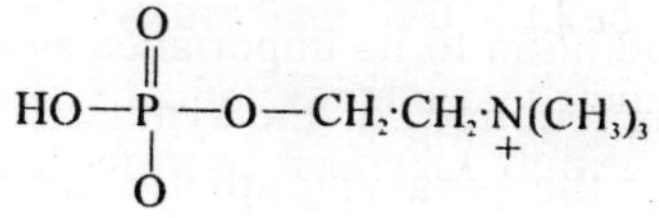

$$HO-\overset{\overset{O}{\|}}{\underset{\underset{O}{|}}{P}}-O-CH_2\cdot CH_2\cdot \underset{+}{N}(CH_3)_3$$

Phosphorylcholine

$$\begin{array}{l} H_2C-OH \\ HO-\overset{|}{C}-H \\ H_2C-O-\overset{\overset{O}{\|}}{\underset{\underset{O^-}{|}}{P}}-O-CH_2\cdot CH_2\cdot \underset{+}{N}(CH_3)_3 \end{array}$$

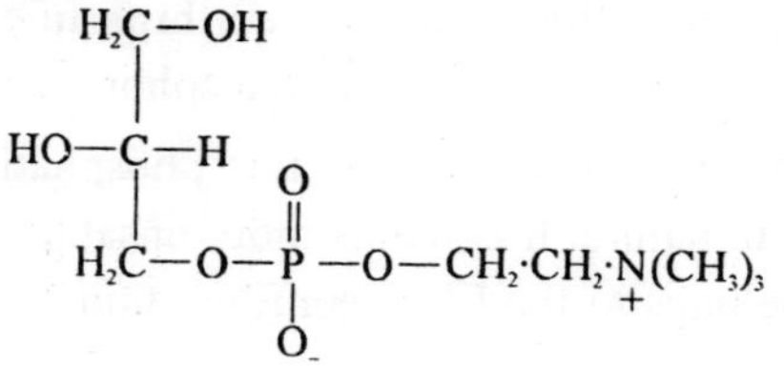

L-α - Glyeerylphosphorylcholine

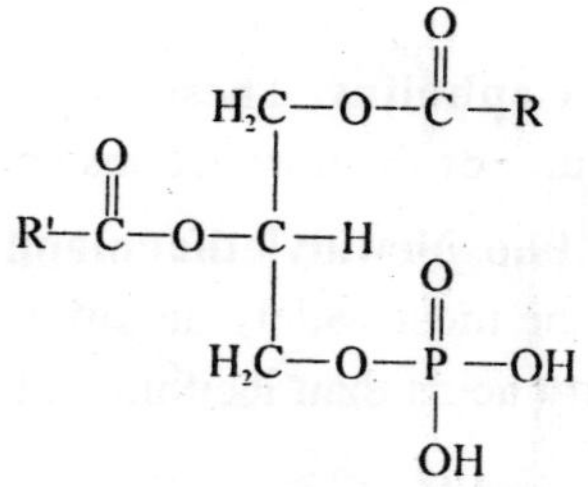

$$\begin{array}{l} \qquad\qquad H_2C-O-\overset{\overset{O}{\|}}{C}-R \\ R'-\overset{\overset{O}{\|}}{C}-O-\overset{|}{C}-H \\ \qquad\qquad H_2C-O-\overset{\overset{O}{\|}}{\underset{\underset{OH}{|}}{P}}-OH \end{array}$$

L-α - Phosphatidic acid

respect to the type of acids attached to the 1-or 2-carbon atom of glycerol. Determination of specific distribution depends on identification of the acids liberated when lecithin is treated with phosphatidase A. Since this enzyme is now believed to hydrolyze the ester bond at the 2-carbon of glycerol, we may determine the exact distribution of fatty acids in the molecule. As a result of these studies, it has been concluded that the 1–carbon atom of glycrol contains saturated acids, whereas the 2-carbon atom is attached to unsaturated acids:

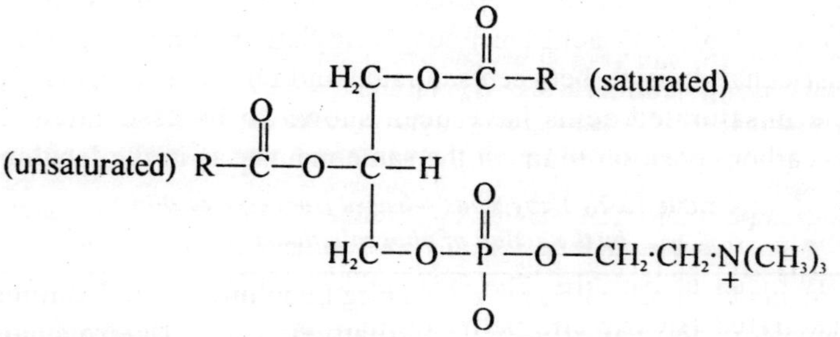

$$\begin{array}{l} \qquad\qquad\qquad\qquad H_2C-O-\overset{\overset{O}{\|}}{C}-R \text{ (saturated)} \\ \text{(unsaturated) } R-\overset{\overset{O}{\|}}{C}-O-\overset{|}{C}-H \\ \qquad\qquad\qquad\qquad H_2C-O-\overset{\overset{O}{\|}}{\underset{\underset{O}{|}}{P}}-O-CH_2\cdot CH_2\cdot \underset{+}{N}(CH_3)_3 \end{array}$$

Choline. This compound is an essential component of the diet of mammals and is therefore included among the vitamins. It is a quaternary ammonium compound with very strong basic properties. Its absence from the diet leads to a fatty infiltration of the liver. In addition to its importance as a part of the lecithin molecule, chlorine is required for the synthesis of acetylcholine, a compound released at the parasympathetic nerve endings when they are stimulated. Acetylchlorine is connected with transmission of the nerve impulse.

$$HO\cdot CH_2\cdot CH_2\cdot \overset{+}{N}(CH_3)_3 \qquad CH_3-\overset{\overset{O}{\|}}{C}-O-CH_2\cdot CH_2\cdot \overset{+}{N}(CH_3)_3$$

Choline Acetylcholine

Cephalins. These compounds differ from the lecithins in the nature of the nitrogenous compounds esterified with phosphoric acid.

Phosphatidylethanolamine. After the lecithins, this phosphatide is the most widely distributed in nature. It contains more unsaturated fatty acids than lecithin and belongs to the L-α-configuration:

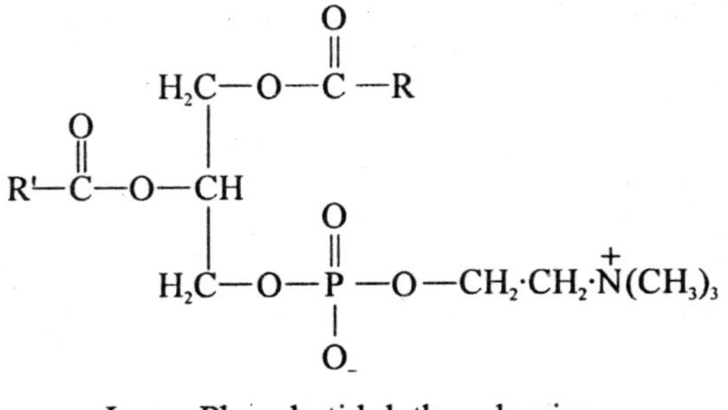

$$\begin{array}{l} \qquad\qquad\quad H_2C-O-\overset{\overset{O}{\|}}{C}-R \\ \qquad\qquad\quad\ \ | \\ R'-\overset{\overset{O}{\|}}{C}-O-CH \\ \qquad\qquad\quad\ \ | \\ \qquad\qquad\quad H_2C-O-\overset{\overset{O}{\|}}{P}-O-CH_2\cdot CH_2\cdot \overset{+}{N}(CH_3)_3 \\ \qquad\qquad\qquad\qquad\quad\ \ | \\ \qquad\qquad\qquad\qquad\quad\ O^- \end{array}$$

L-α- Phosphatidylethanolamine

It is most conveniently prepared from brain, liver, and yeast. The distribution of fatty acids in phosphatidylethanolamine reveals an almost equal division between saturated and unsaturated acids. Most of the unsaturated acids have been shown to be associated with β(2 - carbon) position in much the same manner as in the lecithins.

*Table 12.2 : **Fatty acids released from ego lecithin by the action of phogsphatidase A***

	Egg Lecithin (native)	Egg Lecithin (hydrogenated)
Relative Distribution (moles %)		
Saturated	4	100

unsaturated	96	0
Specific Distribution (moles %)		
Saturated		
C_{16}	3	
C_{18}	1	91
C_{20}		9
Unsaturated		
C'_{16}	0.5	
C'_{18}	51	
C'_{18}	35	
C_{8}	0.5	
C_{20}	1	
C''''_{20}	8	

* The number of prime marks indicates the number of double bonds.

Phosphatidylserine. This compound has been prepared from brain and also appears to belong to the L - α configuration:

$$\begin{array}{l} H_2C-O-\overset{O}{\overset{\|}{C}}-R \\ R'-\overset{O}{\overset{\|}{C}}-O-CH \\ H_2C-O-\overset{O}{\overset{\|}{P}}(O^-)-O\cdot CH_2\cdot CH(\overset{+}{N}H_3)\cdot COO^- \end{array}$$

L - α - Phosphatidylserine

Plasmalogens. These compounds contain a fatty acid residue, an aldehydogenic unit (as a vinyl ether), and glycerylphosphorylethanolamine or the corresponding choline derivative. They are widely distributed in nature in animal tissues. Brain and nerve myelin have relatively large quantities of plasmalogen. It is also present in heart muscle, and in semen. It is now known that the vinyl ether is linked to glycerol at the 1–position.

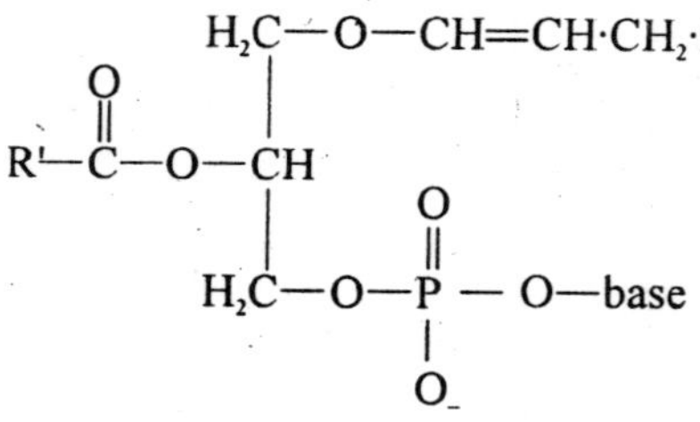

$$
\begin{array}{l}
\qquad\qquad\quad H_2C-O-CH{=}CH{\cdot}CH_2{\cdot}R \\
\qquad\qquad\quad | \\
R'-\overset{\overset{\large O}{\|}}{C}-O-CH \\
\qquad\qquad\quad | \\
\qquad\qquad\quad H_2C-O-\overset{\overset{\large O}{\|}}{\underset{\underset{\large O^-}{|}}{P}}-O-\text{base}
\end{array}
$$

Plasmalogen

Phosphoinositides

Inositol, as a component of phospholipids, has been known for some time, but it is only in recent years that the specific structure of the inositides has been clarified. Inositol is a hexahydroxycyclohexane, of which there are nine possible stereochemical forms. The most important one present in lipids is the optically inactive *myo–inositol.* It has been isolated in a free state:

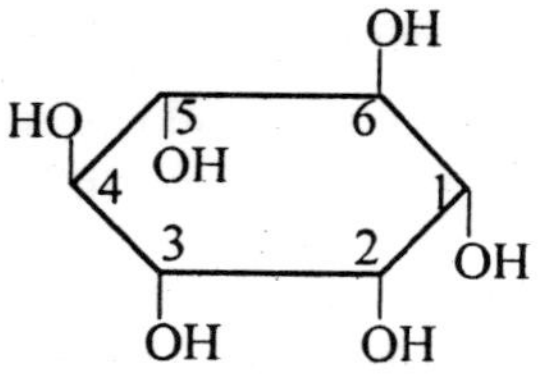

myo-Inositol

Ox and rabbit brain contain at least three inositol lipids: I–phosphatidylinositol, I-phosphatidyl (insoitol-4-phosphate), and I–phosphatidyl (inositol-4,5-diphosphate).

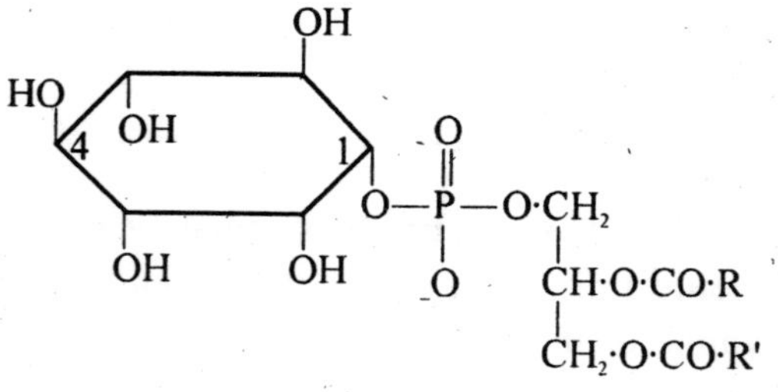

1-Phosphatidylinositol

Sphingolipids

These phospholipids contain a long–chain alcohol–sphingosine.

Recently several additional alcohol have been discovered–dihydrosphingosine, phytosphingosine, and dehydrophtosphingosine. The last two have been found in plant tissues only. The sphingolipids occur in particularly high concentrations in brain and nerve tissue. In the disease *lipidosis*, these compounds accumulate in various organs.

In sphingomyelin, sphingosine or a related base is bound by amide linkage to a long-chain fatty acid and by ester linkage to phosphorylcholine. The fatty acids may be stearic, lignoceric, or nervonic in brain sphingomyelins: or palmitic or lignoceric in sphingomyelins from lung and spleen.

Sulfolipids or sulfatides are sphingolipids that contain sulfur in the form of a sulfate ester of the galactose residue. One such lipid isolated from animal tissues is esterified at the 6-carbon of galactose.

Mucolipids are sphingolipids containing sialic acid (N–acetylneuraminic acid). One of these, *ganglioside* from the ganglion cells, appears to be a cerebroside linked via its galactose residue to a *glucosyl–galactosamine–sialic* acid residue.

Cerebrosides (Glycolipids)

Cerebrosides consist of a nitrogenous base, sphingosine or dihydrosphingosine, a long-chain fatty acid, and a sugar. There are at least four cerebrosides phrenosin, kerasin (cerasin), nervone, and

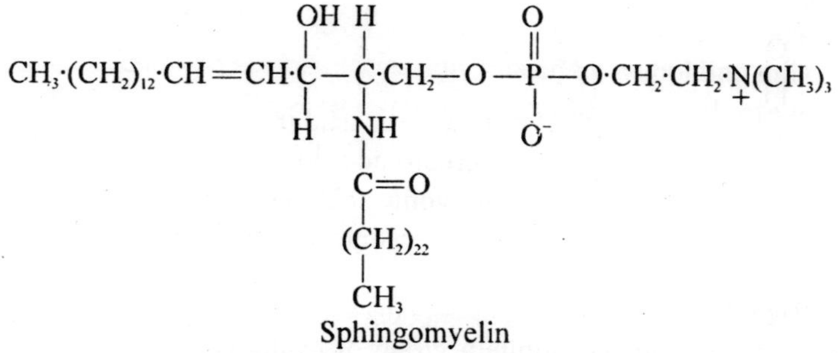

$$CH_3\cdot(CH_2)_{12}\cdot CH{=}CH\cdot \underset{H}{\overset{OH}{C}}-\underset{NH-C(=O)-(CH_2)_{22}-CH_3}{\overset{H}{C}}\cdot CH_2-O-\underset{O^-}{\overset{O}{\overset{\|}{P}}}-O\cdot CH_2\cdot CH_2\cdot \overset{+}{N}(CH_3)_3$$

Sphingomyelin

$$CH_3\cdot(CH_2)_{12}\cdot CH{=}CH\cdot \underset{H}{\overset{OH}{C}}-\underset{NH_2}{\overset{H}{C}}\cdot CH_2OH$$

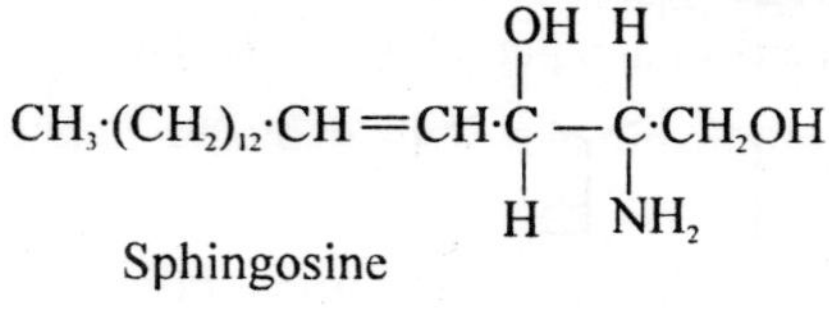

Sphingosine

$$CH_3(CH_2)_{22}\cdot COOH$$

Lignoceric acid

$$C_{14}H_{29}CHOH\cdot CHOH\cdot CHNH_2\cdot CH_2OH$$

Phytosphingosine

oxynervone. They occur in brain, adrenals, kidney, spleen, liver, leukocytes, thymus, lung, retina, egg yolk, and fish sperm. In Gaucher's disease they occur in relatively large quantities in the liver and especially in the spleen.

Phrenosin yields a characteristic hydroxy acid—cerebronic acid—to which the formula $CH_3(CH_2)_7CH_2CH_2(CH_2)_{12}CHOHCOOH$ has been assigned. *Cerasin* contains lignoceric acid, $CH_3(CH_2)_7CH_2CH_2(CH_2)_{12}CH_2COOH$. This acid in nervone is nervonic acid $CH_3(CH_2)_7CH=CH(CH_2)_{12}CH_2COOCH$, which is an unsaturated lignoceric acid. Oxynervonic acid, $CH_3(CH_2)_7CH=CH(CH_2)_{12}CHOHCOOH$, an unsaturated hydroxy lignoceric acid, is the acid that characterizes *oxynervone*.

The general formula for a cerebroside is:

$$CH_3(CH_2)_{12}\cdot CH{=}\underset{H}{\overset{OH}{C}}H\cdot C-\overset{H}{\underset{\underset{\underset{\underset{R}{|}}{C=O}}{|}}{\underset{NH}{C}}}\cdot CH_2-O-C(H)\ [\text{ring: } H\cdot C\cdot OH,\ HO\cdot C\cdot H,\ HO\cdot C\cdot OH,\ H\cdot C-O-,\ CH_2OH]$$

where $R.\overset{O}{\overset{||}{C}}-(OH)=$ lignoceric acid in cerasin
= cerebronic acid in phrenosin
= nervonic acid in nervone
= oxynervonic acid in oxynervone

Steroids

The steroids constitute a group of cyclic compounds that may be considered to be derivatives of a common parent hydrocarbon:

Cyclopentanoperhydrophenanthrene

Sterols

The best known of the sterols is cholesterol. It is present in all animal cells and is particularly abundant in nervous tissue. Varying quantities of this sterol are found in animal but not in vegetable fats.

In the cholesterol nucleus there are eight centers of asymmetry, and, theoretically, something like 240 isomers are possible. Fortunately, only two carbon centers seem to be involved in naturally occurring sterols, those at positions 3 and 5.

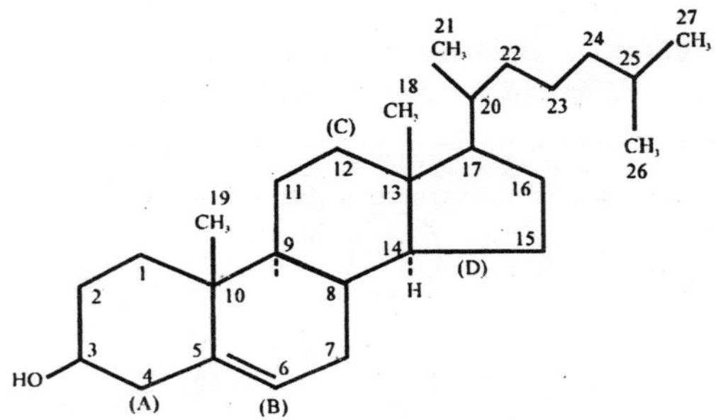

Fig. 12.8 : ***Cholesterol. The four rings are labeled with letters, and the 27 carbons with numbers. The solid lines indicate bonds that project above the plane of the molecules. The dotted lines indicate atoms projecting below the plane.***

Biosynthesis of Cholesterol

In the remainder of this chapter we consider the synthesis of cholesterol and its hormonal derivatives. Although cholesterol can be made in almost all tissues in humans, the principal sites of synthesis are the liver and intestinal mucosa; also important are the adrenal cortex, testes, and ovaries. The enzymes and cofactors required for synthesis are located in the endoplasmic reticulum and the cytosol.

One of the earlist demonstrations of the applicability of isotopic labelling in metabolic studies was the discovery that both carbons of acetate were incorporated into the cholesterol of liver. As these analyses were extended, it was found that the two carbons contributed equally to the cholesterol molecule. When the labeled cholesterol was degraded systematically, the origin of each carbon was shown to be as follows:

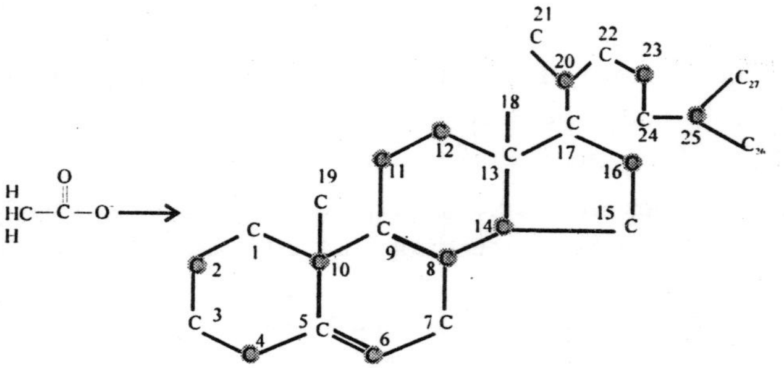

Table: 12.3 ***Metabolic disorders of lipid storage (Ceramide = N-acyl Sphingosine.)***

Disease	Major Sphingolipid Accumulated	Enzyme defect
Niemann-pick	Cer—PChol Sphingomyelin	Sphingomyelinase
Gaucher	Cer —β— Gle Ceramide glucoside (glucocerebroside)	β-Glucosidase
Metachromatic Leukodystrophy	Cer —β— Gle—OSO_3^- Ceramide galactose3 sulfate (sulfatide)	Sulfatidase
Fabry	Cer —β— Gle —β— Gal —β— Gal Ceramide trihexoside	β-Galactosidase
Tay-Sachs	Cer —β— Gle —β— Gal —β— NAcGal; Gal—NAcNA Ganglioside GM_2	Hexosaminidase
Generalized Gangliosidosis	Cer —β— Gle —β— Gal —β— NAcGal —β— Gal; Gal—NAcNA Ganglioside GM_2	β-Galactosidase

The steroids may be grouped into the following classes:

Sterols	Adrenal corticosteroids
Bile acids	Vitamins D
Male sex hormones	Saponins
Female sex hormones	Cardiac glycosides

Cholesterol synthesis begins with two common intermediates in lipid metabolism; acetyl–CoA and acetoacetyl–CoA.

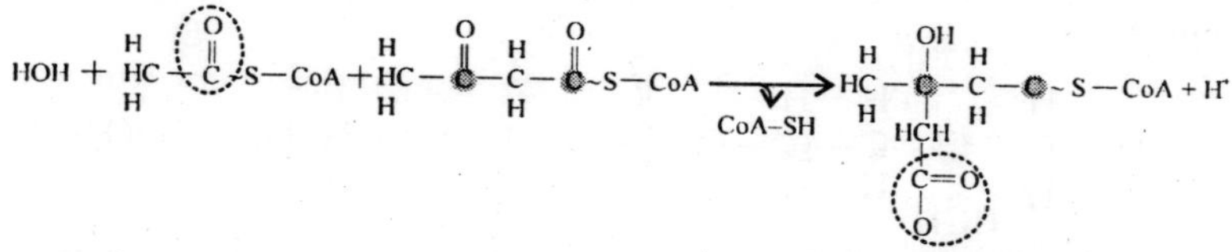

Hydroxymethylglutaryl-CoA

With the addition of four electrons from two NADPH, the thioester end of the glutaryl derivative is reduced to yield L-mevalonate:

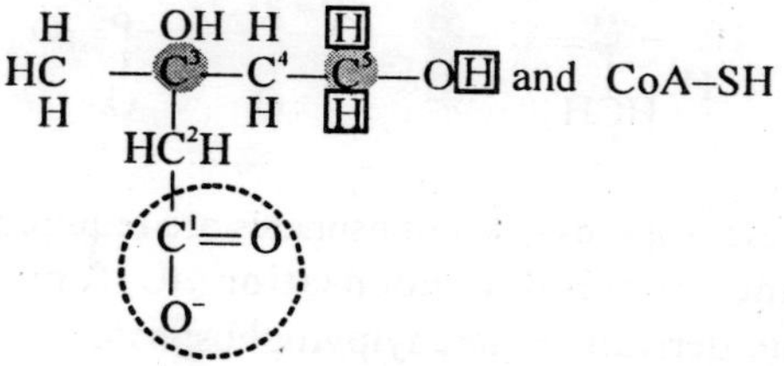

This irreversible reaction is the committed step in cholesterol synthesis and the enzyme that catalyzes the process, 3–hydroxy – 3–methylglutaryl–CoA reductase, is an important control point in the biosynthetic pathway.

By two successive phosphorylations with ATP, the mevalonate is converted to its 5-pyrophosphate form:

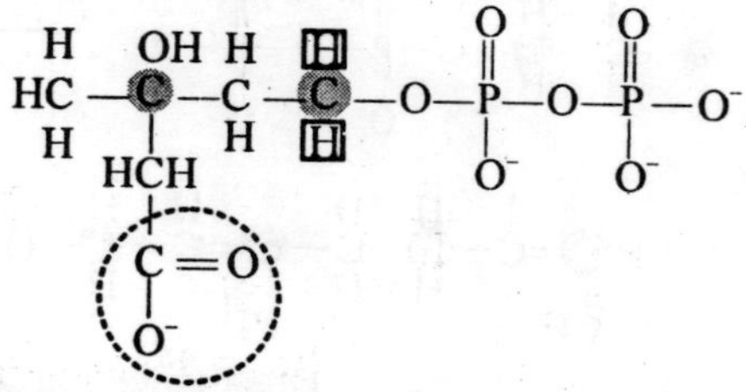

Addition of a third phosphate at C–31 leads to a very unstable intermediate:

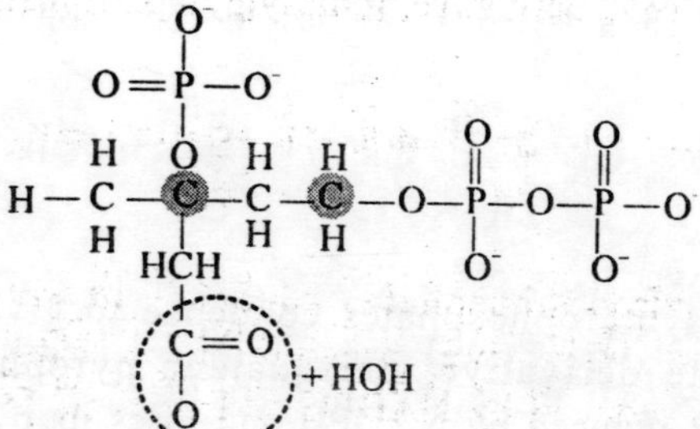

This compound is decarboxylated and loses the phosphate last added:

$$\mathrm{H{-}C(H_2){-}C^3(=CH_2){-}C^2H_2{-}C^1H_2{-}O{-}P(=O)(O^-){-}O{-}P(=O)(O^-){-}O^-} + HCO_3^- + HPO_4^{2-}$$

The isopentenyl unit is isomerized to yield 3,3–dimethylallyl-pyrophosphate:

$$\mathrm{H_3C{-}C(CH_3){=}CH{-}CH_2{-}O{-}P(=O)(O^-){-}O{-}P(=O)(O^-){-}O^-}$$

Both of these isoprenyl pyrophosphates are required for cholesterol synthesis and undergo condensation to form a 10–carbon pyrophosphate derivative, geranylpyrophosphate:

$$\mathrm{H_3C{-}C(CH_3){=}CH{-}CH_2{-}O{-}P(=O)(O^-){-}O{-}P(=O)(O^-){-}O^-} + \mathrm{HC(H_2){-}C{-}CH_2{-}CH_2{-}O{-}P(=O)(O^-){-}O{-}P(=O)(O^-){-}O^-} \longrightarrow \mathrm{HO{-}P(=O)(O^-){-}O{-}P(=O)(O^-){-}O^-}$$

$$\mathrm{H_3C{-}C(CH_3){=}CH{-}CH_2{-}CH_2{-}C(CH_3){=}CH{-}CH_2{-}O{-}P(=O)(O^-){-}O{-}P(=O)(O^-){-}O^-}$$

Geranylpyrophosphate

A second condensation reaction with an isopentenylpyrophosphate yields the 15–carbon derivative, farnesylpyrophosphate:

$$\mathrm{CH_3{-}C(CH_3){=}CH{-}CH_2{-}CH_2{-}C(CH_3){=}CH{-}CH_2{-}CH_2{-}C(CH_3){=}CH{-}CH_2{-}O{-}P(=O)(O^-){-}O{-}P(=O)(O^-){-}O^-}$$

Two farnesylpyrophosphates condense to give a 30–carbon pyrophosphate derivative, presqualene pyrophoshpate. This intermediate is reduced by NADPH and loses its pyrophosphate to become squalene:

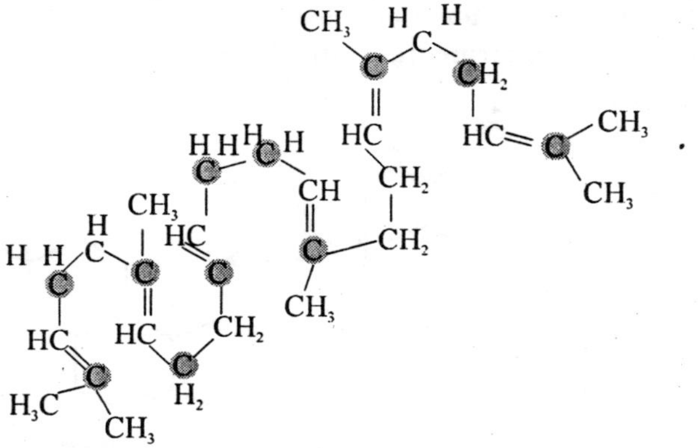

The ring closure of squalene beings with the introduction of the oxygen atom which will ultimately be a part of the hydroxyl group at position 3 of cholesterol. This reaction utilizes molecular $\mathbf{O_2}$ and, like other monooxygenase–catalyzed reactions, requires a reductant in addition to the oxygen:

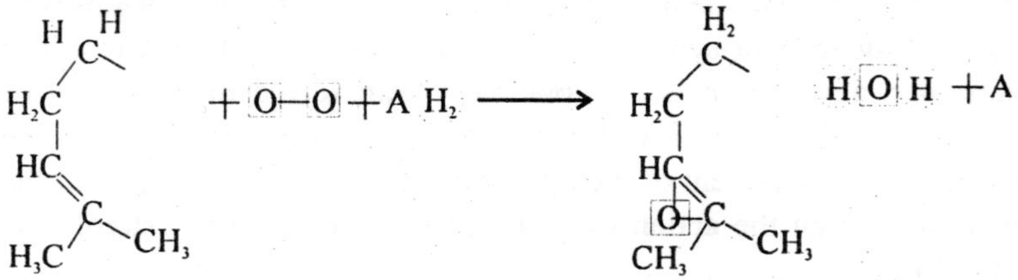

and the epoxide derivative squalene undergoes ring closure to give the lanosterol:

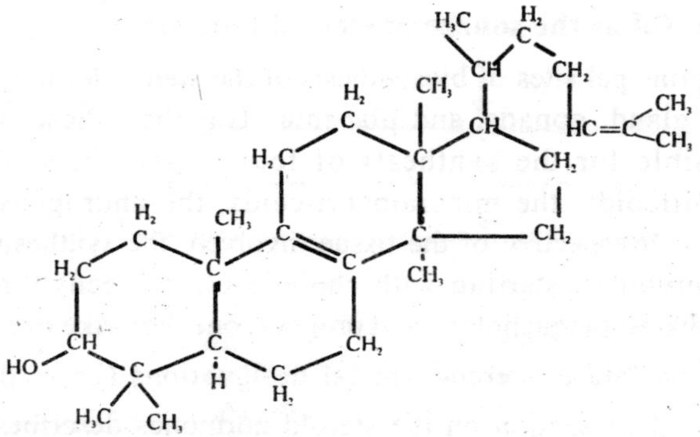

The remaining steps for the completion of cholesterol synthesis

Fig. 12.9 (Contd.)

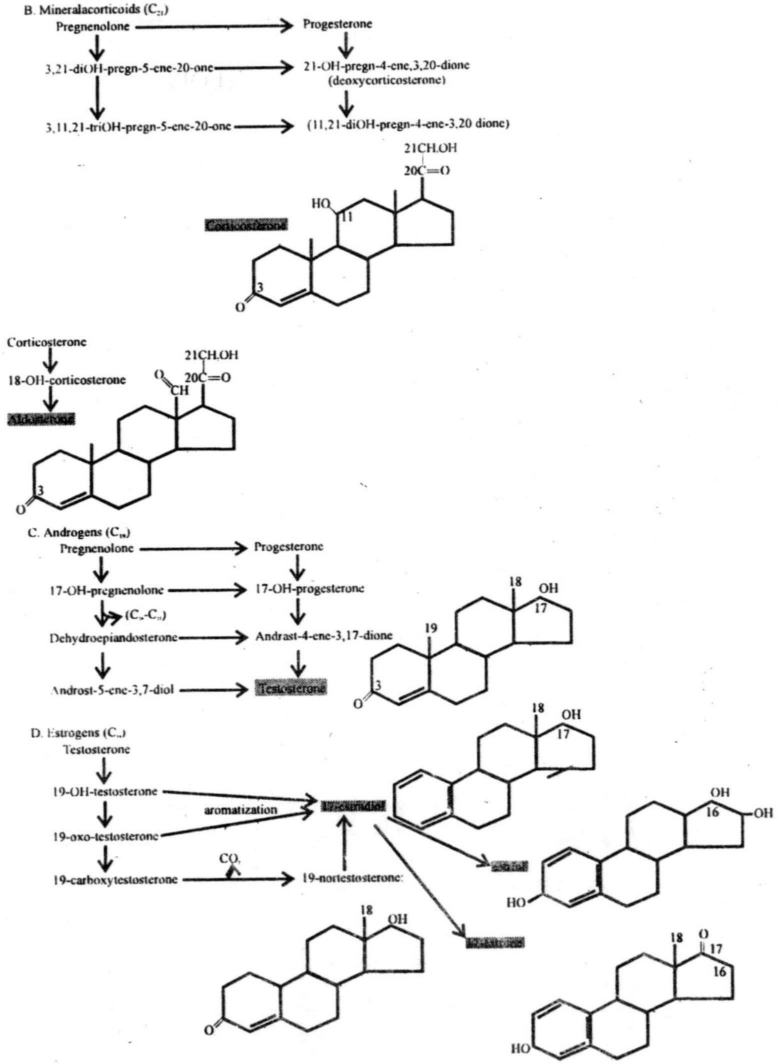

B. Mineralacorticoids (C_{21})
Pregnenolone → Progesterone
3,21-diOH-pregn-5-ene-20-one → 21-OH-pregn-4-ene,3,20-dione (deoxycorticosterone)
3,11,21-triOH-pregn-5-ene-20-one → (11,21-diOH-pregn-4-ene-3,20 dione)
21CH.OH
20C=O
HO 11
Corticosterone
3
O

Corticosterone
18-OH-corticosterone
Aldosterone
21CH.OH
20C=O
O CH
3
O

C. Androgens (C_{19})
Pregnenolone → Progesterone
17-OH-pregnenolone → 17-OH-progesterone
(C_{20}-C_{21})
Dehydroepiandosterone → Andrast-4-ene-3,17-dione
Androst-5-ene-3,7-diol → Testosterone
18 OH 17 19 3 O

D. Estrogens (C_{18})
Testosterone
19-OH-testosterone
19-oxo-testosterone
aromatization
17-estradiol
19-carboxytestosterone
CO_2
19-nortestosterone
Estriol
Estrone
18 OH 17
OH 16 OH
HO
18 OH
O
18 O 17 16
HO

Conversion of cholesterol to pregnenolone

The rate-limiting process in the conversion of cholesterol to the steroid hormones is the initial oxidative removal of the side chain yielding a six-carbon fragment, isocaproic aldehyde, and pregnenolone. The oxidative cleavage is accomplished by two successive mixed–function oxidations at C-20 and C-22:

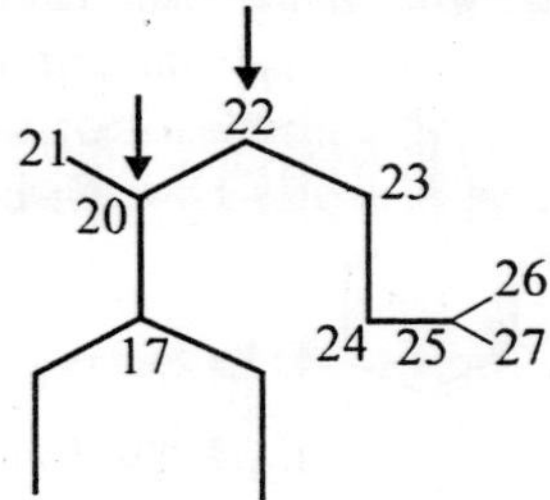

The mitochondrial enzymatic system responsible for the process, called DESMOLASE, occurs in both the adrenal cortex and gonadal tissue. The overall scheme outlined below is typical of mixed-function oxidase or hydroxylation reactions with the exception that an additional electron carrier is interposed between the flavin and cytochrome P_{450}. This carrier, labeled ADRENODOXIN in the adrenal mitochondria and TESTODOXIN in testicular mitochondria, is an iron–sulfur protein. The mechanism of hydroxylation at C-20 of cholesterol, designated below as RH, can be formalized as follows:

(1) R**H** + cyto P_{450} (Fe^{3+}) ⟶ R**H**. Cyto P_{450} (Fe^{3+})

(2) R**H** + cyto P_{450} (Fe^{3+}) + $(Fe^{2+}–S)_I$ ⟶ R**H**. Cyto P_{450} (Fe^{2+}) + $(Fe^{3+}–S)_I$

(3) R**H** + cyto P_{450} (Fe^{2+}). O_2 ⟶ R**H**. Cyto P_{450} (Fe^{2+}) + O_2

(4) R**H** + cyto P_{450} (Fe^{2+}) O_2 + $(Fe^{2+}–S)_{II}$ ⟶ R**H**. Cyto P_{450} (Fe^{2+}) O_2 + $(Fe^{3+}–S)_{II}$

Next, the cytochrome P_{450} must be returned to its Fe^{3+} state and concurrently the hydroxylation of RH occurs:

(5) NADPH + H^+ + FAD ⟶ $NADP^+$ + FAD. 2H

$2(Fe^{3+}—S)_{(I\ and\ II)}$ + R**H**. Cyto P_{450}(Fe^{2+}) O_2 + FAD.2H ⟶

FAD + $2(Fe^{3+}—S)_{(I\ and\ II)}$ + RO**H** + Cyto P_{450}(Fe^{3+}) + HOH

The net reaction becomes

R**H** + NADPH + H^+ + O_2 ⟶ RO**H** + $NADP^+$ + HOH

The hydroxylation of C–22 proceeds by the same mechanism and with an additional oxidation, cleavage occurs between C–20 and C–

22 to yield pregnenolone and isocaproic aldehyde:

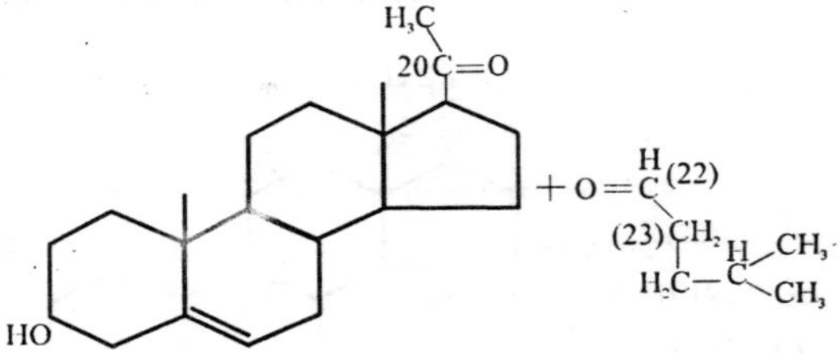

Conversion Of Pregnenolone To Progesterone

Following its synthesis, pregnenolone leaves the mitochondria and is converted to progesterone by two microsomal enzymes, $\Delta^{4,5}$-isomerase and 3-β ol-dehydrogenase:

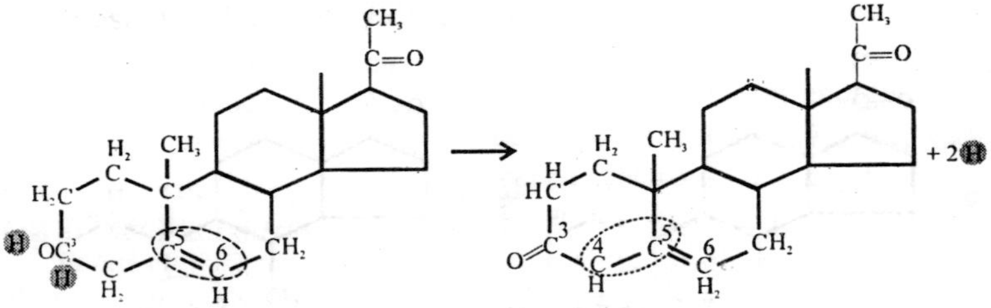

In the corpus luteum of the ovaries there are no reactions beyond this point and progesterone becomes the hormonal product of this tissue.

Conversion Of Progesterone To Cortisol And Cortisone

In considering the synthesis of the mineralo– and glucocorticoids it is useful to recall that the site of their synthesis—the adrenal cortex—has two different types of cell layers. Just beneath the adrenal capsule lies the zona glomerulosa, which produces aldosterone. Deeper in the cortex is the zona fasciculata, the site of cortisol synthesis. The hydroxylation reactions to be described below are divided between the mitochondria and the endoplasmic reticulum and involve the mixed–function type of mechanisms.

The 17α-hydroxylase and the 21–hydroxylase are located in the endoplasmic reticulum and are faster acting than the 11β-hydroxylase, which is in the mitochondria. The bifurcation of the synthetic pathways from progesterone shown above is due primarily to stereochemical restrictions related to C–21. When hydroxylation occurs first at C–21, the presence of the –OH at this position prevents hydroxylation at C–17. As a result, deoxycortisol cannot be formed and the subsequent hydroxylations are routed via DOC and corticosterone to aldosterone. The slow-acting 11 β - hydroxylase can hydroxylate either DOC or deoxycortisol once they have entered the mitochondria.

The sequence of the hydroxylation reactions and the physiological activities of the intermediates may be formulated as follows:

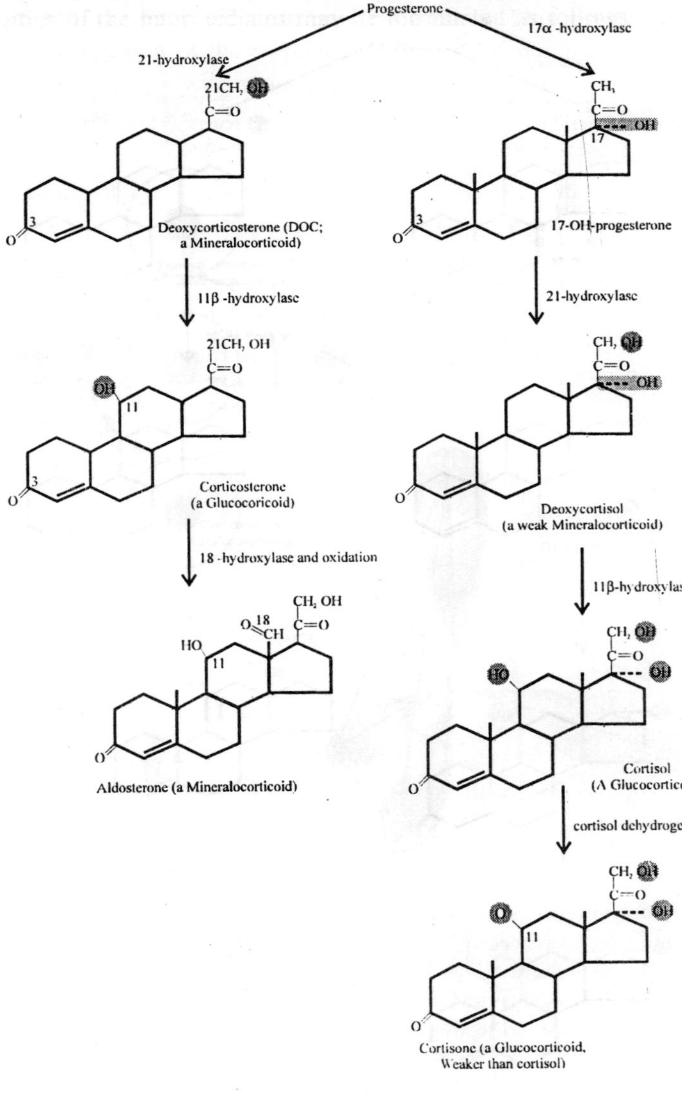

Progesterone

21-hydroxylase

17α -hydroxylase

Deoxycorticosterone (DOC; a Mineralocorticoid)

17-OH-progesterone

11β -hydroxylase

21-hydroxylase

Corticosterone (a Glucocoricoid)

Deoxycortisol (a weak Mineralocorticoid)

18 -hydroxylase and oxidation

11β-hydroxylase

Aldosterone (a Mineralocorticoid)

Cortisol (A Glucocorticoid)

cortisol dehydrogenase

Cortisone (a Glucocorticoid, Weaker than cortisol)

Conversion Of Progesterone to the Sex Hormones in the Gonads.

The synthetic sequence for converting progesterone to the androgens in the endoplasmic reticulum of Leydig cells in the testes begins with hydroxylation of C-17.

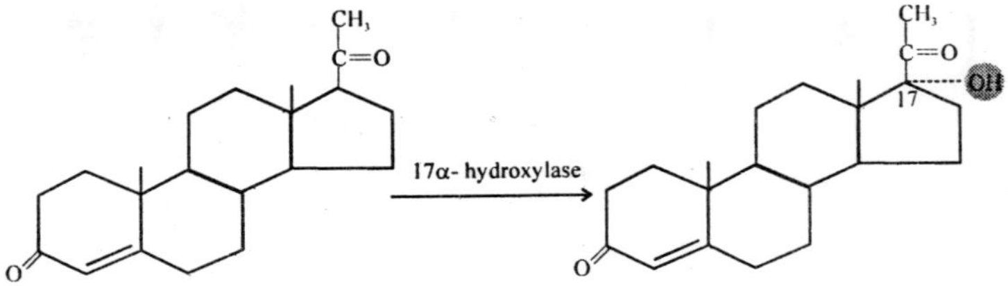

CH_3 C=O → 17α- hydroxylase → CH_3 C=O OH 17

C-20 and C-21 are now removed oxidatively to yield a C–17 ketone, Δ^4 - androstenedione:

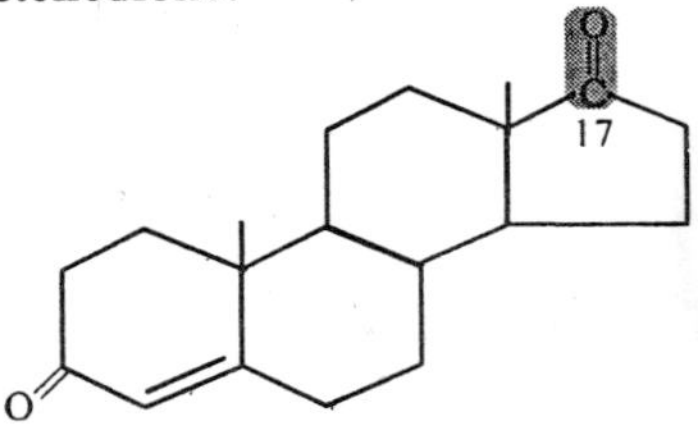

O=C 17

The 17-keto group is capable of a reversible oxidation – reduction and yields testosterone:

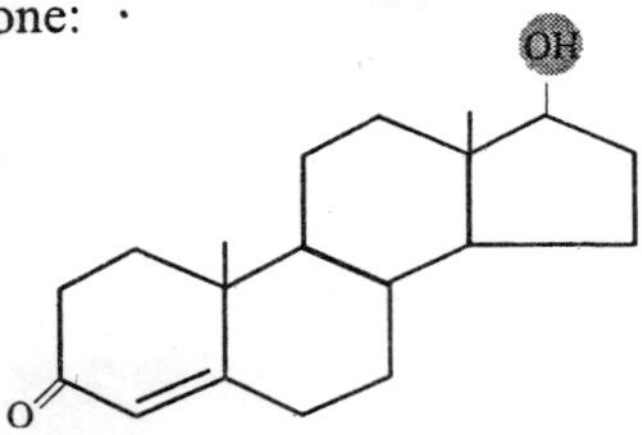

OH

It may be recalled the testosterone can also be formed in the adrenal cortex by the isomer of progesterone, Δ^5 - pregnenolone.

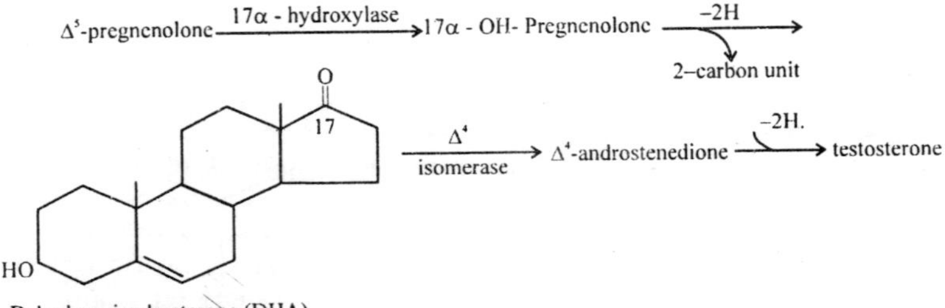

Δ^5-pregnenolone —17α - hydroxylase→ 17α - OH- Pregnenolone —−2H→ (2–carbon unit)

Dehydroepiandrosterone (DHA) —Δ^4 isomerase→ Δ^4-androstenedione —−2H.→ testosterone

Formation of the Female Steroid Hormone from Testosterone

The synthesis of estradiol, the principal estrogen exported by the ovaries, begins with the hydroxylation of C–19 of testosterone, followed by a decarboxylation:

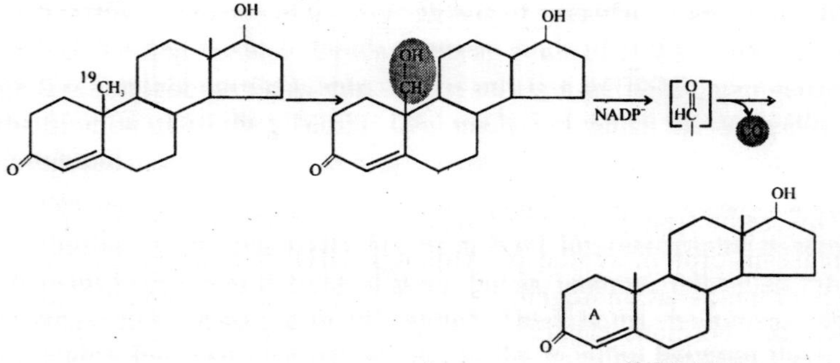

OH OH
19 CH₃ OH CH NADP⁻ HC O

OH
A
19-Nortestosterone

The A ring of the nortestosterone is now aromatized by a mixed–function oxidase system, utilizing NADPH and O_2, and the product is 17 β - estradiol:

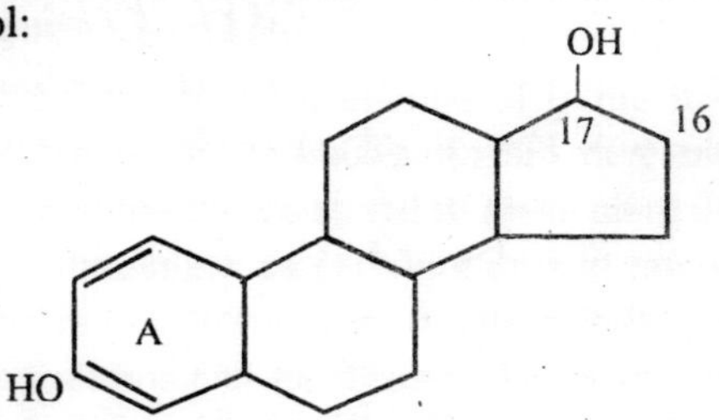

OH
17 16
A
HO

Estradiol can be converted into two less potent hormones either by oxidation of the C-17 hydroxyl or by hydroxylation of C-16:

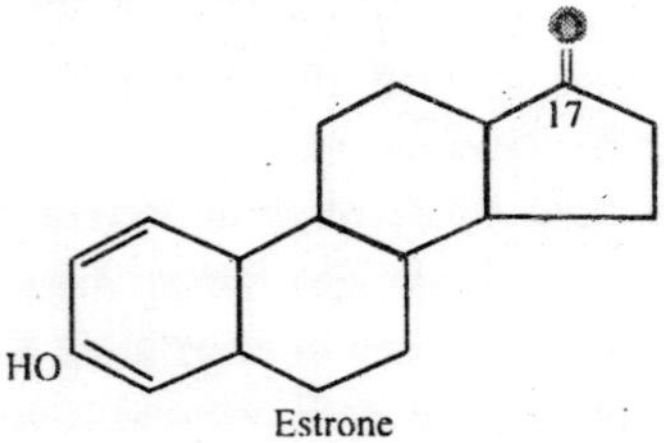

17
HO
Estrone

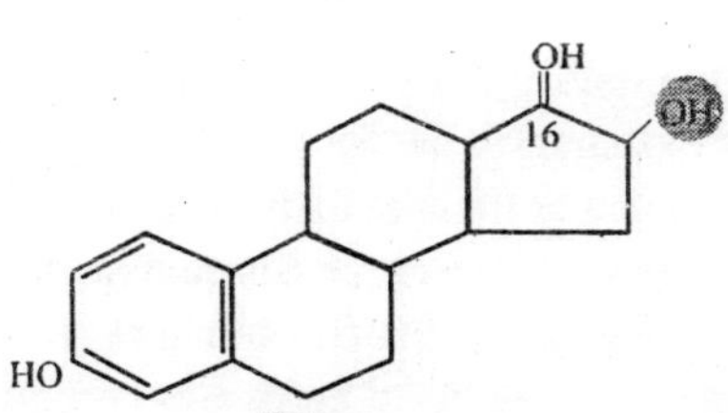

OH
16 OH
HO
Estrone

Formation of Dihydrotestosterone from Testosterone

As the principal circulating androgen, testosterone can enter cells by passive diffusion and be reduced in the endoplasmic reticulum or the nuclear membrane:

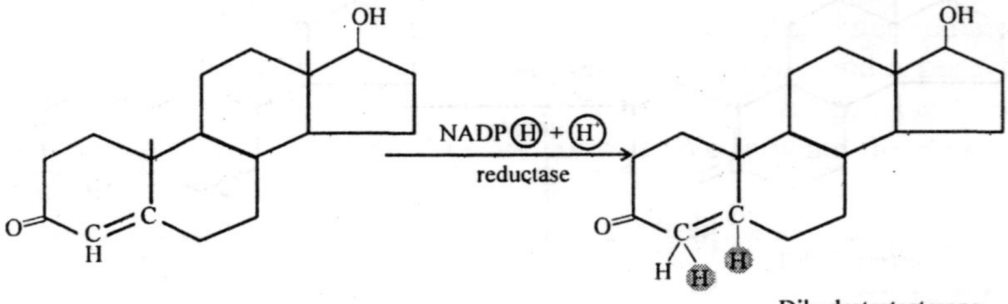

NADPH + H⁺ reductase

Dihydrotestosterone

This reduction product of testosterone (DHT) can effect an increase in protein synthesis. Although DHT is more potent than testosterone as an androgen, its A ring cannot be aromatized and, hence, this hormone cannot serve as a precursor of the estrogen.